THE PSYCHOLOGY OF EXTREMISM

This ground-breaking book introduces a new model of extremism that emphasizes motivational imbalance among individual needs, offering a unique multidisciplinary exploration of extreme behaviors relating to terrorism, dieting, sports, love, addictions, and money.

In popular discourse, the term 'extremism' has come to mean largely 'violent extremism', but this is just one of many different types: extreme sports, extreme diets, political and religious extremisms, extreme self-interest, extreme attitudes, extreme devotion to a cause, addiction to substances, or behavioral addiction (to videogames, shopping, pornography, sex, and work). But do these descriptions have a deeper meaning? Do they reveal a common psychological dynamic? Or are they merely a mode of things about phenomena that have little in common? Bringing together world-leading psychologists from a variety of disciplines, the book uses a brand-new model to examine different expressions of extremism, at different levels of analysis (brain, hormones, and behavior), in order not merely to describe such behaviors but also to explain their occurrence, and the conditions under which they may be likely to emerge.

Also including suggestions for ways in which extremism could be counteracted and to what extent it appears to be harmful to individuals and society, this is essential reading for students and academics in psychology and behavioral sciences.

Arie W. Kruglanski is Distinguished University Professor of Psychology at the University of Maryland, USA. He has received the National Institute of Mental Health Research Scientist Award, the Donald Campbell Award for Outstanding Contributions to Social Psychology, the University of Maryland Regents Award for Scholarship and Creativity, and the Distinguished Scientific Contribution Award from the Society for Experimental Social Psychology. Kruglanski has published over 400 articles, chapters, and books on motivated social cognition; served on NAS panels on the social and behavioral aspects of terrorism; and co-founded the National Center of Excellence for the Study of Terrorism and the Response to Terrorism. He was the PI on a MINERVA grant from the Office of Naval Research on the determinants of radicalization and is presently the PI on a MINERVA grant on Syrian refugees' potential for radicalization.

Catalina Kopetz is Associate Professor of Psychology at Wayne State University, USA. Her research focuses on the mechanisms that underlie multiple goal pursuit and management of goal conflict and their implications for risk-taking. She has published in prestigious journals spanning social and clinical psychology, prevention sciences, psychopharmacology, and behavioral and brain sciences, as well as journals appealing to a broader audience, such as *Perspectives in Psychological Science*, *Current Directions in Psychological Science*, and *Psychological Review*. Her research has been funded by the National Institutes of Health (including NIDA, NCI, and NIAAA).

Ewa Szumowska is a researcher at the Social Psychology Unit in the Institute of Psychology at Jagiellonian University in Krakow, Poland, and a member of the Center for Social Cognitive Studies Krakow, Association for Psychological Science, and the European Association of Social Psychology. She is an author and co-author of scientific publications in journals, such as *Psychological Review*, *Psychological Inquiry*, *Perspectives on Psychological Science*, *Cognition,* and *Personality and Individual Differences*. She studies motivation, information processing, multiple goal pursuit, and extremism.

Frontiers of Social Psychology

Series Editors:
Arie W. Kruglanski, *University of Maryland at College Park*
Joseph P. Forgas, *University of New South Wales*

Frontiers of Social Psychology is a series of domain-specific handbooks. Each volume provides readers with an overview of the most recent theoretical, methodological, and practical developments in a substantive area of social psychology, in greater depth than is possible in general social psychology handbooks. The editors and contributors are all internationally renowned scholars whose work is at the cutting edge of research.

Scholarly, yet accessible, the volumes in the *Frontiers* series are an essential resource for senior undergraduates, postgraduates, researchers, and practitioners and are suitable as texts in advanced courses in specific subareas of social psychology.

Published Titles

Social Judgment and Decision Making, Krueger

Intergroup Conflicts and their Resolution, Bar-Tal

Social Motivation, Dunning

Social Cognition, **Strack & Förster**

Social Psychology of Consumer Behavior, **Wänke**

For continually updated information about published and forthcoming titles in the *Frontiers of Social Psychology* series, please visit: **www.routledge.com/psychology/series/FSP**

THE PSYCHOLOGY OF EXTREMISM

A Motivational Perspective

Edited by
Arie W. Kruglanski, Catalina Kopetz,
and Ewa Szumowska

NEW YORK AND LONDON

First published 2022
by Routledge
605 Third Avenue, New York, NY 10158

and by Routledge
2 Park Square, Milton Park, Abingdon, Oxon, OX14 4RN

Routledge is an imprint of the Taylor & Francis Group, an informa business

© 2022 Taylor & Francis

The right of Arie W. Kruglanski, Catalina Kopetz and Ewa Szumowska to be identified as the authors of the editorial material, and of the authors for their individual chapters, has been asserted in accordance with sections 77 and 78 of the Copyright, Designs and Patents Act 1988.

All rights reserved. No part of this book may be reprinted or reproduced or utilised in any form or by any electronic, mechanical, or other means, now known or hereafter invented, including photocopying and recording, or in any information storage or retrieval system, without permission in writing from the publishers.

Trademark notice: Product or corporate names may be trademarks or registered trademarks, and are used only for identification and explanation without intent to infringe.

Library of Congress Cataloging-in-Publication Data
Names: Kruglanski, Arie W., editor. | Kopetz, Catalina E., editor. | Szumowska, Ewa, 1986– editor.
Title: The psychology of extremism : a motivational perspective / edited by Arie W. Kruglanski, Catalina Kopetz and Ewa Szumowska.
Description: New York, NY : Routledge, 2022. | Includes bibliographical references and index.
Identifiers: LCCN 2021016077 (print) | LCCN 2021016078 (ebook) | ISBN 9780367467623 (hardback) | ISBN 9780367467609 (paperback) | ISBN 9781003030898 (ebook)
Subjects: LCSH: Extreme behavior (Psychology) | Motivation (Psychology)
Classification: LCC BF637.E97 P79 2022 (print) | LCC BF637.E97 (ebook) | DDC 153.1/534—dc23
LC record available at https://lccn.loc.gov/2021016077
LC ebook record available at https://lccn.loc.gov/2021016078

ISBN: 978-0-367-46762-3 (hbk)
ISBN: 978-0-367-46760-9 (pbk)
ISBN: 978-1-003-03089-8 (ebk)

DOI: 10.4324/9781003030898

Typeset in Bembo
by Apex CoVantage, LLC

CONTENTS

List of Contributors   *xiii*

The Psychology of Extremism   1
Arie W. Kruglanski, Ewa Szumowska, and Catalina Kopetz

Extremism as a Motivational Construct 2
  Balance and Moderation 2
  Imbalance and Extremism 2
Consequences of Motivational Imbalance 4
The Present Volume: Motivational Imbalance Across Domains and Levels of Analysis 5
  Part 1: Motivational Imbalance at Different Levels of Analysis 6
  Part 2: Motivational Imbalance Across Domains of Human Endeavor 8

PART 1
Motivational Imbalance at Different Levels of Analysis   13

1   Incentive Salience in Irrational Miswanting and Extreme Motivation   15
Hannah M. Baumgartner, Erin E. Naffziger, David Nguyen, and Kent C. Berridge

Introduction 15
Reward Utilities and 'Wanting' 16
Attribution of Incentive Salience 18

'Irrational Miswanting' and 'Dangerous Desire' 21
Conclusion and Implications in Extreme Aggression 25

2   Attitudinal Extremism   34
    Joseph J. Siev, Richard E. Petty, and Pablo Briñol

    Developing a Model of Attitudinal Extremism 34
    *Candidates for Inclusion in a Model of Attitudinal
       Extremism* 37
       *Conceptualizing Extremism: Motivational and Attitudinal
          Imbalance* 39
    *Properties of Attitudes That Increase Attributions
       of Extremism* 41
    *Processes That Produce Polarized, Confident, and Unusual
       Attitudes* 44
       Determinants of Attitude Polarization 45
    Information Exposure and Processing 46
    Mere Thought 47
    Metacognitive Validation 48
    Normative Influence 49
       Determinants of Attitude Certainty 51
       Determinants of Attitude Unusualness 51
    *Properties of Attitudes That Predict Extreme
       Behavior* 52
    Threat as a Moderator of Compensation Effects 54
    Conclusions 57

3   On Extreme Behavior and Outcomes: The Role
    of Harmonious and Obsessive Passion   66
    Robert J. Vallerand and Virginie Paquette

    A Dualistic Model of Passion 67
       On the Concept of Passion 67
       On Harmonious and Obsessive Passion 68
       Research Methods on Passion 70
    Passion and Extreme Behavior 72
       Passion and Extreme Interpersonal Behavior 72
       Passion and Extreme Intrapersonal Behavior 77
    *Determinants of Harmonious and Obsessive
       Passion* 83
    Conclusions 85

4   The Extreme Group   96
    John M. Levine and Arie W. Kruglanski

    The Present Chapter 96
    Quest for Significance 98
        Activating the Quest for Significance: Significance Loss 99
        Activating the Quest for Significance: Significance Gain 100
        Desire to Reduce Uncertainty 101
        Sacrifice and Identity Fusion 103
        Charismatic and Other Extreme Groups 104
    Narratives 106
        Narratives and Moral Disengagement 108
        Narratives and Group Socialization 109
        Creating a Narrative 111
    Networks 113
        Group Socialization 113
        Group Development 125
    Conclusion 128

5   Masters of Both: Balancing the Extremes of Innovation
    Through Tight–Loose Ambidexterity   140
    Piotr Prokopowicz, Virginia K. Choi, and Michele J. Gelfand

    *Exploration and Exploitation: Understanding the Extremes
        of Innovation* 142
    Cultural Tightness–Looseness 144
    Exploring Looseness, Exploiting Tightness 144
    The Goldilocks Principle of Innovation 147
        Methods 148
        Results 149
    Discussion 151
    Funding 155

6   The Evolution of Extremism   163
    William von Hippel and Nadia Fox

    How Might Extremism Have Evolved? 164
    What Function Might Extremism Serve? 167
    Is Extremism Unique to Humans? 168
    *How Does Morality Attenuate and Exacerbate
        Extremism?* 171
    Implications and Conclusions 173

PART 2
Motivational Imbalance Across Domains of Human Endeavor   **181**

7   The Psychology of Extreme Sports   183
Eric Brymer and Pierre Bouchat

Introduction 183
Traditional Perspectives on the Psychology of Extreme Sports 184
    Sensation Seeking 185
    Edgework 187
    Reversal Theory 188
    Other Psychological Theories Employing the Risk Narrative 189
Beyond the Risk-Taking Narrative 191
Positive Psychology Perspectives 192
    Flow and Peak Experience 192
    Extraordinary and Transcendent Experiences 195

8   The Psychology of Greed   203
Katalin Takacs Haynes

Introduction to Greed Narratives 203
What Is Greed and Why Is It Extreme? 206
History of Greed 208
Greed in the Social Sciences 218
Multilevel Model of Greed as Extremism 222
Mitigating the Problem 224

9   Moral, Extreme, and Positive: What Are the Key Issues for the Study of the Morally Exceptional?   230
William Fleeson, Christian Miller, R. Michael Furr, Angela Knobel, and Eranda Jayawickreme

The Promise of Studying the Morally Exceptional 231
Do Morally Exceptional People Exist? 232
Theoretical, Philosophical, and Theological Accounts of the Morally Exceptional 233
    What Features Are Relevant to Qualifying as Morally Exceptional? 233
    The Desirability and Motivational Balance of the Morally Exceptional 234

*The Role of the Morally Exceptional in Defining and
      Founding the Virtues* 234
   *Empirical Findings: Why Are Some People Extremely
      Moral?* 235
      The Role of Social Structures 237
      Proximate Situational Causes 237
      Developmental History 238
      Personality Traits 238
      Moral Reasoning 239
      Identity as Motivation 240
      *Summary of Empirical Findings: Ordinary People With a
         Deep Concern for Others* 241
   What Is Not Known 244
   *Three Difficulties for the Study of the Morally Exceptional and
      Strategies for Addressing Them* 246
      *Difficulty 1: What Counts as 'Moral' and What Counts as
         'Morally Good'?* 247
      Difficulty 2: What Counts as Exceptional? 249
      *Difficulty 3: What Will Be Added to the Study
         of Morality by Studying the Morally Exceptional in
         Particular?* 252
   Conclusion 253
   Acknowledgments 253

10 The Social Psychology of Violent Extremism   259
   Erica Molinario, Katarzyna Jasko, David Webber,
   and Arie W. Kruglanski

   Introduction 259
   Perceived Efficacy of Violence 260
   Feeling Noticed and Agentic 262
   Ingroup Identification 263
   Culturally Approved Violence 264
   Violence as a Clear Response 265
   Relative Deprivation and Inequality 266
   The 3N Framework: Need, Narrative, and Networks 267
      Needs 267
      Narrative 269
      Network 270
   Conclusion 271

11 Motivational Imbalance in Jihadi Online Recruitment 280
   Gabriel Weimann

   Introduction 280
   Online Recruitment 281
   The Notion of Motivational Imbalance 284
   'Narrowcasting' 286
   Using Motivational Imbalance in Jihadi Campaigns 287
      The Appeals Used 287
      Evidence from Recruits 298
   Conclusions 300

Index 304

CONTRIBUTORS

Hannah M. Baumgartner is a recent doctoral graduate in Psychology at the University of Michigan in Ann Arbor, USA.

Kent C. Berridge is James Olds Distinguished University Professor of Psychology and Neuroscience at the University of Michigan in Ann Arbor, USA.

Pierre Bouchat is Assistant Professor of Social Psychology at the University of Lorraine (Metz) and Associate Researcher at the Universidad del País Vasco (San Sebastian). He completed his doctoral thesis at the Center for Social and Cultural Psychology of the Université libre de Bruxelles and worked as a postdoctoral researcher at the Université catholique de Louvain. His current research focuses on the role of collective memory and collective emotions at the intra- and inter-group levels. Pierre is also a BASE jumper and develops a line of research on risky sports.

Pablo Briñol is Professor of Psychology at the Universidad Autónoma de Madrid (Spain), where he obtained his BA (1995) and PhD (1999) degrees. He is also a regular visiting scholar at Ohio State University. His research interest focuses on the study of the psychological mechanisms underlying attitudes and persuasion, with an emphasis on metacognitive processes and the processes of change. Dr Briñol has published over 200 peer-reviewed publications in top journals of the field (e.g., *Journal of Personality and Social Psychology*, *Perspectives on Psychological Science*, *Psychological Science*, and *Psychological Bulletin*), receiving more than 15,000 citations [Google]. Recently, he was awarded the Kurt Lewin medal by the European Association of Social Psychology which recognizes individuals in mid-career for their scientific excellence. Dr Briñol is also the co-founder of the Sociedad

Científica Española de Psicología Social (SCEPS) where he serves on the executive board. For more information, visit www.pablobrinol.com.

Eric Brymer is Associate Professor at the Australian College of Applied Psychology and Academic Manager. He focuses on investigating extreme sport psychology and the role of the environment for well-being and performance.

Virginia K. Choi is PhD Student in Industrial-Organizational Psychology at the University of Maryland. Her research expertise includes the antecedents and consequences of social norms, organizational cultures, and stigmatization processes. Her work has been published in several scholarly journals, such as *American Psychologist*, *Annual Review of Psychology*, and *Behavioral and Brain Sciences*. She has written several popular press articles for *Harvard Business Review*, *The Conversation*, *Fast Company*, and *Time*.

William Fleeson is Hultquist Family Professor of Psychology at Wake Forest University, received a BA in philosophy from Wisconsin, a PhD degree in psychology from Michigan, and postdoctoral training in human development from Germany. His research interests include the ways people try to be honest and good, the study of those who devote their lives to doing good, the nature of personality, and borderline personality disorder. He takes a personality, social, cognitive, developmental, and philosophical approach to his work. Studies on distributions of behavior and their implications for personality won SPSP's Theoretical Innovation Prize, and the Carol and Ed Diener Award in Personality Psychology.

Nadia Fox is PhD Candidate in the School of Psychology at the University of Queensland. Her dissertation examines the impact of stress on performance and leadership in complex environments through the use of experience sampling and biometric capture.

R. Michael Furr is Professor of Psychology and Wright Faculty Fellow at Wake Forest University. His interests include person-situation interactions, peoples' perceptions of themselves and their social worlds, and psychological measurement. He has been elected a fellow of the Society for Personality and Social Psychology, a fellow of the Association for Psychological Science, and a fellow of both Division 5 (Quantitative and Qualitative Methods) and Division 8 (Social and Personality Psychology) of the American Psychological Association. In addition, he received the Award for Distinguished Contributions to the Literature in Personality Assessment from the Society for Personality Assessment, and he received the Award for Excellence in Research from Wake Forest University. He earned a BA at the College of William and Mary, an MS at Villanova University, and a PhD at the University of California at Riverside.

Michele J. Gelfand is Distinguished University Professor of Psychology at the University of Maryland, College Park. Gelfand uses field, experimental, computational, and neuroscience methods to understand the evolution of culture and its multilevel consequences. Her work has been published in outlets, such as *Science*, the *Proceedings of the National Academy of Sciences*, *Psychological Science*, *Nature Human Behavior*, the *Journal of Personality and Social Psychology*, *Journal of Applied Psychology*, *Academy of Management Journal*, among others. Gelfand is the founding co-editor of the Advances in Culture and Psychology series (Oxford University Press). Her book *Rule Makers, Rule Breakers: How Tight and Loose Cultures Wire the World* was published by Scribner in 2018. She is the Past President of the International Association for Conflict Management and co-founder of the Society for the Study of Cultural Evolution. She received the 2016 Diener award from SPSP, the 2017 Outstanding International Psychologist Award from the APA, the 2019 Outstanding Cultural Psychology Award from SPSP, the 2020 Rubin Theory to Practice award from IACM, and the Annaliese Research Award from the Humboldt Foundation. Gelfand was elected to the American Academy of Arts and Sciences in 2019. She is joining the Stanford Graduate School of Business in Fall 2021.

Katalin Takacs Haynes is an award-winning Associate Professor of Strategic Management and Leadership. Her research intersects corporate governance and the dark side of management, and includes studies on executive compensation, corruption, greed, and hubris. Her work has been published in *Administrative Science Quarterly*, *Journal of Management*, *Journal of Management Studies*, and *Strategic Management Journal*. Her studies on executive greed and hubris have been cited in the *Washington Post* and the *London Business School's Business Review* for executives and deposited in the Vault of the Daedalus Trust.

William von Hippel is Social Psychologist at the University of Queensland. He conducts research in evolutionary psychology on topics ranging from self-deception and social intelligence to innovation and leadership. He recently published his first book, *The Social Leap*, which traces our evolutionary history over the last 6 million years and received the 2019 Society for Personality and Social Psychology Book Award.

Katarzyna Jasko (PhD, Jagiellonian University) works at the Institute of Psychology, Jagiellonian University. Her research focuses on issues related to political engagement, such as motivation to participate in collective action, political extremism and ideology, and political violence.

Eranda Jayawickreme is Associate Professor of Psychology and Senior Research Fellow in the Program for Leadership and Character and at Wake Forest University. His research focuses on posttraumatic growth, moral personality, wisdom, and well-being. His awards include the 2018 Faculty Excellence in Research Award from Wake Forest, the 2015 Rising Star Award from the Association for Psychological Science, and multiple grants from the John Templeton Foundation, Templeton Religion Trust, Templeton World Charity Foundation, the European Association for Personality Psychology and the Asia Foundation/USAID.

Angela Knobel is Associate Professor of Philosophy at the University of Dallas. She has published extensively on medieval and contemporary ethics. Her work focuses primarily on Aquinas' theory of infused virtue, virtue ethics, and applied ethics. She has published widely in journals, such as *The Thomist, American Catholic Philosophical Quarterly, Nova et Vetera, International Philosophical Quarterly*, and *The Journal of Moral Theology*. Her book *Aquinas and the Infused Moral Virtues* is forthcoming from the University of Notre Dame Press. Dr Knobel received her BA degree from The Catholic University of America, her MA degree in mathematics from the University of Maryland, and her PhD degree in philosophy from the University of Notre Dame.

Catalina Kopetz is Associate Professor of Psychology at Wayne State University, USA. Her research focuses on the mechanisms that underlie multiple goal pursuit and management of goal conflict and their implications for risk-taking. She has published in prestigious journals spanning social and clinical psychology, prevention sciences, psychopharmacology, behavioral and brain sciences, as well as journals appealing to a broader audience, such as *Perspectives in Psychological Science, Current Directions in Psychological Science*, and *Psychological Review*. Her research has been funded by the National Institutes of Health (including NIDA, NCI, and NIAAA).

Arie W. Kruglanski is Distinguished University Professor of Psychology at the University of Maryland, USA. He has received the National Institute of Mental Health Research Scientist Award, the Donald Campbell Award for Outstanding Contributions to Social Psychology, the University of Maryland Regents Award for Scholarship and Creativity, and the Distinguished Scientific Contribution Award from the Society for Experimental Social Psychology. Kruglanski has published over 400 articles, chapters, and books on motivated social cognition; served on NAS panels on the social and behavioral aspects of terrorism; and co-founded the National Center of Excellence for the Study of Terrorism and the Response to Terrorism. He was the PI on a MINERVA grant from the Office of Naval Research on the determinants of radicalization and is presently the PI on a MINERVA grant on Syrian refugees' potential for radicalization.

John M. Levine (PhD, University of Wisconsin) is Professor Emeritus of Psychology at the University of Pittsburgh. He has published widely on various facets of group behavior, including conformity and reaction to deviance, group socialization, and loyalty/disloyalty. He served as Editor of the *Journal of Experimental Social Psychology* and Chair of the Society of Experimental Social Psychology. Dr Levine was co-recipient of the McGrath Award for Lifetime Achievement in the Study of Groups from the Interdisciplinary Network for Group Research. He has been an Honorary Professor of Psychology at the University of Kent and received an Honorary Doctorate from the University of Lausanne.

Christian B. Miller is the A. C. Reid Professor of Philosophy at Wake Forest University. He is currently the Director of the Honesty Project, funded by a $4.4 million grant from the John Templeton Foundation, and was the Philosophy Director of the Beacon Project and the Director of the Character Project. He is the author of over 90 academic papers and editor of five volumes, as well as author of *Moral Psychology* with Cambridge University Press (2021) and four books with Oxford University Press, *Moral Character: An Empirical Theory* (2013), *Character and Moral Psychology* (2014), *The Character Gap: How Good Are We?* (2017), and *Honesty: The Philosophy and Psychology of a Neglected Virtue* (2021). He is a science contributor for *Forbes*, and his writings have also appeared in *The New York Times*, *Wall Street Journal*, *Dallas Morning News*, *Slate*, *The Conversation*, *Newsweek*, *Aeon*, and *Christianity Today*.

Erica Molinario (PhD, Sapienza University of Rome) is Postdoctoral Research Associate at the University of Maryland, College Park. Her main research interests focus on human motivation as a framework to understand behaviors related to current social issues, such as pro-environmental behaviors, political extremism and radicalization, and intergroup conflicts.

Erin E. Naffziger is PhD candidate in Psychology at the University of Michigan.

David Nguyen is Postdoctoral Research Fellow in the Psychology Department at the University of Michigan in Ann Arbor, USA.

Virginie Paquette holds BSc in Psychology from Université de Montréal. She is a PhD candidate at the Université du Québec à Montréal in the Research Laboratory on Social Behavior with professor Robert Vallerand. Her research interests are related to motivation and passion.

Richard E. Petty is Distinguished University Professor of Psychology at Ohio State University. He received his BA degree from the University of Virginia and his PhD degree from Ohio State. Petty's research, funded by NSF and NIMH, focuses on social influence and the factors responsible for changes in attitudes and behaviors. He has published eight books and over 400 articles which have been cited over 100,000 times. Petty is a fellow of the American Academy of Arts and Sciences, the American Association for the Advancement of Science, the Association for Psychological Science, the Society for Consumer Psychology, and the American Psychological Association. His honors include receiving Distinguished Scientific Contribution Awards from the Societies of Experimental Social Psychology (SESP), Personality and Social Psychology (SPSP), and Consumer Psychology (SCP). His professional service includes being past Editor of the *Personality and Social Psychology Bulletin*, Associate Editor of *Emotion*, and President of SPSP, the Midwestern Psychological Association, and the Foundation for Personality and Social Psychology. For more information, visit www.richardepetty.com.

Piotr Prokopowicz is Assistant Professor in the Department of Sociology of Economy, Education, and Social Research Methods, a researcher at the Center for Evaluation and Analysis of Public Policies, and a Postdoctoral Fellow at the University of Maryland's Culture Lab. He specializes in organizational culture, leadership, innovation, and evidence-based management. He holds a PhD degree in sociology and a double MA degree in psychology and sociology. He is a co-founder at Freenovation, speaker, author, and talent management consultant. A Ryoichi Sasakawa Young Leaders Fellowship Fund and Kosciuszko Fellow, he has also cooperated with numerous international institutions and businesses. In his research and consulting, he focuses on helping organizations in reaching their maximum potential through participation, transparency, and evidence-based management.

Joseph J. Siev is PhD candidate in social psychology at Ohio State University. His research examines the effects of attitude strength on social judgment and behavior, with a focus on implications for social problems, such as extremism, political polarization, and prejudice. To learn more about his research, which has been published in leading journals, including the *Journal of Personality and Social Psychology* and the *Journal of Consumer Psychology*, visit www.joesiev.com.

Ewa Szumowska (PhD) is Researcher at the Social Psychology Unit in the Institute of Psychology at Jagiellonian University in Krakow, Poland, and a member of the Center for Social Cognitive Studies Krakow, Association for Psychological Science and the European Association of Social Psychology. She is an author and co-author of scientific publications in journals, such as *Psychological Review*, *Psychological Inquiry*, *Perspectives on Psychological Science*, *Cognition*, and *Personality and Individual Differences*. She studies motivation, information processing, multiple goal pursuit, and extremism.

Robert J. Vallerand is Full Professor of Psychology at the Université du Québec à Montréal. He has published eight books and well over 350 scientific articles and book chapters. His research focuses on motivational processes and especially passion for activities, including sport and exercise. He has served as President of the Canadian Psychological Association and the International Positive Psychology Association. He is a fellow of more than a dozen associations, including the American Psychological Association (and APA Division 15 Educational Psychology), the Association for Psychological Science, and many others. He has received numerous awards, including the Christopher Peterson Gold Medal Award from the International Positive Psychology Association, the William James Award from the American Psychological Association for his 2015 book, *The Psychology of Passion* with Oxford University Press, and the Sport Science Award from the International Olympic Committee.

David Webber (PhD, University of Alberta) is Assistant Professor of Homeland Security and Emergency Preparedness at the L. Douglas Wilder School of Government of Public Affairs at Virginia Commonwealth University. His research focuses on understanding the radicalization and deradicalization processes of violent extremists. He co-authored *The Radical's Journey* (2019) on far-right extremists in Germany.

Prof. Gabriel Weimann (PhD) is Full Professor of Communication (Emeritus) at the Department of Communication at Haifa University, Full Professor at the Interdisciplinary Center in Herzelia, Israel, and Visiting Professor at the University of Maryland, USA. His research interests include the study of political campaigns, persuasion and influence, modern terrorism and the mass media, online terrorism, and cyberterrorism. He published nine books: *Communicating Unreality; The Influentials: People Who Influence People; The Theater of Terror; Hate on Trial; The Singaporean Enigma; Terror in the Internet: The New Arena, the New Challenges; Freedom and Terror; Social Research in Israel,* and *Terrorism in Cyberspace*. His 190 papers have been published in scientific journals such as *Journal of Communication, Public Opinion Quarterly, Communication Research, Journal of Broadcasting and Electronic Media,* and *American Sociological Review*. He received numerous grants and awards from international foundations, including the Fulbright Foundation, the Canadian-Israel Foundation, the Alexander von Humboldt-Stiftung, the German National Research Foundation, the Sasakawa Foundation, the United States Institute for Peace, The Woodrow Wilson Center, the Australian Research Council, and others.

THE PSYCHOLOGY OF EXTREMISM

*Arie W. Kruglanski, Ewa Szumowska,
and Catalina Kopetz*

The term extremism and its linguistic 'relatives' (extremists and extreme) are used often in the popular language. People talk of extreme sports, extreme diets, political and religious extremisms, violent extremism, and extreme self-interest. One could be extremely devoted to a cause (e.g., saving the environment) or an activity, such as playing basketball or chess. One could be an extreme supporter of a political party or an extreme sports fan. One can hold extreme attitudes or be extremely in love—with a person or a public figure. People can be prepossessed by their extreme craving for a drug or experience extreme urges of various sorts. But do these expressions have a deeper meaning? Do they reveal a common psychological dynamic? Or are they merely ways of talking about phenomena that have little if anything in common? This is the question we are asking in this volume. And we propose an answer, a hypothesis, and a model described in the remainder of this chapter and explored in those that follow.

Specifically, we submit that despite the striking differences in specifics, extreme behaviors share a deeper psychological dynamic. They are the result of a motivational imbalance, wherein a given need overrides other basic concerns. It means the focusing of one's mental resources on a given prioritized need and their proportionate withdrawal from other needs. As a consequence, the constraints that the latter exert on behavior are relaxed, thereby allowing behaviors that are inconsistent with those needs and would not be otherwise considered. This has important consequences for motivation, cognition, behavior, affect, and sociality.

The aim of this chapter is to present the motivational imbalance model of extremism and to lay the theoretical grounds for chapters in this volume. We start by presenting the imbalance model and its consequences (for a more detailed discussion, see Kruglanski, Szumowska, Kopetz, Vallerand, & Pierro, 2021; Kruglanski, Szumowska, & Kopetz, 2021). This is followed by an overview of this volume's

DOI: 10.4324/9781003030898-1

contents in terms of the levels of analysis being applied to the analysis of extremism and the specific domains where striking instances of extremism have been observed.

Extremism as a Motivational Construct

The concept of *extremism* stands in contrast to that of *moderation*. We define the two in terms of relations—balance or imbalance—between people's basic needs. Psychological theorists agree that humans have a set of basic needs whose fulfillment is indispensable to their well-being. Some needs are biological in nature (the needs for nutrition, hydration, rest, etc.), and others are psychogenic (e.g., autonomy, relatedness, control, understanding, and significance) (Deci & Ryan, 2000; Fiske, 2010; Higgins, 2012; Maslow, 1943). Note that the very concept of *basic* needs suggests that people try to satisfy them all, so that their nonfulfillment foments distress. The balanced attempt to satisfy all one's basic needs defines the state of *moderation* whereas the dominance of one need over others defines the state of *extremism*.

Balance and Moderation

The several needs exercise constraints on each other such that behaviors that gratify one need while undermining others tend to be avoided. For instance, one's hunger may coexist with concerns about taste and health. Therefore, people try to eat foods that are both healthy and tasty, while avoiding foods that are foul-tasting and unhealthy (Kopetz, Faber, Fishbach, & Kruglanski, 2011). One's need for intimacy and relatedness may temper one's need for achievement, thus promoting a work-family balance (as opposed to focusing entirely on work, for example, as in workaholism).

Importantly, moderation does not imply 'moderation' in the magnitude of one's needs, as some individuals may experience all their needs more intensely than others. However, as long as one of the needs (or goals originating from these needs) does not dominate the behavior, motivational balance may exist also between several high-level needs. This is nicely seen in the distinction between harmonious (balanced) and obsessive (imbalanced) passion (Vallerand, 2015; see also Vallerand & Paquette, this volume). Both harmoniously and obsessively passionate individuals are 'passionate', that is, highly motivated to pursue a given activity. Yet the harmoniously passionate individuals are also highly motivated to pursue other activities that satisfy their other needs. In contrast, the obsessively passionate individuals are highly motivated to pursue a given focal activity and are relatively unmotivated by much else.

Imbalance and Extremism

At times the balanced attempt to satisfy all one's basic needs can get upset and one need can gain dominance over others. This happens when one of the needs is aroused at an exaggerated magnitude (via deprivation or incentivization). When that happens, the person's other needs are crowded out via a goal-shielding mechanism (Shah, Friedman, & Kruglanski, 2002). Alternatively, a motivational imbalance may be created when the saliency of the alternative needs subsides (often because of declined expectancy of satisfying them). In both cases, the person's mental resources are then channelled disproportionately toward the satisfaction of the focal need.

Moreover, the suppressed alternative needs are less able to exert constraints upon behavior. As a result, actions that would normally be constrained by those needs and hence avoided are now permitted. Referring to the hunger example once more—whereas under moderate hunger, one may choose foods that are nutritious, tasty, and healthy, under extreme hunger one may eat whatever becomes available, even if it is unhealthy, tasteless, or disgusting. When one's need for a given intoxicating substance (e.g., alcohol, crack, and heroin) is particularly intense, it may prompt behaviors aimed at obtaining that substance, even when such behaviors are destructive to self and others (e.g., neglecting one's work obligations, foregoing healthy nutrition, and stealing or engaging in other criminal activities). Similarly, an extremely self-interested person, driven by greed and an unstoppable drive to succeed, may sacrifice her or his personal relations, health, and the interest of others, society, or the environment (see Haynes, this volume). Such behaviors denote a deviation from a motivational balance. And because people in general strive to maintain balance, extreme behavior is infrequent.

The constraints on behavior imposed in a motivationally balanced state and their relaxation under motivational imbalance were investigated by Kopetz et al. (2011). In one of their studies, researchers examined the foods that students, concerned both with *eating enjoyment* and with *weight control*, were considering for lunch. In one group, commitment to the goal of food enjoyment was experimentally increased. In this group, unlike in the control group where the eating enjoyment and weight control goals were in balance, participants listed more foods they would consider having, which reflected an expansion of the number of means to the dominant goal under motivational imbalance. Interestingly too, the additional foods listed were rated higher on taste but also on caloric content (at odds with the weight control objective), whereas in the control (balanced) condition, the fewer foods listed were equal on taste but lower on caloric content. Subsequent studies using different goal commitment manipulations replicated this finding and showed that the motivational imbalance explained it; specifically, the increase in the number of means to the dominant goal was mediated by the *inhibition* of the dieting (weight control) goal through focusing on eating enjoyment.

These findings attest that imbalance in one's needs leads to actions which, while gratifying the main need, undermine others—illustrating the essence of extreme behavior. Importantly, temporary imbalances do not necessarily denote extreme pursuits. Indeed, momentary spikes in magnitudes of given needs may occur often, resulting in important concerns being temporarily put aside in favor of other salient needs. Moreover, such shorter and weaker states of imbalance are often necessary for goal progress—one needs to focus on one task and shield it against distraction in order to bring it to completion. However, people typically gratify the dominant need and then relatively quickly proceed to restore a motivational balance by taking care of their alternative, temporarily neglected, needs.

By contrast, the term *extremism* refers to enduring states of motivational imbalance. As Maslow (1943) argued, an unsatisfied need dominates the organism 'if it is extreme enough and chronic enough' (p. 376). It is a long-lasting substance addiction rather than occasional use, an enduring obsession with a given activity (e.g., playing video games), or a long-lasting commitment to violent pursuits (e.g., terrorism), which is typically viewed as extreme. Because people generally strive to satisfy all of their basic needs, such prolonged obsession is rare as noted earlier.

In summary, we suggest that extremism is the result of the psychological processes underlying goal prioritization and goal pursuit. These processes enable people (and other organisms) to channel their resources toward a single, dominant purpose and to broaden their strategic considerations in order to fulfill momentarily important goals and to cope with the pressing current challenges. This typically produces a momentary goal prioritization soon followed by a restoration of balance among needs. The basic capacity to develop a motivational imbalance may be prolonged in some circumstances, permitting the emergence of different forms of extreme behaviors.

Consequences of Motivational Imbalance

Motivational imbalance has important consequences for motivation, cognition, behavior, affect, and sociality. Perhaps the most important consequence is the legitimation of behaviors that undermine other basic needs. For example, free solo climbers ascend cliffs without a rope or other aids as protection, extreme mountaineers venture above the death zone (8000 m) where their bodies are tested to their limits, and extreme skiers navigate sheer cliffs where a slip would likely result in an uncontrollable tumble (see Brymer & Bouchat, this volume). Morally exceptional individuals risk their lives to save the lives of other people (e.g., Holocaust rescuers, Fleeson et al., this volume).

At the level of motivation, the imbalance manifests itself in differential concern for the focal versus the alternative needs. This translates into increased relative preferences for activities seen as serving the focal need compared with those serving the alternative needs. It also translates into greater explicit value being attached to the focal versus the alternative needs, and greater readiness to sacrifice the alternative for the focal need. For instance, studies on workaholism showed a negative correlation between commitment to work and satisfaction with other life areas. Burke (1999) showed that managers and professionals reporting greater workaholism also reported less satisfaction in domains external to work, such as family, friends, and community. Similar exaggerated wanting of an object (a drug), which is detrimental to other needs, is common in addictions and can take such striking forms as wanting to experience a hurtful stimulus (such as an electric shock, see Baumgartner et al., this volume).

At the cognitive level, the greater the magnitude of the motivational imbalance, the stronger the tendency to allocate attention to the focal need and to withdraw attention from the alternative needs (e.g., focusing one's attention solely on a passionate activity, Vallerand & Paquette, this volume). Also, prolonging motivational imbalance promotes a pronounced rumination about the object(s) of one's desire and promotes the tendency to project the focal versus the alternative needs onto others—greedy managers behave as if others were greedy too (see Haynes, this volume) and obsessive passion (OP) for sexuality leads one to experience intrusive cognitive sexual thoughts and biased processing of stimuli, such as seeing sex-related words where there are none (Philippe, Vallerand, Bernard-Desrosiers, Guilbault, & Rajotte, 2017).

Motivational imbalance also affects one's emotionality. When fully committed to one goal, a person puts 'all (of their) eggs in one basket', that is, invests all their resources into a single motivational concern. The person is thus highly emotionally dependent on gratification of his or her dominant need. Therefore, satisfaction of the dominant need engenders a highly positive affect and its frustration, a highly negative one. Linville's (1985) self-complexity model suggests that people whose self-worth depends on the satisfaction of fewer, less distinct goals (lower in self-complexity) are more susceptible to greater swings in affect in response to success and failure. For example, obsessively (hence motivationally imbalanced) passionate recreational golfers exhibited particularly high levels of positive affect after success, and high levels of negative affect after failure (Verner-Filion, Schellenberg, Rapaport, Belanger, & Vallerand, 2018).

The Present Volume: Motivational Imbalance Across Domains and Levels of Analysis

An intense and frequent form of motivational imbalance, which defines extremism, can be observed in a variety of domains and at different levels of analysis. This diversity around a common core is represented in the present volume. In this vein, Baumgartner, Naffziger, Nguyen, and Berridge (this volume) discuss how motivational imbalance can emerge at the neural level. Siev, Petty, and Briñol (this volume) discuss the role individual factors, such as extreme attitudes, in contributing to extreme behavior. Vallerand and Paquette (this volume) discuss the role passion plays in extreme behaviors at individual and interindividual level. Levine and Kruglanski (this volume) analyze extreme behaviors at the level of groups, and Prokopowicz, Choi, and Gelfand (this volume) discuss extreme phenomena at the level of nations and cultures. Even broader scope of analysis is offered by von Hippel and Fox (this volume) who, while analyzing the evolution of extremism, examine it across levels of phylogeny and across time (from appearance of the *Homo sapiens*, 200,000–300,000 years ago to the present).

Similar variety characterizes the domains of human pursuits in which extreme behaviors are studied. Several of those important domains are represented in this volume. Thus, Brymer and Bouchat (this volume) discuss extreme sports, Haynes (this volume) focuses on extreme self-interest or greed, and Fleeson, Miller, Furr, Knobel, and Jayawickreme (this volume) explain extreme humanitarianism. Molinario, Jasko, Webber, and Kruglanski (this volume) elaborate on the social psychology of violent extremism, and Weimann (this volume) explains how violent extremists are recruited via the social media. A brief overview of these chapters is given in what follows.

Part 1: Motivational Imbalance at Different Levels of Analysis

As with all behavioral phenomena, extremism can be approached at different levels of analysis. Accordingly, the present volume features such strikingly distinct approaches to extremism, ranging all the way from the neural to the cultural perspectives.

In Chapter 1, Baumgartner, Naffziger, Nguyen, and Berridge discuss the role incentive salience plays in irrational wanting (or 'miswanting', using the authors' term), which might contribute to motivational imbalance. They propose that extreme behaviors arise due to brain mechanisms amplifying incentive salience processes, thus leading to imbalances in reward utility subtypes (Kahneman, Wakker, & Sarin, 1997). Specifically, 'irrational miswanting' occurs when decision utility is high, whereas remembered utility, predicted utility, and experienced utility are low. This happens via enhancements in incentive salience, mediated neurobiologically by the mesolimbic dopamine system: increases in dopamine interact with reward cues to produce irrational elevations in decision utility. Such cues become motivational magnets and elicit intense and narrowly focused incentive motivation to pursue a given target, at the expense of alternative targets, and despite deleterious consequences. The authors explain how these neural mechanisms may be responsible for addiction, relapse, and 'wanting for what hurts you', and how they contribute to motivational imbalance in general.

In Chapter 2, Siev, Petty, and Briñol demonstrate how attitudes can contribute to the understanding of extreme behavior. Building on the motivational imbalance model of extremism (Kruglanski et al., 2021; this volume) and concepts of attitude strength (Petty & Krosnick, 1995), they offer a conceptualization of attitudinal extremism including the identification of antecedents of both 'authentic' and 'perceived' (attributed) extremism. In so doing, they go beyond the attitude literature's reliance on polarization as the sole defining feature of attitudinal extremism and expand on other attitude properties, such as attitude strength (certainty), unusualness (infrequency), and social disapproval associated with holding extreme beliefs. They describe the key determinants of extreme attitudes, paying particular attention to processes that can mediate the effects of motivation on their formation and also address the question under what conditions attitudes are more or less likely to result in extreme behavior. The authors describe how motivational imbalance is mirrored by *attitudinal imbalance* represented by discrepancies between properties of attitudes that are relevant versus irrelevant to the dominant need.

In Chapter 3, Vallerand and Paquette analyze the role that passion, a particularly intense form of motivation, plays in extreme behaviors. They present the Dualistic Model of Passion (Vallerand, 2015) and argue that obsessive passion presents motivational imbalance and leads to extreme behaviors whereas harmonious passion does not. The latter passion represents the case in which beyond the focal need that is served by the 'passionate' activity, alternative needs are not suppressed or neglected but rather are served sequentially in their proper turn. As Vallerand and Paquette demonstrate, passion plays a role in a variety of extreme behaviors at both the intrapersonal and interpersonal levels. On the intrapersonal level, obsessive (but not harmonious) passion foments addiction, burnout, and poorer physical and mental health. On the interpersonal level, obsessive (but not harmonious) passion leads to negative outcomes, such as dehumanization, immoral behaviors, aggression, stalking, and road rage. The authors also demonstrate how harmonious passion leads to overall well-being and obsessive passion to overall malaise, thus emphasizing the difference between balanced and imbalanced pursuit of one's passionate activities.

In Chapter 4, Levine and Kruglanski (this volume) analyze how extremism operates at the level of groups. They analyze the psychological and social processes that underlie group extremism, understood as willful collective behavior that substantially violates the norms of expected conduct in a given context (e.g., cults, religious orders, and single-issue political groups). They base their analysis on Significance Quest Theory (Kruglanski et al., 2014, in press; Kruglanski, Jasko, Webber, Chernikova, & Molinario, 2018; Kruglanski, Bélanger, & Gunaratna, 2019), which assumes that the dominant need underlying extremism is desire for personal significance. This need is assumed to be linked to action by an ideological *narrative* that justifies extreme means for achieving group goals; and a social *network* which provides validation for the narrative and dispenses rewards (in the form of respect) for engaging in narrative-consistent behavior. In particular, the authors highlight the crucial role that networks play in extreme groups by providing the *means* by which groups engage in narrative-specified behavior designed to satisfy their quest for significance.

In Chapter 5, Prokopowicz, Choi, and Gelfand (this volume) offer an analysis of extremism at the level of cultures. They analyze how nations handle innovation depending on where they lie on the tightness-looseness continuum referring to the extent to which a culture can be characterized by clear and strong norms (Gelfand et al., 2011). They show that loose cultures are better at exploration and tight cultures at exploitation and provide evidence for the Goldilocks principle of innovation—such that cultures that balance tightness and looseness are higher on innovation. Such ambidextrous cultures enable the right balance between freedom and structure and boost creativity while simultaneously driving implementation. That is, only balance between exploration and exploitation, rather than their extremes, promotes innovation and pushes nations and cultures forward.

Finally, in Chapter 6, von Hippel and Fox offer a broad perspective on how extremism evolves and what functions it serves. They discuss the role that genetic and environmental factors (such as evolved tribalism and intergroup conflict) play in the evolution of extremism. The authors also analyze evolutionary pressures toward extremism, such as virtue and commitment-signaling as well as leadership aspirations. And they discuss the forms of extremism that are unique (and non unique) to human beings and reflect on the role morality plays in extreme behaviors.

Part 2: Motivational Imbalance Across Domains of Human Endeavor

Extremism spans across domains of human endeavor. Depending on which need gains dominance over other needs and what behavioral means are identified as best fitting to serve the dominant need, different extreme activities would be pursued. In this part of the volume, we present important representatives of the 'extreme behaviors' category.

Thus, in Chapter 7, Brymer and Bouchat discuss the fascinating domain of extreme sports. Activities such as BASE jumping, proximity flying, big-wave surfing, or rope-free solo are prime examples of extreme behaviors in that they, per definition, involve risk—they are activities where a mismanaged mistake or accident is most likely to result in death (Brymer, 2005). And whereas the traditional accounts have mainly focused on this component of risk, the authors propose to go beyond the risk-focused definitions. Instead, they highlight the positive psychological experiences that participation in extreme sports brings, such as the opportunity to experience the extraordinary or transcend the everyday life. They argue that this contemporary view is more nuanced and imbedded in the lived experience of extreme sports participants. They also compare and contrast the relatively balanced and imbalanced participation in sports generally labeled as extreme, attesting to the continuum of imbalance and hence degrees of extremism (Kruglanski et al., 2021).

In Chapter 8, Haynes analyzes the psychology of greed. She makes the case for greed as a form of extremism. Specifically, she portrays greed as a state of motivational imbalance where the extreme desire for material wealth (or other physical or nonmaterial objects) is unconstrained by other needs. Greed is extreme self-interest, superseding the person's other needs and consideration for the needs of other individuals and groups. Intriguingly, the latter notion implies the phenomenon of interpersonal or societal imbalance. Greedy pursuits serve only the needs of the person and are misaligned with the needs of society. Therefore, greed creates not only a state of intrapersonal imbalance but also an imbalance in the social environment of the greedy individual. As an illustration, Haynes and colleagues analyze the greedy behavior of leaders (CEOs, top managers, and others in leadership positions) and show that there is a tipping point at which rational self-interested behavior in managers turns into the extreme pursuit of wealth. Then managerial actions stop benefiting the greater good and begin to harm it.

The opposite of extreme self-interest, namely extreme humanitarianism, is presented in Chapter 9, where Fleeson, Miller, Furr, Knobel, and Jayawickreme focus on positive moral exceptionality. They ask the question of why ordinary people do extraordinary things, such as donating organs to strangers, rescuing Jews during the Holocaust, whistleblowing on a powerful organization or crusading for social justice. Such unusual moral acts share main characteristics with other extreme behaviors: there is a dominant need to be morally good or virtuous; this need motivates actions that involve sacrifices and severe (sometimes mortal) risks. However, rather than using the term 'imbalance' (mainly due to its negative connotations), the authors prefer to view behaviors of the morally exceptional as balanced and conducive to full and meaningful life. In considering the morally exceptional behaviors, it is important to remember, however, that motivational imbalance is a matter of degree. Indeed, some of the morally exceptional behaviors discussed by the authors (e.g., hiding persecuted Jews in the countryside where the likelihood of being discovered is much reduced) occupy an intermediate point on the extremism continuum in which the risks and potential sacrifices are mitigated.

In Chapter 10, Molinario, Jasko, Webber, and Kruglanski concentrate on violent extremism and the appeal of violence. They review many reasons why people select violence as a way to attain their political goals—such as economic grievances, political discrimination, social marginalization, and affront to their social identity—and propose that they all hark back to one motivational force, the need for significance (Kruglanski, Chen, Dechesne, Fishman, & Orehek, 2009; Kruglanski et al., 2013, 2014; 2019, in press; Kruglanski, Jasko, Chernikova, Dugas, & Webber, 2017). When strong enough, this need can dominate other basic needs and if inserted in a certain ideological and social milieu, it can guide one toward violence. Specifically, when a given ideology, or *narrative*, justifies violence and bestows significance on people who act violently on behalf of a cherished cause, and when a given group, or *network*, offers respect and admiration for the extremist's devotion and sacrifice, people may embark on violent behavior in their quest for significance. The three factors—the need, narrative, and network—thus constitute the 3Ns framework that the authors use to explain violent extremism.

The topic of violent extremism is further addressed in Chapter 11, where Weimann analyzes Jihadi online recruitment process. He explains how, driven by the principles of motivational imbalance and narrowcasting (focusing persuasive communication on selected target audiences), terrorist organizations attract, seduce, and radicalize new members. Based on a rare database on online terrorism and testimonies of the recruits themselves, the author demonstrates that motivational imbalance, wherein one need—the quest for significance—is overemphasized to the point of overriding others—can lend key persuasive appeal to arguments advanced in terrorist online recruitment. Among other points, Weimann also explains how the new communication tools in the social media contribute to the unmistakable growth in the participation of women and youth in terrorist activity.

We believe that the panoply of content domains and analytic perspectives on extremism that the present volume offers make a strong case for the concept of extremism as a meaningful psychological phenomenon, whose dynamics are shared by behaviors that have been rarely thought of in common terms. And we hope that the fascinating phenomena that the present chapters describe will motivate the authors as well as the readers of this volume to continue and extend the empirical and conceptual work on the human capacity for extremism and its constructive and destructive potentials.

Arie W. Kruglanski, Ewa Szumowska, and Catalina Kopetz
University of Maryland, College Park
Institute of Psychology, Jagiellonian University
Wayne State University

References

Brymer, E. (2005). *Extreme dude: A phenomenological exploration into the extreme sport experience* (Doctoral dissertation). University of Wollongong, Wollongong. Retrieved from http://ro.uow.edu.au/theses/379

Burke, R. J. (1999). Workaholism and extra-work satisfactions. *The International Journal of Organizational Analysis*, 7, 352–364.

Deci, E. L., & Ryan, R. M. (2000). The 'what' and 'why' of goal pursuits: Human needs and the self-determination of behavior. *Psychological Inquiry*, 11, 227–268.

Fiske, S. T. (2010). *Social beings: Core motives in social psychology* (2nd ed.). Hoboken, NJ: Wiley Blackwell.

Gelfand, M. J., Raver, J. L., Nishii, L., Leslie, L. M., Lun, J., Lim, B. C., . . . Aycan, Z. (2011). Differences between tight and loose cultures: A 33-nation study. *Science*, 332(6033), 1100–1104.

Higgins, E. T. (2012). *Beyond pleasure and pain: How motivation works*. Oxford: Oxford University Press.

Kahneman, D., Wakker, P. P., & Sarin, R. (1997). Explorations of experienced utility. *The Quarterly Journal of Economics*, 112(2), 375–406.

Kopetz, C., Faber, T., Fishbach, A., & Kruglanski, A. W. (2011). The multifinality constraints effect: How goal multiplicity narrows the means set to a focal end. *Journal of Personality and Social Psychology*, 100, 810–826.

Kruglanski, A. W., Bélanger, J. J., Gelfand, M., Gunaratna, R., Hettiarachchi, M., Reinares, F., . . . Sharvit, K. (2013). Terrorism—A (self) love story: Redirecting the significance quest can end violence. *American Psychologist*, 68, 559–575.

Kruglanski, A. W., Bélanger, J. J., & Gunaratna, R. (2019). *The three pillars of radicalization: Needs, narratives, and networks.* Oxford: Oxford University Press.

Kruglanski, A. W., Chen, X., Dechesne, M., Fishman, S., & Orehek, E. (2009). Fully committed: Suicide bombers' motivation and the quest for personal significance. *Political Psychology, 30*, 331–557.

Kruglanski, A. W., Gelfand, M. J., Bélanger, J. J., Sheveland, A., Hetiarachchi, M., & Gunaratna, R. (2014). The psychology of radicalization and deradicalization: How significance quest impacts violent extremism. *Political Psychology, 35*, 69–93.

Kruglanski, A. W., Jasko, K., Chernikova, M., Dugas, M., & Webber, D. (2017). To the fringe and back: Violent extremism and the psychology of deviance. *American Psychologist, 72*(3), 217–230.

Kruglanski, A. W., Jasko, K., Webber, D., Chernikova, M., & Molinario, E. (2018). The making of violent extremists. *Review of General Psychology, 22*(1), 107–120.

Kruglanski, A. W., Molinario, E., Jasko, K., Webber, D., Leander, P., & Pierro, A. (in press). Significance quest theory. *Perspectives on Psychological Science.*

Kruglanski, A. W., Szumowska, E., & Kopetz, C. H. (2021). The call of the wild: How extremism happens. *Current Directions in Psychological Science, 30*(2), 181–185.

Kruglanski, A. W., Szumowska, E., Kopetz, C. H., Vallerand, R. J., & Pierro, A. (2021). On the psychology of extremism: How motivational imbalance breeds intemperance. *Psychological Review, 128*(2), 264–289.

Linville, P. W. (1985). Self-complexity and affective extremity: Don't put all of your eggs in one cognitive basket. *Social Cognition, 3*(1), 94–120.

Maslow, A. H. (1943). A theory of human motivation. *Psychological Review, 50*(4), 370–396.

Petty, R. E., & Krosnick, J. A. (Eds.). (1995). *Attitude strength: Antecedents and consequences.* Hillsdale, NJ: Erlbaum.

Philippe, F. L., Vallerand, R. J., Bernard-Desrosiers, L., Guilbault, V., & Rajotte, G. (2017). Understanding the cognitive and motivational underpinnings of sexual passion from a dualistic model. *Journal of Personality and Social Psychology, 113*(5), 769–785.

Shah, J. Y., Friedman, R., & Kruglanski, A. W. (2002). Forgetting all else: On the antecedents and consequences of goal shielding. *Journal of Personality and Social Psychology, 83*, 1261–1280.

Vallerand, R. J. (2015). *The psychology of passion: A dualistic model.* New York, NY: Oxford University Press.

Verner-Filion, J., Schellenberg, B. J. I, Rapaport, M., Belanger, J. J., & Vallerand, R. J. (2018). "The thrill of victory . . . and the agony of defeat": Passion and emotional reactions to success and failure among recreational golfers. *Journal of Sport Exercise Psychology, 40*(5), 280–283.

PART 1
Motivational Imbalance at Different Levels of Analysis

1
INCENTIVE SALIENCE IN IRRATIONAL MISWANTING AND EXTREME MOTIVATION

Hannah M. Baumgartner, Erin E. Naffziger, David Nguyen, and Kent C. Berridge

Introduction

Motivated behaviors, including extreme and infrequent ones, are influenced by both subjective cognitive decisions and more basic motivational processes, such as incentive salience, that can occur either consciously or unconsciously (Anselme & Robinson, 2016; Kahneman, Wakker, & Sarin, 1997; Kruglanski, Chernikova, Rosenzweig, & Kopetz, 2014; Kruglanski, Fernandez, Factor, & Szumowska, 2019; Higgins, 2012; Komissarouk, Chernikova, Kruglanski, & Higgins, 2018; Winkielman & Gogolushko, 2018; Berridge, 2018; Wilson et al., 2014). We describe a form of extreme behavior motivated by incentive salience that we call 'irrational miswanting', which occurs when the decision utility of an act is extremely high, while predicted, experienced, and remembered utilities of its outcome are all low or even negative. Throughout this chapter, we use the term 'irrational miswanting' as shorthand for this category of extreme behavior which can be demonstrated in rodent neuroscience experiments and may also underlie some human extreme behaviors.

For example, why would a rat choose to compulsively seek out and self-administer electric shocks that it has previously experienced repeatedly? Why do some individuals addicted to crack cocaine 'chase ghosts' or compulsively search for white specks on the ground to smoke, even if they know that the only specks available are merely small pebbles, or grains of sugar or salt? And why would addicts continue to pursue drugs if they were no longer in withdrawal, knew that their lives would be ruined by any further taking of drugs, and no longer even liked the drugs anymore? We propose that these extreme behaviors arise due to brain mechanisms amplifying incentive salience motivation processes, thus leading to these imbalances in reward utility subtypes. In this chapter, we will describe the reward utility types that become detached during 'irrational miswanting' and explain how that may happen psychologically through enhancements in incentive salience, mediated neurobiologically by mesolimbic brain systems. We describe examples of 'irrational miswanting' and suggest a theoretical framework for how it might conceivably extend even to extreme human aggression.

Reward Utilities and 'Wanting'

A useful way of thinking about motivation for rewards is offered by the different subtypes of reward utility proposed by Kahneman et al. (1997), namely, *predicted utility, experienced utility, decision utility*, and *remembered utility*. Predicted utility refers to the cognitive or associative *expectation* of value that a future outcome will have (i.e., how much it is predicted to be liked or disliked). Experienced utility is the affective experience of how much the outcome is *actually liked* when it is obtained (i.e., the pleasure experienced). Remembered utility is reconstructed *memory* of a previously experienced outcome's value (i.e., memory of how much it was liked in the past). Decision utility refers to the final *decision* or motivation to choose or pursue the outcome, visibly manifested in behavior (Kahneman et al., 1997).

In most cases, decision utility, predicted utility, remembered utility, and experienced utility all cohere together. For example, an individual may be motivated to pursue a chocolate dessert (high decision utility), informed by memories of previous positive experience (high remembered utility), which guides prediction that the dessert will be enjoyed in the future (high predicted utility), which is ultimately confirmed when they consume the treat that is then indeed enjoyed (high experienced utility). In typical cases where people pursue goals, they choose so that all these different types of reward utility are maximized, and so make rational choices.

However, exceptions occur when 'miswanting' an outcome (high decision utility) that turns out not to be liked (low experienced utility) (Gilbert & Wilson, 2000; Fredrickson & Kahneman, 1993). Most cases of 'miswanting' turns out to be based simply on wrong expectations (incorrect high predicted utility), due to either wrong memories about what was liked in the past or wrong theories about liking for future outcomes that were never before experienced (Gilbert & Wilson, 2000; Fredrickson & Kahneman, 1993). As an example of false remembered utility that drives wrong expectations, people can be induced to choose the most painful of two aversive options, based on their false memory that it gave less pain than the other (because it ended with a decrement that distorted their memory of how much pain had been endured); consequently, they pick the 'wrong' more painful option when offered the choice again (Kahneman, Fredrickson, Schreiber, & Redelmeier, 1993). Other times, wrong predictions are based on false theories or overthinking (Gilbert & Wilson, 2000; Wilson & Schooler, 1991). For example, a wrong theory might include believing that winning the lottery will make a person enduringly happy, then coming to find out that the experience of winning the lottery is followed by other unanticipated obstacles that make life more difficult (Gilbert & Wilson, 2000). In an example of overthinking, college students asked to pick the best of several brands of strawberry jam did best at matching experts' picks when they chose based on immediate 'gut reaction', but worse when they introspectively analyzed and explained in detail why they felt the way they did (Wilson & Schooler, 1991). In these cases of 'miswanting' based on false expectations, predicted utility, remembered utility, and decision utility all still cohere together and all are high. Only experienced utility stands apart as disappointingly low in these standard cases of 'miswanting'.

However, we suggest there are other situations where a more truly 'irrational decision' may exist, in which experienced utility coheres with remembered and predicted utilities, but decision utility detaches and stands alone. For example, this could mean wanting and pursuing a bad outcome, which is remembered to be bad, expected to be bad, and actually is bad when experienced—but is nonetheless positively wanted as an incentive. Irrational incentive motivation in 'wanting' a bad outcome on its own is different from rationally choosing a bad outcome simply to avoid a worse outcome, such as preferring an uncomfortable visit to the dentist today over the worse pain of a toothache tomorrow. It is also different from suppressing recognition of an outcome's bad consequences to justify its choice, such as succumbing to the temptation of eating high-calorie cake by momentarily suppressing recognition that it will lead to unwanted weight gain. In what we will call an 'irrational' case of 'miswanting' a bad outcome becomes 'wanted' on its own, not to avoid a worse alternative, yet its bad consequences are known from past experiences and are accurately predicted for future. In such 'irrational miswanting', remembered utility and predicted utility align with experienced utility and are low, so that predictions and memories are accurate for a disliked outcome. Yet decision utility remains high, and so becomes the outlier that is detached from all other types of utility. In these cases, one may 'want' (high decision utility) an outcome that is accurately remembered to be disliked (low remembered utility), accurately predicted to be disliked in future (low predicted utility), and which is actually disliked when obtained (low experienced utility).

What could cause decision utility to detach and soar above remembered, predicted, and experienced utilities for an outcome? A key to this dissociation in some cases, we propose, is mesolimbic incentive salience, a basic psychological process of motivation that can powerfully and specifically influence decision utility. We refer to incentive salience as '*wanting*' in quotation marks to distinguish this mesolimbically mediated motivational process, which can give urgency to conscious desires but also can sometimes occur unconsciously, from the unmodified term *wanting* which always means a conscious, cognitive desire mediated by more cortically weighted brain systems (Berridge & Aldrige, 2008; Anselme & Robinson, 2016; Berridge, 2018; Winkielman & Gogolushko, 2018).

Attribution of Incentive Salience

For most goals, a high level of 'wanting' usually tracks a high predicted utility for a cognitively wanted goal, and both contribute to decision utility. That is, usually one both wants and 'wants' what is expected to be 'liked'. However, in some special cases to be described here, high 'wanting' that leads to high decision utility can be directed toward an outcome that is not predicted to be 'liked' nor 'liked' when actually received. In such cases, one may irrationally come to 'want' an outcome that is accurately expected to be not 'liked', and cognitively might not be wanted. It is even possible to 'want what hurts you': a bad outcome that is remembered to be painful, predicted to be painful, and experienced as painful can still become positively 'wanted' with high decision utility in certain circumstances, as will be shown here.

Similar to 'wanting', we use 'liking' in quotations to describe the basic hedonic impact that is derived from consuming a pleasant reward. 'Liking' typically corresponds to conscious pleasure, as an input to brain circuitry underlying subjective hedonic experience, but can also in some circumstances occur as a basic hedonic reaction that is not necessarily consciously experienced (Winkielman & Gogolushko, 2018; Berridge, 2018; Berridge & Winkielman, 2003; Anselme & Robinson, 2016). 'Liking' is enhanced by a set of brain anatomical hedonic hotspots operating together as an integrated hedonic circuit. Hedonic hotspots are small, specialized subregions in orbitofrontal cortex, insular cortex, nucleus accumbens, and ventral pallidum with unique capacity to amplify hedonic impact in response to neurochemical opioid, endocannabinoid, and orexin signals. Brain mechanisms of 'liking' are separable from those of 'wanting', which is mediated by a larger mesocorticolimbic circuitry that additionally includes the mesolimbic dopamine system (Castro & Berridge, 2017; Berridge & Kringelbach, 2015).

Incentive salience operates by a signature set of psychological rules, involving Pavlovian learning that enables reward cues to trigger motivations, dynamic brain modulations that can amplify the motivational power of particular cues, and major individual differences in susceptibility to attribute extreme levels of incentive salience to particular cues (Bindra, 1978; Toates, 1986; Colaizzi et al., 2020; Berridge, 2018). Incentive salience is typically triggered by reward-associated Pavlovian cues (i.e., cues that were previously paired with actual rewards as 'wants' to consume the relevant reward) (Bindra, 1978; Toates, 1986). In addition, reward cues become attractive in their own right, so that a cue elicits approach, which is sometimes called sign-tracking (Flagel, Watson, Robinson, & Akil, 2007), and may even elicit attempts to 'consume' the cue similarly to its associated reward.

When the cue is encountered again, surges in mesolimbic dopamine transmission are an important step in imbuing the cue itself with incentive salience and momentarily enable the cue to take on some motivational properties of its associated 'wanted' reward (Robinson & Berridge, 2013; Berridge, 2012). At a given moment, the intensity of incentive salience triggered by a particular cue can be powerfully amplified by relevant physiological appetites, stress or emotional excitement states, by drug intoxication, or by direct brain manipulations of 'wanting' circuitry—including heightening mesolimbic dopamine reactivity to reward cues (Robinson & Berridge, 2013; Berridge, 2012). Thus, encountering the same reward cue can trigger varying degrees of incentive salience in different situations, in some cases leading to surprisingly high levels of 'wanting'—a phenomenon that can contribute to relapse in addiction.

The mesolimbic dopamine system is an important contributor to dynamic amplification of incentive salience. Increases in dopamine can interact with reward cues to produce irrational elevations in decision utility. For example, in early studies of cue-triggered 'wanting' in our laboratory, rats were first trained to work for sucrose pellets by pressing a lever (Wyvell & Berridge, 2000). In separate training sessions, the same rats learned that Pavlovian auditory cues predicted free sucrose rewards, without having to work for them (i.e., Pavlovian pairings of conditioned stimulus [CS, auditory cue] that predicted reward unconditioned stimulus [sucrose]). Instrumental lever pressing behavior was then tested in what is called a Pavlovian to instrumental transfer (PIT) test, during an extinction session wherein pressing the lever no longer led to sucrose reward. Although they decrease their responding on the lever the rats learned that engaging with the lever no longer led to reward (lowered experienced utility); therefore, in future encounters with the lever, the rats should gradually come to expect that engaging with the lever will not produce a rewarding outcome (lowered predicted utility). Although gradually decreasing their levels of motivated pressing for sucrose that no longer came during the PIT extinction session (lowered predicted utility), presentations of the Pavlovian auditory cue elicited sudden spurts of higher 'wanting' and lever presses that were several times higher than their current extinguished level of responding, each spurt lasting about a minute. Some rats were then given a painless brain microinjection of a tiny droplet of amphetamine into the shell region of the nucleus accumbens, which increased levels of dopamine release for the next half hour (Daberkow et al., 2013), or else microinjection of drug-free saline as a control comparison. Rats with amphetamine-induced elevated dopamine dramatically amplified each minute-long cue-triggered peak of 'wanting' manifest in lever pressing whenever the tone cue was presented but did not raise their plateaus of lesser effort in between reward cues. That is, the amphetamine in nucleus accumbens selectively amplified cue-triggered 'wanting' as short spurts of higher decision utility right after Pavlovian CS presentations, without creating any stable expectation or predicted utility of higher reward, that would last after the temporary incentive salience spurt faded. It is important to note that the Pavlovian CS cue had never been presented before while rats ever pressed instrumentally to earn sucrose, so the rats had no reason to expect that the Pavlovian cue would make their lever pressing more successful—and it didn't. That consideration also supports the conclusion that the Pavlovian cue did not raise the predicted utility of the act of pressing during the PIT test. Also, the amphetamine failed to enhance the rats' 'liking' reactions to the sweetness of sucrose when they received it in a separate test of affective taste reactions, indicating that dopamine did not increase the experienced utility of its hedonic impact. These findings suggested that elevated levels of dopamine release in the nucleus accumbens shell enhanced cue-triggered 'wanting' but not 'liking', selectively magnifying cue-triggered spurts of decision utility as incentive salience that could detach from both experienced utility and predicted utility of the same reward. Further studies from our laboratory using this and related tasks have since demonstrated robust mesolimbic amplification of 'wanting' using various forms of dopamine stimulations, as well as brain stimulations of opioid and corticotropin-releasing factor (CRF) neurochemical and neural components of the nucleus accumbens and amygdala (Baumgartner, Cole, Olney, & Berridge, 2020; Castro & Berridge, 2014; Cole, Robinson, & Berridge, 2018; DiFeliceantonio & Berridge, 2012; Mahler & Berridge, 2012; Peciña & Berridge, 2013; Smith & Berridge, 2007).

More long-lasting amplifications of incentive salience may occur in particularly vulnerable individuals through the phenomenon of incentive sensitization, which can be produced by drugs of abuse to cause drug addiction (Robinson & Berridge, 1993; Berridge & Robinson, 2016). Psychological incentive sensitization is due to neural dopamine sensitization, which refers to near-permanent hyper-reactivity to drug cues in mesolimbic dopamine systems, induced in vulnerable individuals by binge-like exposures to addictive drugs (individual vulnerability is determined largely by genetic factors as well as by a few experiential factors) (Horger, Shelton, & Schenk, 1990; Kalivas & Stewart, 1991; Mendrek, Blaha, & Phillips, 1998; Robinson et al., 2015; Robinson, Fischer, Ahuja, Lesser, & Maniates, 2016; Vezina, Lorrain, Arnold, Austin, & Suto, 2002).

For example, repeated exposures to addictive drugs are known to induce a state of enduring sensitization in some individuals that can last years and are characterized by increased responsiveness of dopamine neural circuitry, such that subsequent drug may persistently trigger even higher levels of dopamine release, and postsynaptic neurons become more responsive to dopamine and corticolimbic glutamate release that also carries signals about reward cues to neurons that receive dopamine (Boileau et al., 2007; Cox et al., 2009; Paulson & Robinson, 1995; Robinson & Kolb, 2004; Steketee & Kalivas, 2011; Singer et al., 2009; Wolf, 2016). Therefore, in the case of drug addiction, neural sensitization of dopamine-related brain systems in these susceptible individuals is posited to cause addiction and relapse by causing the drug and its associated cues to take on exaggerated incentive salience (Robinson & Berridge, 1993). As such, when a recovering drug addict encounters a drug-associated cue, they may experience high levels of 'wanting' impulses that compel them to seek out the drug. Sensitized cue-triggered 'wanting' can cause relapse (decision utility) in addicts who have long been drug-abstinent, even despite their cognitive desire to avoid the drug due to knowledge of the negative consequences that accompany drug use (predicted utility). Incentive sensitization involving similar changes in brain dopamine systems may also possibly occur spontaneously, even without drugs, in some very vulnerable individuals to produce behavioral food addiction, sex addiction, gambling addiction, and so on (Olney, Warlow, Naffziger, & Berridge, 2018; Devoto et al., 2018; Stice & Yokum, 2016; Davis & Carter, 2009; Gearhardt et al., 2011; Hartston, 2012; Imperato, Angelucci, Casolini, Zocchi, & Puglisi-Allegra, 1992; Linnet et al., 2012; O'Sullivan et al., 2011; Pfaus et al., 1990; Ray et al., 2012; Voon et al., 2014; Zeeb, Li, Fisher, Zack, & Fletcher, 2017).

'Irrational Miswanting' and 'Dangerous Desire'

Examples of 'irrational miswanting' due to incentive salience can explain certain extreme motivated behaviors that happen when decision utility is high, but predicted and experienced utility are low. For example, incentive salience may explain why someone may relapse back into drug taking repeatedly, even if they truly want to give up drugs, even when they are no longer suffering withdrawal, and perhaps even reporting that they don't like the drugs anymore. While relapse in addiction can occur for a number of reasons, in incentive sensitization an excessive 'want' is triggered by previously paired drug-related cues, and the amplitude of the incentive salience peak is heightened especially in moments of stress or emotion. That may be because stressors cause CRF, the brain's master stress signal, to be released in several brain structures including the nucleus accumbens and amygdala. CRF microinjections in nucleus accumbens of rats can enhance cue-triggered incentive salience as effectively as dopamine-releasing amphetamine microinjections, and CRF neural activation in amygdala can also increase incentive motivation to pursue rewards such as sucrose (Peciña, Schulkin, & Berridge, 2006; Lemos et al., 2012; Baumgartner, Schulkin, & Berridge, 2021). CRF in nucleus accumbens and related structures also amplifies the responsiveness of mesolimbic dopamine systems in moments of stress.

Similar sensitization of incentive salience in addicts, creating mesolimbic hyper-reactivity to drug cues, could therefore drive higher decision utility to seek out the drug once again (Robinson & Berridge, 1993). For sensitized individuals, drug-related cues become powerful *motivational magnets* able to cause intense 'wanting' to take drugs, even in seemingly irrational ways that may be related to 'irrational miswanting'. For example, heightened dopamine response to drug cues, making the cues become 'wanted' themselves, could explain why some crack cocaine addicts have reported to irrationally 'chase ghosts': that is, to compulsively and intensely search for small white specks on the ground to smoke even when the individual cognitively knows the specks are merely small pebbles or sugar granules, and not cocaine (Rosse et al., 1993; Rosse, Fay-McCarthy, Collins, Alim, & Deutsch, 1994). Furthermore, incentive sensitization may also explain why Parkinson's patients who have been treated with medications that are dopamine direct agonists, which mimic dopamine at particular D2-type receptors to give an abnormal 'fake dopamine' stimulation, sometimes develop behavioral addictions induced by the medications, such as compulsive gambling, compulsive sex, compulsive shopping, and so on (Warren, O'Gorman, Lehn, & Siskind, 2017; Bostwick, Hecksel, Stevens, Bower, & Ahlskog, 2009; Vela et al., 2016; Voon et al., 2017). These compulsive Parkinson's patients have been reported to have a sensitized dopamine system, in the sense that they actually release more dopamine in nucleus accumbens when given a pharmacological L-dopa dose than nonaddicted Parkinson's patients given the same dose, due to their medication-induced dopamine hyper-reactivity combined with their individual vulnerability (O'Sullivan et al., 2011). Individual differences in susceptibility to incentive sensitization may explain why some Parkinson's patients develop such medication-induced behavioral addictions while other patients do not.

Recent animal neuroscience studies have explored related individual differences in incentive salience features, such as the tendency to assign high levels of incentive salience to reward cues in ways that make the cue itself become 'irrationally miswanted', by examining the neural basis of individual differences between 'sign-tracking' tendencies versus 'goal-tracking' tendencies (Flagel et al., 2007; Robinson & Flagel, 2009; Pitchers, Phillips, Jones, Robinson, & Sarter, 2017; Phillips & Sarter, 2020; Roughley & Killcross, 2019). This is done using a Pavlovian autoshaping paradigm, where sudden protrusion of a lever through a wall into the chamber serves as a Pavlovian cue for a sucrose pellet reward. Some individual rats are 'sign-trackers': attracted by the Pavlovian lever cue, eagerly approaching, nibbling, and biting the reward cue lever, even though doing so earns nothing more, and sometimes even at the expense of eating the reward. Other rats are 'goal-trackers': drawn toward the goal dish that delivers the actual sucrose reward as soon as the lever cue appears, taking the cue as a predictor of sucrose but not as an incentive target of desire (Flagel et al., 2007; Robinson & Flagel, 2009). Sign-trackers attribute higher incentive salience to the reward cue, so that the lever becomes attractive and 'wanted' itself and postmortem analysis of sign-tracker brain tissue reveals increased expression of dopamine receptors in mesolimbic regions distinct from goal-trackers along with enhanced patterns of c-fos in mesocorticolimbic regions after cue presentations (Flagel et al., 2007, 2011). This sign-tracking phenomenon also underlies the long-known history of 'misbehaving animals', such as pigs initially eager to retrieve coins to deposit in a wooden bank for a food reward when trained to perform the task for television commercials, soon beginning to 'dilly-dally', repetitively picking up and dropping coins as though rooting in the dirt, even though doing so substantially delayed their reward (Breland & Breland, 1961; Bolles, 1972). In such cases, the Pavlovian cue or CS, such as the pigs' coin, took on incentive qualities of the reward itself and became so attractively tempting as a food-like Pavlovian incentive that disrupted the ability of the animal to perform the instrumental task, preventing earning of actual rewards (Bindra, 1978; Toates, 1986). Related individual propensities to sign-track reward cues also have been reported in human children performing a Lego operant task, as well as in human adults playing a monetary laboratory task, and may share similar underlying neural activity in mesolimbic 'wanting' circuits (Joyner, Gearhardt, & Flagel, 2018; Garofalo & di Pellegrino, 2015; Schad et al., 2020).

Perhaps the strongest example of 'irrational miswanting', namely 'wanting' something that is correctly remembered and predicted to be painful, was recently demonstrated in our laboratory using optogenetic brain stimulation techniques in rats to make them seek out and touch an object that delivers electric shocks (Warlow, Naffziger, & Berridge, 2020). People will sometimes voluntarily give themselves an unpleasant electric shock or two when left alone with nothing else to do, apparently out of boredom, even if they've previously said they would rather avoid the shock (Wilson et al., 2014). But it seems very different to suggest that one could be powerfully motivated to compulsively seek out repeated electric shocks in the same way one would be powerfully motivated to consume sugar or cocaine.

Yet our studies in the laboratory have shown it is possible to induce 'wanting for what hurts you' (Warlow et al., 2020). By giving painless laser-induced stimulations of the amygdala that recruits mesolimbic incentive circuitry, appropriately timed to activate and assign incentive salience to a paired painful target, 'irrational miswanting' can be formed. In this example, high decision utility was given to that target, detaching it from the low predicted utility, remembered utility, and experienced utility of the painful outcome. In short, the rat can be made to 'irrationally want' or 'want what hurts it', causing it to voluntarily subject itself to a painful outcome repeatedly. Using the same brain stimulation technique paired instead with earning sugar, rather than with the painful target, another rat can be made into a 'sugar addict' that intensely pursues and consumes only sugar while ignoring freely available intravenous cocaine. A third rat can be made into a 'cocaine addict' that pursues cocaine and ignores sugar—all at the whim of the experimenter who controls the brain's assignment of incentive salience by pairing the brain stimulation with these respective targets.

The majority of humans, rats, and animals will avoid something they've learned will deliver pain. To convert that into high decision utility leading to pursuit, our laboratory studies used an electrified small 'shock rod' as the painful target (Warlow et al., 2020). The rod protruded a few inches into the chamber, which gave an electric shock to the paws or snout whenever a rat *voluntarily* touched it. We paired stimulations of the subcortical 'wanting' system with voluntary encounters in which a rat approached and usually touched the shock rod, resulting in irrational decision utility for the shock rod that led the rat to subject itself to pain. To stimulate the amygdala and therefore recruit the subcortical 'wanting' system, we used optogenetic techniques, in which a laser light is shone painlessly into the brain via implanted optic fiber, to excite particular neurons to fire. The amygdala neurons respond to light, due to photoreceptor molecules that have sprouted on their outer membrane because a droplet of virus solution, containing a gene for the photoreceptor molecule, was previously microinjected into the amygdala when the rats were surgically anesthetized a few weeks earlier.

Several seconds of laser activation of the amygdala were paired repeatedly with encounters with the shock-delivering rod, each time the rats approached and touched it. After a few pairings, rats began to compulsively approach and touch the rod, spending most of their time hovering closely over the rod and occasionally even nibbling and chewing the rod, subjecting themselves to multiple unnecessary shocks to the mouth and paws (Warlow et al., 2020). This was not simply due to diminished experienced utility impact of shock, such as feeling less pain. The laser did not seem to reliably reduce pain, instead in other situations it seemed to increase pain and fear. For example, when the same amygdala laser activation was paired with an auditory tone predicting an uncontrollable and unlocalizable footshock in a Pavlovian fear learning situation, the amygdala activation intensified the fearful *freezing* conditioned response elicited by the tone, as though the shock had become more painful. Similarly, with the 'irrationally miswanted' shock rod, rats still reliably emitted reflexive brief withdrawals each time they touched the shock rod, suggesting that its shock was still perceived as painful. In other words, aspects of the situation controlled whether the amygdala activation increased incentive desire versus fear. In the Pavlovian fear learning situation, shock was unavoidable and inescapable, and was delivered throughout the entire floor of the chamber. By contrast, in the shock rod situation, rats could control whether they received a shock, which was localized to the shock rod object, whereas the rest of the chamber constituted a safe space.

Furthermore, it became clear that rats had an accurate remembered utility of the rod as painful, and accurate predicted utility that touching the rod would be painful in the future. In fact, the painful memory and prediction that the rod would give a shock was essential to the rat's maladaptive attraction to it. For example, the rats never became attracted to a different 'dummy rod' that gave no shocks, when encounters with the dummy rod were paired with similar amygdala activations. The pain of the shock rod when voluntarily touched, combined with the brain activation, was needed together to produce the maladaptive attraction. The necessity of that combination was further evident when we stopped delivering the laser that activated the amygdala. Within minutes, the rats flipped from attraction to fear of the rod, avoiding it and defensively kicking bits of cage litter toward it just as ordinary rats did who had never had amygdala activation.

Further psychological analysis indicated that high incentive salience of cues mediated the maladaptive attraction (Warlow et al., 2020). The rod object itself became an irresistibly attractive cue due to amygdala pairings, and other cues associated with its shock also were made similarly and abnormally attractive. For example, another cue was a distinct auditory sound (either a pure-frequency tone or a mixed-frequency white noise; counterbalanced), which had been heard every time the rats approached their rod and received amygdala stimulation and electric shock. This sound became a Pavlovian CS+ for the electric shock associated with shock rod encounters, a sound which normal rats would never wish to hear. However, the amygdala-stimulated rats attracted to the rod did want to hear their shock-predicting CS+ sound: they learned to poke their noses into a hole in order to hear brief presentations of the CS+ sound and worked hard to hear their pain-paired sound many times in a session (i.e., instrumental conditioned reinforcement). This illustrates that the pain-related sound cue had become imbued with abnormally high incentive value.

Subsequent brain analyses of neural activation traces revealed that the laser-stimulation recruited neural activation in mesolimbic 'wanting' systems at the moment when rats were attracted to their shock rod, providing an explanation that incentive salience underlay their maladaptive attraction and 'irrational miswanting' (i.e., elevated Fos expression in recently activated neurons) (Warlow et al., 2020). This laboratory result provides evidence that extreme 'irrational miswanting', even for a painfully unpleasant outcome, can be generated when decision utility becomes hijacked by mesolimbic brain systems, even when the remembered utility and predicted utility are accurately known to be a negative outcome that delivers only pain as experienced utility.

Similar compulsive 'wanting' can be turned to other objects, depending on what the experimenter pairs with amygdala stimulation. For example, pairing laser stimulation with a sugar reward, when rats were given the choice between earning sugar or intravenous cocaine rewards, caused rats to become 'sugar addicts', intensely pursuing and consuming only sugar, while ignoring the cocaine. Alternatively, when other rats given the same choice had their amygdala activation paired with earning cocaine and not sugar, those rats became 'cocaine addicts' which intensely pursued and consumed only cocaine, while ignoring sugar. Those cocaine-laser rats also began to bite and nibble the inedible metal cues associated with the cocaine, such as the instrumental nose-poke porthole or lever they had to manipulate to earn a cocaine infusion, which had been paired with amygdala activation (Warlow, Robinson, & Berridge, 2017).

We suggest that these extreme laboratory examples of 'irrational miswanting' arise because of the attribution of incentive salience to cues for the stimulation-paired target, even the shock rod, thus selectively driving up its decision utility. Incentive salience attributed to cues causes them to become motivational magnets, thereby eliciting intense and narrowly focused incentive motivation to pursue that target, at the expense of other alternative targets and even despite deleterious consequences. Our laboratory results suggest that extreme incentive salience, mediated by brain mesocorticolimbic systems, can produce such distortions of decision utility without needing to change predicted or experienced utility of an adverse outcome.

Conclusion and Implications in Extreme Aggression

Does incentive salience and the induction of extreme 'irrational wanting' have implications for understanding extreme motives in humans, such as motivated acts of aggression? That is a question we must leave for social psychologists to answer conclusively, as we can only discuss it as a theoretical possibility. Still, the maladaptive 'wanting' described previously does bear similarities to human extreme behavior, which has been defined psychologically as 'a motivational imbalance in which one need rises in saliency and magnitude to the point of dominating and "crowding out" other basic needs' (Kruglanski et al., 2019). What we have done here is to show how extreme incentive salience or 'irrational miswanting' can create a narrowly focused addictive-type motivation, for example, causing a rat to even 'want what hurts' it. In humans, we suggest that incentive sensitization in addicts causes similar excessive incentive salience that can produce recurrent relapse, as in the case of an addict who wants to quit but cannot maintain abstinence despite knowing the negative consequences. These cases can arise through a selective distortion of raising decision utility, even if predicted utility, experienced utility, and remembered utility all remain low. Although many extreme behaviors may be entirely 'rational' in the sense that the outcome carries high predicted utility for the individual (Kruglanski et al., 2019), it may still be that high predicted utility is not always necessary for extreme behavior to occur, if mediated by high decision utility via mesocorticolimbic incentive salience mechanisms described here. If so, an individual may 'want' to perform an extreme behavior that lacks high predicted utility, even one with aversive consequences, leading to 'irrational miswanting'. Such seemingly irrational choices are caused by mesolimbic dopamine systems involving the nucleus accumbens, recruited by amygdala-related systems, as parts of a larger mesocorticolimbic circuitry that assigns incentive salience to particular targets.

In principle, it is conceivable that extreme 'wanting' could contribute in some cases to extreme behaviors such as intense violence and aggression. There is evidence from animal studies that acts of aggression can become appetitive or guided by positively valenced incentive motivation processes (Golden, Aleyasin et al., 2017; Golden, Heins et al., 2017). Aggressive acts may even take on addictive-like features for some individuals, such as being prone to relapse for aggression-seeking even when facing aversive consequences (Golden, Heins et al., 2017; Golden, Jin, & Shaham, 2019; Fish, De Bold, & Miczek, 2002). Furthermore, the mesolimbic dopamine system is integral to appetitive aggression in animals (Golden, Jin, Heins et al., 2019), and increases in nucleus accumbens volume and connectivity have been related to chronic aggression (Schiffer et al., 2011). Therefore, aggressive behaviors may be driven at least in part through this mesolimbic 'wanting' system.

In certain cases of human repeat-violent offenders, previously neutral cues may become more salient and lead to continued aggressive behavior. For example, males identified as highly aggressive and chronically violent men show increased amygdala reactivity to angry and even neutral facial cues (Coccaro, McCloskey, Fitzgerald, & Phan, 2007; Pawliczek et al., 2013; Pardini & Phillips, 2010) and this is linked to higher impulsivity (Pawliczek et al., 2013). Furthermore, provoking violent offenders cause hyper-reactivity in both the nucleus accumbens and amygdala, which coincides with increased aggressive behavior (da Cunha-Bang et al., 2017). Therefore, it is possible that especially salient cues may elicit and magnify aggressive behaviors, at least in some individuals (da Cunha-Bang, Fisher, Hjordt, Holst, & Knudsen, 2019). If incentive salience as mesolimbic 'wanting' contributed to the attractiveness of extreme behaviors in such cases, it is conceivable that such cue reactivity may engage violent acts via high decision utility, even in situations where the individual knows the acts will lead to undesirable outcomes that are contrary to the individual's cognitive goals (low predicted utility). This could cause extreme behavior through a narrowing of focus on one motivational goal over other potential deterrents via incentive salience mechanisms (Kruglanski et al., 2019).

It is of course difficult to know whether individuals actually are ever motivated to 'want' to continue violent behaviors, without feeling pleasure or 'liking' for the outcome nor expecting it to achieve a goal. It is not our purpose to assert so here. We simply raise the theoretical possibility and offer descriptions of how extreme 'wanting' works, together with explanation of how it may be generated by underlying mesocorticolimbic neural circuits, which might contribute to a motivational imbalance leading to extreme behaviors (Kruglanski, Jasko, Chernikova, Dugas, & Webber, 2017; Kruglanski et al., 2019). If so, understanding how extreme 'wanting' can become 'irrational miswanting' may be helpful to providing better treatments and interventions to manage pathologically extreme behaviors.

References

Anselme, P., & Robinson, M. J. (2016). "Wanting," "liking," and their relation to consciousness. *Journal of Experimental Psychology: Animal Learning and Cognition, 42*(2), 123–140.

Baumgartner, H. M., Cole, S. C., Olney, J. J., & Berridge, K. C. (2020). Desire or dread from nucleus accumbens inhibitions: Reversed by same-site optogenetic excitations. *Journal of Neuroscience, 40*(13), 2732–2752.

Baumgartner, H. M., Schulkin, J., & Berridge, K. C. (2021). Activating corticotropin-releasing factor systems in the nucleus accumbens, amygdala, and bed nucleus of stria terminalis: Incentive motivation or aversive motivation? *Biological Psychiatry, 89*(12), 1162–1175.

Berridge, K. C. (2012). From prediction error to incentive salience: Mesolimbic computation of reward motivation. *European Journal of Neuroscience, 35*(7), 1124–1143.

Berridge, K. C. (2018). Evolving concepts of emotion and motivation. *Frontiers in Psychology, 9*(1647), 1–20.

Berridge, K. C., & Aldrige, J. W. (2008). Decision utility, the brain, and the pursuit of hedonic goals. *Social Cognition, 26*(5), 621–646.

Berridge, K. C., & Kringelbach, M. L. (2015). Pleasure systems in the brain. *Neuron, 86*(3), 646–664.

Berridge, K. C., & Robinson, T. E. (2016). Liking, wanting, and the incentive-sensitization theory of addiction. *American Psychologist, 71*(8), 670–679.

Berridge, K. C., & Winkielman, P. (2003). What is an unconscious emotion? (The case for unconscious "liking"). *Cognition & Emotion, 17*(2), 181–211.

Bindra, D. (1978). How adaptive behavior is produced: A perceptual-motivational alternative to response-reinforcement. *Behavioral and Brain Sciences, 1*(1), 4191. https://doi.org/10.1017/S0140525X00059380

Boileau, I., Dagher, A., Leyton, M., Welfeld, K., Booij, L., Diksic, M., & Benkelfat, C. (2007). Conditioned dopamine release in humans: A positron emission tomography [11C]raclopride study with amphetamine. *Journal of Neuroscience, 27*(15), 3998–4003. https://doi.org/10.1523/JNEUROSCI.4370-06.2007

Bolles, R. C. (1972). Reinforcement, expectancy, and learning. *Psychological Review, 79*(5), 394–409.

Bostwick, J. M., Hecksel, K. A., Stevens, S. R., Bower, J. H., & Ahlskog, J. E. (2009). Frequency of new-onset pathologic compulsive gambling and hypersexuality after drug treatment of idiopathic Parkinson disease. *Mayo Clinic Proceedings, 84*(4), 310–316.

Breland, K., & Breland, M. (1961). The misbehavior of organisms. *American Psychologist, 16*(11), 681–684.

Castro, D., & Berridge, K. C. (2014). Opioid hedonic hotspot in nucleus accumbens shell: Mu, delta, and kappa maps for enhancement of sweetness "liking" and "wanting". *Journal of Neuroscience, 34*(12), 4239–4250.

Castro, D., & Berridge, K. C. (2017). Opioid and orexin hedonic hotspots in rat orbitofrontal cortex and insula. *Proceedings of the National Academy of Sciences, 114*(43), E9124–E9134.

Coccaro, E. F., McCloskey, M. S., Fitzgerald, D. A., & Phan, K. L. (2007). Amygdala and orbitofrontal reactivity to social threat in individuals with impulsive aggression. *Biological Psychiatry, 62*(2), 168–178. https://doi.org/10.1016/j.biopsych.2006.08.024

Colaizzi, J. M., Flagel, S. B., Joyner, M. A., Gearhardt, A. N., Stewart, J. L., & Paulus, M. P. (2020). Mapping sign-tracking and goal-tracking onto human behaviors. *Neuroscience & Biobehavioral Reviews, 111*, 84–94.

Cole, S. L., Robinson, M. F., & Berridge, K. C. (2018). Optogenetic self-stimulation in the nucleus accumbens: D1 reward versus D2 ambivalence. *PLoS One, 13*(11), e0207694.

Cox, S. M., Benkelfat, C., Dagher, A., Delaney, J. S., Durand, F., McKenzie, S. A., . . . Leyton, M. (2009). Striatal dopamine responses to intranasal cocaine self-administration in humans. *Biological Psychiatry, 65*(10), 846–850. https://doi.org/10.1016/j.biopsych.2009.01.021

da Cunha-Bang, S., Fisher, P. M., Hjordt, L. V., Holst, K., & Knudsen, G. M. (2019). Amygdala reactivity to fearful faces correlates positively with impulsive aggression. *Social Neuroscience, 14*(2), 162–172. https://doi.org/10.1080/17470919.2017.1421262

da Cunha-Bang, S., Fisher, P. M., Hjordt, L. V., Perfalk, E., Persson Skibsted, A., Bock, C., . . . Knudsen, G. M. (2017). Violent offenders respond to provocations with high amygdala and striatal reactivity. *Social Cognitive and Affective Neuroscience, 12*(5), 802–810. https://doi.org/10.1093/scan/nsx006

Daberkow, D. P., Brown, H. D., Bunner, K. D., Kraniotis, S. A., Doellman, M. A., Ragozzino, M. E., . . . Roitman, M. F. (2013). Amphetamine paradoxically augments exocytotic dopamine release and phasic dopamine signals. *The Journal of Neuroscience: The Official Journal of the Society for Neuroscience, 33*(2), 452–463. https://doi.org/10.1523/JNEUROSCI.2136-12.2013

Davis, C., & Carter, J. C. (2009). Compulsive overeating as an addiction disorder. A review of theory and evidence. *Appetite, 53*(1), 1–8.

Devoto, F., Zapparoli, L., Bonandrini, R., Berlinger, M., Ferrulli, A., Luzi, L., . . . Paulesu, E. (2018). Hungry brains: A meta-analytical review of brain activation imaging studies on food perception and appetite in obese individuals. *Neuroscience & Biobehavioral Reviews*, *94*, 271–285.

DiFeliceantonio, A. G., & Berridge, K. C. (2012). Which cue to 'want'? Opioid stimulation of central amygdala makes goal-trackers show stronger goal-tracking, just as sign-trackers show stronger sign-tracking. *Behavioral Brain Research*, *230*(2), 399–408.

Fish, E. W., De Bold, J. F., & Miczek, K. A. (2002). Aggressive behavior as a reinforcer in mice: Activation by allopregnanolone. *Psychopharmacology*, *163*(3–4), 459–466. https://doi.org/10.1007/s00213-002-1211-2

Flagel, S. B., Cameron, C. M., Pickup, K. N., Watson, S. J., Akil, H., & Robinson, T. E. (2011). A food predictive cue must be attributed with incentive salience for it to induce c-fos mRNA expression in cortico-striatal-thalamic brain regions. *Neuroscience*, *196*, 80–96. https://doi.org/10.1016/j.neuroscience.2011.09.004

Flagel, S. B., Watson, S. J., Robinson, T. E., & Akil, H. (2007). Individual differences in the propensity to approach signals vs goals promote different adaptations in the dopamine system of rats. *Psychopharmacology*, *191*(3), 599–607. https://doi.org/10.1007/s00213-006-0535-8

Fredrickson, B. L., & Kahneman, D. (1993). Duration neglect in retrospective evaluations of affective episodes. *Journal of Personality and Social Psychology*, *65*(1), 45–55. https://doi.org/10.1037/0022-3514.65.1.45

Garofalo, S., & di Pellegrino, G. (2015). Individual differences in the influence of task-irrelevant Pavlovian cues on human behavior. *Frontiers in Behavioral Neuroscience*, *24*, 163.

Gearhardt, A. N., Yokum, S., Orr, P. T., Stice, E., Corbin, W. R., & Brownell, K. D. (2011). Neural correlates of food addiction. *Arch Gen Psychiatry*, *68*(8), 808–816.

Gilbert, D. T., & Wilson, T. D. (2000). Miswanting: Some problems in the forecasting of future affective states. In J. P. Forgas (Ed.), *Thinking and feeling: The role of affect in social cognition* (pp. 178–197). Cambridge: Cambridge University Press.

Golden, S. A., Aleyasin, H., Heins, R., Flanigan, M., Heshmati, M., Takahashi, A., . . . Shaham, Y. (2017). Persistent conditioned place preference to aggression experience in adult male sexually-experienced CD-1 mice. *Genes Brain Behavior*, *16*, 44–55.

Golden, S. A., Heins, C., Venniro, M., Caprioli, D., Zhang, M., Epstein, D. H., & Shaham, Y. (2017). Compulsive addiction-like aggressive behavior in mice. *Biological Psychiatry*, *82*(4), 239–248. https://doi.org/10.1016/j.biopsych.2017.03.004

Golden, S. A., Jin, M., Heins, C., Venniro, M., Michaelides, M., & Shaham, Y. (2019). Nucleus Accumbens Drd1-expressing neurons control aggression self-administration and aggression seeking in mice. *The Journal of Neuroscience: The Official Journal of the Society for Neuroscience*, *39*(13), 2482–2496. https://doi.org/10.1523/JNEUROSCI.2409-18.2019

Golden, S. A., Jin, M., & Shaham, Y. (2019). Animal models of (or for) aggression reward, addiction, and relapse: Behavior and circuits. *The Journal of Neuroscience*, *39*(21), 3996–4008. https://doi.org/10.1523/JNEUROSCI.0151-19.2019

Hartston, H. (2012). The case for compulsive shopping as an addiction. *Journal of Psychoactive Drugs*, *44*(1), 64–67.

Higgins, E. T. (2012). *Beyond pleasure and pain—how motivation works*. Oxford: Oxford University Press.

Horger, B. A., Shelton, K., & Schenk, S. (1990). Preexposure sensitizes rats to the rewarding effects of cocaine. *Pharmacology, Biochemistry and Behavior*, *37*(4), 707–711. https://doi.org/10.1016/0091-3057(90)90552-S

Imperato, A., Angelucci, L., Casolini, P., Zocchi, A., & Puglisi-Allegra, S. (1992). Repeated stressful experiences differently affect limbic dopamine release during and following stress. *Brain Research, 577*(2), 194–199.

Joyner, M. A., Gearhardt, A. N., & Flagel, S. B. (2018). A translational model to assess sign-tracking and goal-tracking behavior in children. *Neuropsychopharmacology: Official Publication of the American College of Neuropsychopharmacology, 43*(1), 228–229. https://doi.org/10.1038/npp.2017.196

Kahneman, D., Fredrickson, B. L., Schreiber, C. A., & Redelmeier, D. A. (1993). When more pain is preferred to less: Adding a better end. *Psychological Science, 4*(6), 401–405. https://doi.org/10.1111/j.1467-9280.1993.tb00589.x

Kahneman, D., Wakker, P. P., & Sarin, R. (1997). Explorations of experienced utility. *The Quarterly Journal of Economics, 112*(2), 375–406.

Kalivas, P. W., & Stewart, J. (1991). Dopamine transmission in the initiation and expression of drug- and stress-induced sensitization of motor activity. *Brain Research Reviews, 16*(3), 223–244.

Komissarouk, S., Chernikova, M., Kruglanski, A. W., & Higgins, E. T. (2018). Who is most likely to wear rose-colored glasses? How regulatory mode moderates self-flattery. *Personality and Social Psychology Bulletin, 45*(3), 327–341.

Kruglanski, A. W., Chernikova, M., Rosenzweig, E., & Kopetz, C. (2014). On motivational readiness. *Psychological Review, 121*(3), 367–388. https://doi.org/10.1037/a0037013

Kruglanski, A. W., Fernandez, J. R., Factor, A. R., & Szumowska, E. (2019). Cognitive mechanisms in violent extremism. *Cognition, 188*, 116–123. https://doi.org/10.1016/j.cognition.2018.11.008.

Kruglanski, A. W., Jasko, K., Chernikova, M., Dugas, M., & Webber, D. (2017). To the fringe and back: Violent extremism and the psychology of deviance. *The American Psychologist, 72*(3), 217–230.

Lemos, J. C., Wanat, M. J., Smith, J. S., Reyes, B. A., Hollon, N. G., Van Bockstaele, E. J., . . . Phillips, P. E. (2012). Severe stress switches CRF action in the nucleus accumbens from appetitive to aversive. *Nature, 490*(7420), 402–406. https://doi.org/10.1038/nature11436

Linnet, J., Mouridsen, K., Peterson, E., Møller, A., Doudet, D. J., & Gjedde, A. (2012). Striatal dopamine release codes uncertainty in pathological gambling. *Psychiatry Research: Neuroimaging, 204*(1), 55–60.

Mahler, S. V., & Berridge, K. C. (2012). What and when to "want"? Amygdala-based focusing of incentive salience upon sugar and sex. *Psychopharmacology (Berl), 221*(3), 407–426.

Mendrek, A., Blaha, C. D., & Phillips, A. G. (1998). Pre-exposure of rats to amphetamine sensitizes self-administration of this drug under a progressive ratio schedule. *Psychopharmacology, 135*(4), 416–422. https://doi.org/10.1007/s002130050530

Olney, J. J., Warlow, S. M., Naffziger, E. E., & Berridge, K. C. (2018). Current perspectives on incentive salience and applications to clinical disorders. *Current Opinions in Behavioral Sciences, 22*, 59–69.

O'Sullivan, S. S., Wu, K., Politis, M., Lawrence, A. D., Evans, A. H., Bose, S. K., . . . Piccini, P. (2011). Cue-induced striatal dopamine release in Parkinson's disease-associated impulsive-compulsive behaviours. *Brain, 134*(Pt 4), 969–978.

Pardini, D. A., & Phillips, M. (2010). Neural responses to emotional and neutral facial expressions in chronically violent men. *Journal of Psychiatry & Neuroscience: JPN, 35*(6), 390–398. https://doi.org/10.1503/jpn.100037

Paulson, P. E., & Robinson, T. E. (1995). Amphetamine-induced time-dependent sensitization of dopamine neurotransmission in the dorsal and ventral striatum: A microdialysis study in behaving rats. *Synapse*, *19*(1), 56–65. https://doi.org/10.1002/syn.890190108

Pawliczek, C. M., Derntl, B., Kellermann, T., Kohn, N., Gur, R. C., & Habel, U. (2013). Inhibitory control and trait aggression: Neural and behavioral insights using the emotional stop signal task. *NeuroImage*, *79*, 264–274. https://doi.org/10.1016/j.neuroimage.2013.04.104

Peciña, S., & Berridge, K. C. (2013). Dopamine or opioid stimulation of nucleus accumbens similarly amplify cue-triggered 'wanting' for reward: Entire core and medial shell mapped as substrates for PIT enhancement. *European Journal of Neuroscience*, *37*(9), 1529–1540.

Peciña, S., Schulkin, J., & Berridge, K. C. (2006). Nucleus accumbens corticotropin-releasing factor increases cue-triggered motivation for sucrose reward: Paradoxical positive incentive effects in stress? *BMC Biology*, *4*, 8. https://doi.org/10.1186/1741-7007-4-8

Pfaus, J. G., Damsma, G., Nomikos, G. G., Wenkstern, D. G., Blaha, C. D., Phillips, A. G., & Fibiger, H. C. (1990). Sexual behavior enhances central dopamine transmission in the male rat. *Brain Research*, *530*(2), 345–348.

Phillips, K. B., & Sarter, M. (2020). Addiction vulnerability and the processing of significant cues: Sign-, but not goal-, tracker perceptual sensitivity relies on cue salience. *Behavioral Neuroscience*, *134*(2), 133–143.

Pitchers, K. K., Phillips, K. B., Jones, J. L., Robinson, T. E., & Sarter, M. (2017). Diverse roads to relapse: A discriminative cue signaling cocaine availability is more effective in renewing cocaine seeking in goal trackers than sign trackers and depends on basal forebrain cholinergic activity. *Journal of Neuroscience*, *37*(30), 7198–7208.

Ray, N., Miyasaki, J. M., Zurowski, M., Ko, J. H., Cho, S. S., Pellecchia, G., . . . Strafella, A. P. (2012). Extrastriatal dopaminergic abnormalities of DA homeostasis in Parkinson's patients with medication-induced pathological gambling: A [11C] FLB-457 and PET study. *Neurobiology of Disease*, *48*(3), 519–525.

Robinson, M. J., & Berridge, K. C. (2013). Instant transformation of learned repulsion into motivational "wanting". *Current Biology*, *23*(4) 282–289.

Robinson, M. J. F., Burghardt, P. R., Patterson, C. M., Nobile, C. W., Akil, H., Watson, S. J., . . . Ferrario, C. R. (2015). Individual differences in cue-induced motivation and striatal systems in rats susceptible to diet-induced obesity. *Neuropsychopharmacology*, *40*(9), 2113–2123. https://doi.org/10.1038/npp.2015.71

Robinson, M. J. F., Fischer, A. M., Ahuja, A., Lesser, E. N., & Maniates, H. (2016). Roles of "wanting" and "liking" in motivating behavior: Gambling, food, and drug addictions. *Current Topics in Behavioral Neurosciences*, *27*, 105–136.

Robinson, T. E., & Berridge, K. C. (1993). The neural basis of drug craving: And incentive-sensitization theory of addiction. *Brain Research Reviews*, *18*(3), 247–291.

Robinson, T. E., & Flagel, S. B. (2009). Dissociating the predictive and incentive motivational properties of reward-related cues through the study of individual differences. *Biological Psychiatry*, *65*(10), 869–873.

Robinson, T. E., & Kolb, B. (2004). Structural plasticity associated with exposure to drugs of abuse. *Neuropharmacology*, *47*(Suppl 1), 33–46. https://doi.org/10.1016/j.neuropharm.2004.06.025

Rosse, R. B., Fay-McCarthy, M., Collins, J. P., Alim, T. N., & Deutsch, S. I. (1994). The relationship between cocaine-induced paranoia and compulsive foraging: A preliminary report. *Addiction*, *89*(9), 1097–1104.

Rosse, R. B., Fay-McCarthy, M., Collins, J. P., Risher-Flowers, D., Alim, T. N., & Deutsch, S. I. (1993). Transient compulsive foraging behavior associated with crack cocaine use. *The American Journal of Psychiatry, 150*(1), 155–156.

Roughley, S., & Killcross, S. (2019). Differential involvement of dopamine receptor subtypes in the acquisition of Pavlovian sign-tracking and goal-tracking responses. *Psychopharmacology, 236*, 1853–1862.

Schad, D. J., Rapp, M. A., Garbusow, M., Nebe, S., Sebold, M., Obst, E., . . . Huys, Q. J. M. (2020). Dissociating neural learning signals in human sign- and goal-trackers. *Nature Human Behaviour, 4*(2), 201–214. https://doi.org/10.1038/s41562-019-0765-5

Schiffer, B., Müller, B. W., Scherbaum, N., Hodgins, S., Forsting, M., Wiltfang, J., . . . Leygraf, N. (2011). Disentangling structural brain alterations associated with violent behavior from those associated with substance use disorders. *Archives of General Psychiatry, 68*(10), 1039–1049. https://doi.org/10.1001/archgenpsychiatry.2011.61

Singer, B. F., Tanabe, L. M., Gorny, G., Jake-Matthews, C., Li, Y., Kolb, B., & Vezina, P. (2009). Amphetamine-induced changes in dendritic morphology in rat forebrain correspond to associative drug conditioning rather than non-associative drug sensitization. *Biological Psychiatry, 65*(10), 835–840. https://doi.org/doi:10.1016/j.biopsych.2008.12.020

Smith, K. S., & Berridge, K. C. (2007). Opioid limbic circuit for reward: Interaction between hedonic hotspots of nucleus accumbens and ventral pallidum. *Journal of Neuroscience, 27*(7), 1594–1605.

Steketee, J. D., & Kalivas, P. W. (2011). Drug wanting: Behavioral sensitization and relapse to drug-seeking behavior. *Pharmacological Reviews, 63*(2), 348–365. https://doi.org/10.1124/pr.109.001933

Stice, E., & Yokum, S. (2016). Neural vulnerability factors that increase risk for future weight gain. *Psychological Bulletin, 142*(5), 447–471

Toates, F. M. (1986). *Motivational systems.* Cambridge [Cambridgeshire] and New York: Cambridge University Press.

Vela, L., Martinez Castrillo, J. C., Garcia Ruiz, P., Gasca-Salas, C., Macias Macias, Y., Perez Fernandez, E., . . . Marasescu, R. (2016). The high prevalence of impulse control behaviors in patients with early-onset Parkinson's disease: A cross-sectional multicenter study. *Journal of the Neurological Sciences, 368*, 150–154.

Vezina, P., Lorrain, D. S., Arnold, G. M., Austin, J. D., & Suto, N. (2002). Sensitization of midbrain dopamine neuron reactivity promotes the pursuit of amphetamine. *Journal of Neuroscience, 22*(11), 4654–4662. https://doi.org/10.1523/jneurosci.22-11-04654.2002

Voon, V., Mole, T. B., Banca, P., Porter, L., Morris, L., Mitchell, S., . . . Irvine, M. (2014). Neural correlates of sexual cue reactivity in individuals with and without compulsive sexual behaviours. *PLoS One, 9*(7), e102419.

Voon, V., Napier, T. C., Frank, M. J., Sgambato-Faure, V., Grace, A. A., Rodriguez-Oroz, M., . . . Fernagut, P. O. (2017). Impulse control disorders and levodopa-induced dyskinesias in Parkinson's disease: An update. *The Lancet Neurology, 16*(3), 238–250.

Warlow, S. M., Naffziger, E. E., & Berridge, K. C. (2020). The central amygdala recruits mesolimbic circuitry for pursuit of sugar, cocaine, or pain. *Nature Communications, 11*, 2716.

Warlow, S. M., Robinson, M. J. F., & Berridge, K. C. (2017). Optogenetic central amygdala stimulation intensifies and narrows motivation for cocaine. *Journal of Neuroscience, 37*(5), 8330–8348.

Warren, N., O'Gorman, C., Lehn, A., & Siskind, D. (2017). Dopamine dysregulation syndrome in Parkinson's disease: A systematic review of published cases. *Journal of Neurology, Neurosurgery, & Psychiatry*, *88*(12), 1060–1064.

Wilson, T. D., Reinhard, D. A., Westgate, E. C., Gilbert, D. T., Ellerbeck, N., Hahn, C., . . . Shaked, A. (2014). Just think: The challenges of the disengaged mind. *Science*, *345*(6192), 75–77.

Wilson, T. D., & Schooler, J. W. (1991). Thinking too much: Introspection can reduce the quality of preferences and decisions. *Journal of Personality and Social Psychology*, *60*(2), 181–192. https://doi.org/10.1037/0022-3514.60.2.181

Winkielman, P., & Gogolushko, Y. (2018). Influence of suboptimally and optimally presented affective pictures and words on consumption-related behavior. *Frontiers in Psychology*, *8*(2261), 2261

Wolf, M. E. (2016). Synaptic mechanisms underlying persistent cocaine craving. *Nature Reviews Neuroscience*, *17*(6), 351–365.

Wyvell, C. L., & Berridge, K. C. (2000). Intra-accumbens amphetamine increases the conditioned incentive salience of sucrose reward: Enhancement of reward "wanting" without enhanced "liking" or response reinforcement. *Journal of Neuroscience, 20*(21), 8122–8130.

Zeeb, F. D., Li, Z., Fisher, D. C., Zack, M. H., & Fletcher, P. J. (2017). Uncertainty exposure causes behavioural sensitization and increases risky decision-making in male rats: Toward modelling gambling disorder. *Journal of Psychiatry and Neuroscience*, *42*(6), 404–413. https://doi.org/10.1503/jpn.170003

2
ATTITUDINAL EXTREMISM

Joseph J. Siev, Richard E. Petty, and Pablo Briñol

The extensive social psychological literature on *attitudes* (people's overall evaluations of issues, objects, and people; Petty & Cacioppo, 1981) has featured prominently in attempts to understand a diversity of behavior from consumer purchases to voting patterns (e.g., Fishbein & Ajzen, 1975). Thus, it might be surprising that with respect to understanding the determinants of radical or extreme behavior, with only a few exceptions (e.g., McCauley & Moskalenko, 2017; Doosje, van den Bos, Loseman, Feddes, & Mann, 2012), research on attitudes as a determinant has been largely ignored with more attention placed on other variables and nonattitudinal processes. To be sure, a variety of factors can motivate radical behavior, and no single approach is likely to provide a complete model. Nevertheless, the theme of this chapter is that the accumulated literature over the last 70 years on the topic of attitudes can contribute to an understanding of extreme behavior. We offer a conceptualization of attitudinal extremism including the identification of some antecedents of both 'authentic' and 'perceived' extremism, and we also address when attitudes are more or less likely to result in extreme behavior.

Developing a Model of Attitudinal Extremism

Prior research documents several factors that can result in extreme attitudes and actions, and it is important to examine how or if these findings fit together. Researchers have operationalized attitudinal extremism (and closely related constructs like ideological extremity) in very different ways. For example, the most widely used general measure of attitude extremity is to compute the discrepancy between a person's overall evaluative rating (positive or negative) and neutrality (Abelson, 1995; see further discussion in the following). This can be adapted to

DOI: 10.4324/9781003030898-4

politics, for example, by calculating the distance from self-report ratings of political orientation (liberal-conservative) to having no political leaning (Brandt, Evans, & Crawford, 2015; Frimer, Brandt, Melton, & Motyl, 2019). Other approaches within politics (where attitudinal extremism is most studied) include assessing the degree of identification with a political party (Zmigrod, Rentfrow, & Robbins, 2020) and the ideologies of the social media accounts that an individual follows (Sterling, Jost, & Bonneau, 2020). Still other work has focused on more general individual differences such as overall intolerance of other views (van Prooijen & Krouwel, 2019; Rollwage, Dolan, & Fleming, 2018) and unwillingness to compromise (Webber et al., 2018).

Although these are all plausible approaches, it may be useful to identify with more conceptual clarity what are the core features of a person being an attitudinal extremist. In the view (e.g., Kruglanski, Jasko, Chernikova, Dugas, & Webber, 2017; Kruglanski, Szumowska, Kopetz, Vallerand, & Pierro, 2021) elaborated in many chapters in this volume, extremism is a general psychological phenomenon rooted in the dynamics of motivation (discussed further shortly). Given the need for a conceptual framework to help organize the growing literature on attitudinal extremism, we take this motivational perspective as our point of departure and proceed to explore its implications for attitudes. In adopting this approach, our goal is to advance the study of both attitudes and extremism in an integrative and reciprocally generative manner.

We begin our analysis with three critical issues that remain unresolved. First, research has not yet established what properties of attitudes should be present before labeling a person as an attitudinal extremist. We offer some suggestions before turning to the empirical questions of: (1) whether these factors are indeed considered by others in labeling people as extremists and (2) whether these factors are implicated in the link between attitudes and extreme behavior. In addition to literal attitude extremity or *polarization* (i.e., deviation from neutrality), we consider the attitude's strength and in particular the *certainty* with which it is held, its *unusualness*, and the *social disapproval* associated with it. We propose these as reasonable contributors to attitudinal extremism based on the prior literature on behavioral extremism (e.g., Kruglanski et al., 2021) and attitude strength (Petty & Krosnick, 1995).

After considering these dimensions, we turn to people's perceptions of others as extremists and assessing whether the factors just identified play a role in such attributions. It is important to note that the judgment that someone else is an extremist (*perceived extremism*) might or might not reflect or be based on all of the factors that actually lead people to engage in radical actions (*authentic extremism*). We detail potential similarities and differences between authentic and perceived extremism in the upcoming section on determinants of attributing extremism to others. We also report the results of two experiments that assess targets' attitude polarization, certainty, and unusualness as determinants of attributions of extremism. Although our aim is simply to identify key variables and delineate a few central considerations in this area, our framework can ultimately guide future research on the degree to which people accurately attribute extremism to others versus over-attributing it (to nonextremists) or under-attributing it (i.e., failing to recognize it in authentic extremists).

Following our consideration of attributions of extremism, we aim to clarify the conditions under which attitudes are likely to result in extreme behavior. The dearth of prior research on this topic echoes the difficulties researchers encountered in early attempts to use attitudes to predict *any* behavior (Wicker, 1969). Although considerable progress has been made in understanding when and for whom attitudes predict so-called normal behaviors (e.g., whether attitudes toward ice cream predict ice cream purchases; Fazio, 1990; Petty, Haugtvedt, & Smith, 1995), some have argued that attitudes have not proven to be very helpful in predicting extreme behaviors (McCauley & Moskalenko, 2017). Thus, we conclude this chapter by reporting the results of a new study suggesting that attitudes may be more predictive of tendencies toward extreme action once other factors are taken into account.

In sum, this chapter proceeds in four sections. First, we propose going beyond the attitude literature's near-exclusive reliance on polarization as the sole defining feature of attitudinal extremism and expand on other attitude properties that might be needed. In weighing these additional variables, we incorporate considerations raised by the literature on attitude strength (Petty & Krosnick, 1995) such as attitude certainty, and especially Kruglanski et al.'s motivational account of extremism (this volume; Kruglanski, Bélanger, & Gunaratna, 2019; Kruglanski et al., 2021). In the motivational account, the concept of extremism connotes intensity, defined as the amount of a characteristic (as in extreme heat) as well as the characteristic's frequency or unusualness (as in an extreme situation). From an attitudinal perspective, the notion of intensity can be conceptualized as encompassing both the amount of positivity or negativity an attitude has (i.e., its polarization) as well as the strength associated with that amount of positivity or negativity (e.g., the certainty with which the particular attitude is held, its importance, or one's commitment to it). We consider both attitude polarization and attitude strength, as well as unusualness (infrequency) and social disapproval, in our discussion of the characteristics associated with attitudinal extremism. Although attitude polarization can be related to attitude strength (e.g., more polarized attitudes tend to be held with greater certainty than more moderate ones) and unusualness is related to disapproval (i.e., attitudes that are unique in society tend to be approved of less than those that are common), we argue that each variable is sufficiently distinct that considering their independent contributions can further clarify the psychology of extremism.

The second section turns to an impression formation context, examining determinants of people's proclivities to attribute extremism to others. No prior research of which we are aware has examined this issue. In the third section, we describe key determinants of polarized, certain, and unusual attitudes, paying particular attention to processes that can mediate effects of motivation on their formation. Finally, the fourth section takes up the question of how attitudes are linked to extreme behaviors, and new data are reported that begin to address this important issue.

Candidates for Inclusion in a Model of Attitudinal Extremism

From the perspective of the research literature on attitudes, the most obvious candidate for inclusion in a conceptualization of attitudinal extremism might simply be holding polarized attitudes (cf., Abelson, 1995). Conceptually, *polarization* represents the degree (magnitude) of liking or disliking for an attitude object (i.e., the thing being evaluated), and as noted earlier, is typically operationalized as the distance on an evaluative scale between the endorsed position and the scale neutral point. Thus, on a *bipolar* scale (e.g., −4 to +4), where low values reflect strong disliking or negativity (e.g., extremely bad) and high values represent strong liking or positivity (e.g., extremely good), polarization is operationalized as the distance from the selected position to the neutral (0) midpoint (e.g., Tesser, 1976). Alternatively, when a *unipolar* scale (e.g., 1 = not good, 7 = extremely good) measures a single evaluative dimension, more polarization is reflected in higher values. Although attitude polarization has often been identified in research by the term attitude extremity (Abelson, 1995), we use the term polarization rather than extremity to avoid confusion with the broader construct of attitudinal *extremism*, which as we argue shortly, should include additional criteria.

We argue that polarization is a key component of attitudinal extremism because although a person can value a *moderate* position (e.g., political centrism), such a view would not be considered extreme unless the person adopts a polarized attitude toward that position (i.e., liking or disliking centrism a great deal). In such a case, this polarized attitude could lead to extreme behavior with respect to centrism.[1] In contrast, a mildly positive attitude toward a radical position (e.g., white supremacy) would not be polarized (even though the attitude object itself is extreme). A mild favoring of an extreme attitude object on its own would be unlikely to produce extreme behavior. It is also possible for a person to be very favorable toward both sides of an issue (i.e., be extremely ambivalent; Priester & Petty, 1996). On a bipolar scale, this person might appear to be neutral (e.g., see Kaplan, 1972) when in fact the person's attitude is highly polarized in *both* positive and negative directions. Such a person would therefore be capable of engaging in extremely positive or negative behavior depending on which aspect of the attitude is salient (Luttrell, Petty, & Briñol, 2016). Or, if both evaluations are simultaneously salient, the level of felt conflict could be extreme (Newby-Clark, McGregor, & Zanna, 2002), which could be paralyzing in the short term (Durso, Briñol, & Petty, 2016), but potentially provoke extreme action in the longer term as the person aims to resolve the conflict.

Related to but distinct from polarization, another candidate feature of attitudinal extremism is attitude *strength* (Petty & Krosnick, 1995), and in particular, the *certainty* or confidence with which attitudes are held. Certainty largely reflects the extent to which people believe their attitudes are valid (Petrocelli, Tormala, & Rucker, 2007; Rucker, Tormala, Petty, & Briñol, 2014). Attitude certainty could be relevant to extremism in multiple ways, and we discuss its role both as a possible criterion for identifying someone as an attitudinal extremist and as a moderator of the effect of attitudes on extreme behavior. Certainty is a key example of a host of other attitude strength indicators (e.g., attitude importance and accessibility), but because it is the one that has garnered the most research attention, we focus on it in this chapter. Importantly, attitudes can vary in certainty (and other strength features) regardless of their degree of polarization, so just as a highly polarized attitude can be held with uncertainty or doubt (e.g., doubting one's intense love for one's spouse), a moderate or neutral attitude can be held with considerable conviction (e.g., being confident in one's apathy about politics). Importantly, attitudes held with certainty are generally more predictive of behavior than those held with some doubt, although research shows that sometimes uncertainty can be highly motivating (e.g., McGregor, 2003), and attitudes that are held with doubt can inspire efforts to increase certainty and might therefore result in extreme behavior. We return to this issue in the section linking attitudes to extreme behavior.

Although attitude polarization and certainty (strength) are plausibly important components of attitudinal extremism, we also argue that they might be insufficient to define the construct. This is because some attitudes that are both polarized and strong seem unlikely to result in extremist action. For example, passionate and confidently held love of commonly liked attitude objects such as ice cream hardly seems to merit the designation of extremism and would be unlikely to predict who would stock a garage freezer full of the treat or kill to be the first to obtain a special flavor (i.e., engage in extremist ice cream behavior). Similarly, confident hatred of commonly disliked objects such as parking tickets likely would not predict who would be willing to go to jail in protest or set the local traffic court on fire. Thus, in accord with other approaches to extremism (e.g., Kruglanski et al., 2021), we suggest that another potentially important dimension is an attitude's *unusualness* in the sense of a particular attitude being different from other people's attitudes or one's own other attitudes. As was the case with polarization and strength, unusualness refers to the evaluation and not the attitude object.

Attitudes can be unusual in at least two ways. First, they can be unusually polarized such as when someone likes or dislikes some object much more or less than most other people do (e.g., I hate Mary but everyone else loves her) or when a person likes or dislikes some object much more or less than other objects (I hate Mary but I love all other people). Indeed, considering not only a person's attitude toward one object (e.g., a political candidate) but also the attitude toward other relevant objects (e.g., the alternative candidates) improves prediction of relevant behavior (Fishbein, 1980; Petty & Cacioppo, 1981; Ajzen & Kruglanski, 2019). Second, attitudes can be unusually strong or weak such as when someone has much more or less confidence in a particular attitude than most other people, or when the individual's attitude toward a given object is held with much more or less certainty than the person's other attitudes.[2] Unusualness is conceptually independent of polarization and strength, and it is possible that even neutral or weak attitudes could reflect extremism if they are incredibly deviant. Nevertheless, most examples of attitudinal extremism, from religious fanatics to rabid sports fans to extreme dieters, reflect high degrees of polarization and/or strength.

A final and related property of attitudes that might relate to extremism is *social disapproval* or holding an attitude that is rejected by the mainstream. Other people can disapprove of a person's attitude because of the particular position taken (i.e., holding an attitude that others see as inappropriately positive or negative in degree) or because of the attitude's strength (e.g., holding attitudes that others consider inappropriately confident or important). Although social disapproval can be related to unusualness, these dimensions need not go together. For example, attitudes can be very unusual but not socially disapproved (e.g., a person who greatly likes a very unique flavor of ice cream but most others, though vehemently disagreeing, do not disapprove of liking unusual flavors). Some unusual attitudes might even be highly admired, such as caring a great deal about alleviating poverty. Similarly, even very popular attitudes (e.g., liking for junk food) might not be very high in social approval (i.e., people might think this is the wrong attitude to have; DeMarree, Wheeler, Briñol, & Petty, 2014). Nevertheless, the unusualness of the position taken and social disapproval are likely correlated across many attitude objects because people generally take socially acceptable positions (Cialdini, 2003). Because of this, a key question is whether unusualness alongside polarization and strength is sufficient to characterize attitudinal extremism or whether the additional element of social disapproval adds explanatory power.

Conceptualizing Extremism: Motivational and Attitudinal Imbalance

As noted, the polarization, strength (e.g., certainty), and unusualness criteria on their own do not require attitudes or behavior to be antisocial or negative in order to be extremist. The issue of whether they are diagnostic of attitudinal extremism in the absence of social disapproval turns on whether very unusual but socially approved (or at least not disapproved) attitudes can be meaningfully classified as extremist. According to Kruglanski et al., unusualness is inherently linked to extremism, with or without social disapproval, through the dynamics of motivation (Kruglanski et al., 2017; this volume). That is, core motivations such as the needs for understanding and social approval (Murray, 1955) ordinarily maintain a relative state of balance with each other, and people prefer *multifinal* means (i.e., behaviors that satisfy multiple goals or motivations; Kruglanski et al., 2002). This constrains the range of behaviors likely to be appealing as a means of satisfying any given motive because single-minded pursuit of one motive can undermine the satisfaction of others. As a consequence, most people's behaviors reflect trade-offs between motivations, whereas behaviors aimed at maximally satisfying a single motivation even at the expense of others are comparatively rare. It is only when an individual is deprived of satisfaction of an important motivation, or when it is enhanced through incentivization, or when competing motives become less influential that a state of *motivational imbalance* can arise such that some motives are neglected in favor of single-minded pursuit of a particular motive that has become dominant (Kruglanski et al., 2021). The pursuit of this dominant motive can give rise to extreme behaviors.

The same reasoning can be applied to attitudes, which are often linked to core motives (Briñol & Petty, 2005), with several key implications for attitudinal extremism. First, since motivational imbalance is a relatively unusual state, the attitudes associated with imbalanced motives are likely to be uncommon too. For example, political partisans can be motivated to (a) see their party succeed relative to the opposing party and also to (b) view themselves as relatively unbiased consumers of political information. Most of the time, these two motives should stay roughly balanced, producing partisans who support their side but are nevertheless somewhat skeptical of information that is excessively favorable to it. When the motive to support one's party becomes dominant, however, people might develop unusual political attitudes and beliefs (e.g., endorsing conspiracy theories) and even their more conventional opinions might become unusually polarized, while they refuse to entertain uncertainty (van Prooijen & Krouwel, 2019) and make minimal efforts to correct for bias (Wegener & Petty, 1997). We give further consideration to how motivational imbalance can produce unusual as well as polarized and certain attitudes in a subsequent section.

Second and related, motivational imbalance would often involve *attitudinal imbalance* resulting in discrepancies between properties of attitudes that are relevant to the dominant need versus properties of non-need relevant attitudes (Kruglanski et al., 2021). That is, as noted earlier, extremist attitudes can differ not only from other people's attitudes toward relevant objects (e.g., video game addicts like video games much more than other people do) but also from the person's own attitudes toward other objects (e.g., video game addicts like video games much more than they like social interaction). Moreover, we argue that these within-person discrepancies can involve not only relative polarization (liking or disliking need-relevant objects much more than nonrelevant objects) but also relative strength and unusualness. For example, someone with a single-minded motivation to pursue their political goals might dislike people who disagree with them about politics more than they dislike anyone for any non-political reasons (relative polarization), insist upon the superiority of their political ideology despite having epistemic humility about non-political information (relative certainty), and endorse outlandish political ideas while having conventional opinions about non-political topics (relative unusualness). That is, even attitudes that are polarized, strong, and unusual might not reflect extremism if they are not *especially* so for that individual because people can possess many attitudes with these features, and this would not reflect motivational or attitudinal imbalance. Thus, an individual who holds uniformly strange political and non-political views, believes all their opinions to be infallible, and is contemptuous of anyone who disagrees with them about any topic would not be a *political* extremist.

Third, the notion of attitudinal imbalance, and especially relative polarization, is also helpful in predicting attitude-consistent behavior because relative liking for a future state of need satisfaction over alternative states produces a desire for that state and can catalyze the formation of a goal to attain it (Ajzen & Kruglanski, 2019). Similarly, relative liking for a different attitude from one's current view can produce a desire for that attitude and relevant behaviors to achieve it (DeMarree et al., 2014). Thus, attitudes toward need-relevant objects under conditions of motivational imbalance should be more predictive of behavior than the same attitudes under conditions of balance because such attitudes are especially likely to translate into desire and goal formation. Although we address this only briefly later in this chapter, we consider it a promising avenue for future research. Next, we consider the factors that contribute to people's attributions of extremism to others.

Properties of Attitudes That Increase Attributions of Extremism

What do people mean when they refer to others as *extremists*? We argued that authentic attitudinal extremism involves holding attitudes that are polarized and/ or strong as well as unusual in some way. But what about perceptions? We expect that observers would rely on these same dimensions in judging extremism in other people. There are also reasons unique to the domain of attribution (i.e., that do not apply to the question of extremist motivations) to expect unusualness to play a role. In particular, a long tradition of research on attribution (how people explain behavior in terms of personal and/or situational causes; Kelley, 1967) demonstrates that common behaviors are often discounted (i.e., perceived as nondiagnostic of the actor's traits), whereas unusual behaviors need an explanation and are often explained with reference to the actor's traits (e.g., Skowronski & Carlston, 1989). In other words, uncommon attitudes are seen as reflecting more on the person who holds them than common ones. As such, if polarized attitudes are more likely than neutral ones to be perceived as extremist, then that should especially be so when those attitudes are also unusual in some way.

We conducted two studies to address the prediction that polarized attitudes (and, in the second study, polarized and confident attitudes) are more likely to be perceived as extremist when they are also unusual. We focused on how unusual attitudes were with respect to other people's attitudes rather than the person's own attitudes since observers are more likely to be aware of the former than the latter. But, according to attribution theories, both sources of unusualness should produce similar effects (Kelley, 1967). We did not consider social disapproval in these studies, although the unusual attitude in the second study was likely seen by many participants as inappropriate as well as unusual.

The studies used similar designs with several key differences. Both involved participants reading brief summaries of the attitudes of four targets who took either a polarized or more moderate position in favor of or against a particular issue. After receiving the relevant information, participants judged the extent to which each target was an extremist on a seven-point Likert scale. The first study used a fictional issue, and the second study used a real and presumably more engaging issue.

Participants in the first study (101 undergraduates) were asked to consider a situation in which a society was contemplating a potential change to an unspecified law—called 'Proposition 6'. They were told that if a majority voted in favor of the proposition, the change would be implemented. Participants were randomly assigned to receive information that 'the majority of voters' in society took one of four positions with respect to Proposition 6: slight support, strong support, slight opposition, or strong opposition. In this way, the majority position valence (support versus opposition) and polarization (slight support/opposition versus strong support/opposition) were manipulated. Regardless of which of the four societal positions they received, participants were then exposed to a series of four individuals, each taking one of these four positions. All participants provided attributions of extremism for each of the four targets to which they were exposed.

For analysis, we recoded the variables into a 2 (target polarization: slight/strong position) × 2 (societal polarization: slight/strong position) × 2 (target position: support/oppose Proposition 6) × (societal position: support/oppose Proposition 6) mixed factorial design. In addition to a significant main effect of target polarization (i.e., polarized targets were perceived as more extremist than nonpolarized targets ($p < .001$) and a significant two-way interaction between target position and societal position (i.e., disagreement with society was seen as more extremist than agreement with society, $p < .001$), a four-way interaction emerged among the variables ($p < .001$). Because the results were not further moderated by the particular position that targets took, for ease of interpretation, the two position factors are recoded in Figure 2.1 into a single target-society agreement factor. As can be seen in the figure, target polarization increased attributions of extremism except when the target was in agreement with a polarized societal position. Thus, this study shows that both the degree to which a target expresses a polarized position and the extent to which that position deviates from society (it is unusual) contribute to perceptions of extremism.

The second study was designed to extend these results to a real-world issue. The study focused on the issue of social distancing as a method of preventing the spread of COVID-19 (data were collected in May, 2020, during the pandemic).

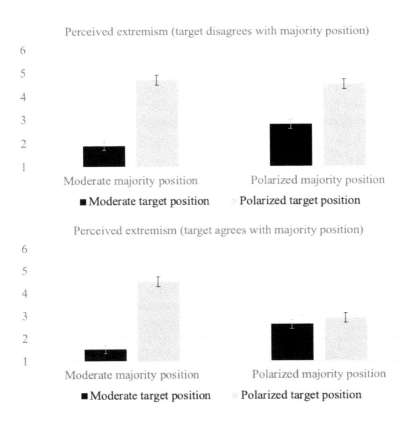

FIGURE 2.1 Attributions of extremism as a function of target and societal position, polarization, and target agreement with society in Study 1.

Because we expected this to be a very familiar issue at the time the study was run, it seemed unrealistic to manipulate perceptions of the normative position in society, though it was likely that participants would realize that a clear majority of the public favored social distancing (Kaiser Family Foundation, 2020). Thus, we dropped the normative position manipulation and added a within-subject manipulation of the target's certainty about the position expressed as a potential contributor to attributions of extremism. Using social distancing as the topic also suggested measuring participants' own prior attitudes in order to determine how those might affect judgments of others.

Participants in Study 2 (328 Amazon MTurk workers) first reported their own attitudes about social distancing using three 9-point scales (e.g., −4 *Dislike* and +4 *Like*). They next rated the extremism of each of four groups of people who took the following positions on social distancing: confident support, doubtful support, confident opposition, and doubtful opposition. Furthermore, participants were randomly assigned to read these four positions expressed in either a polarized or a moderate way. For example, in the polarized (moderate) confident support condition, the position taken was described as follows: 'People in Group A strongly (slightly) support social distancing, are completely certain about their opinion and have no doubts about it'. In the polarized (moderate) doubtful support condition, it was said that group members 'strongly (slightly) support social distancing but are very uncertain about their opinion and have clear doubts about it'. The opposition to social distancing conditions was identically worded with 'oppose' replacing 'support'.

Attitudes about social distancing among participants were highly favorable with less than 5% of the sample on the negative side of the neutral point. We thus excluded the 16 with negative attitudes. A 2 (target polarization: moderate/polarized, between subjects) × 2 (target certainty: certainty/doubt, within subjects) × 2 (target position: support/oppose social distancing, within subjects) mixed analysis of variance was performed. Although the main effect of target polarization was not significant, there was a main effect of target position. That is, targets who opposed social distancing and thus held an unusual (and perhaps socially disapproved) position (i.e., in disagreement with the majority of society) were viewed as more extremist than supporters, $p < .001$. In addition, targets expressing certainty in their attitudes were viewed as more extremist than those expressing doubt, $p < .001$. Furthermore, an interaction was present among the three variables ($p = .02$). Polarization increased attributions of extremism only when targets were *certain* about opposing social distancing (i.e., the target disagreed with participants' views and presumably societal views as well; Figure 2.2A). Viewed differently, expressing doubt in one's unusual position attenuated the normal effect of polarization on perceived extremism. When targets were on the same side of the issue as the subject and society (in support of social distancing), neither certainty nor polarization affected extremism (Figure 2.2B).

In sum, consistent with our argument about the roles of attitude polarization, certainty, and unusualness in producing attributions of extremism, participants in the first study perceived polarized attitudes as indicative of extremism, especially when the position and degree of polarization were unusual (nonnormative). In the second study, polarization increased attributions of extremism only when targets held a disagreeable position that was also counter to society (i.e., unusual) and certainty rather than doubt in that view was expressed. More research is needed to understand these dynamics fully, but our findings are consistent with the notion that polarization, unusualness, and certainty all contribute to perceptions of extremism in others.

Processes That Produce Polarized, Confident, and Unusual Attitudes

So far, we have focused on identifying core features of extremism, which we approached by incorporating insights from prior work on attitudes, attitude

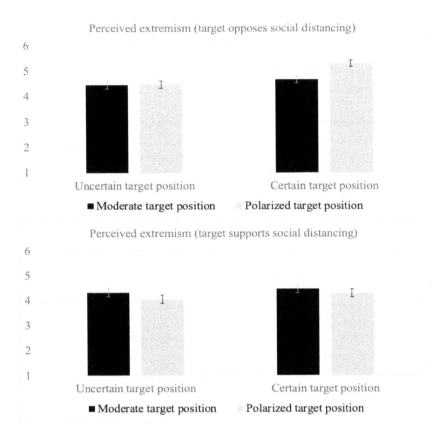

FIGURE 2.2 Attributions of extremism as a function of target position, polarization, and certainty in Study 2.

strength, and motivation. We also reported new evidence for the impact of three attitude features on perceptions of extremism—polarization, certainty, and unusualness. In this section, we continue our analysis by briefly considering the determinants of each of these contributors to extremism.

Determinants of Attitude Polarization

Producing polarized attitudes can be viewed as producing extreme initial attitudes or extreme amounts of attitude change (i.e., moving people with neutral or somewhat moderate attitudes to more polarized ones in either direction, or converting someone from one extreme position to its opposite). Thus, the processes involved in producing polarized attitudes should be compatible with those involved in forming or changing attitudes in general. We group these processes into four broad categories: *information exposure and processing*, *mere thought*, *metacognitive validation*, and *normative influence*.

Common models of attitude formation and change suggest that first, a person is exposed to and attends to some sort of information (McGuire, 1985). Second, they might elaborate upon this information, generating new thoughts and/or feelings and integrating those with content already stored in memory (Petty & Cacioppo, 1986; Petty & Briñol, 2012). People can also engage in mere thought where they contemplate an issue in the absence of any external information (Tesser, 1978). Finally, they may engage in metacognition (Petty, Briñol, Tormala, & Wegener, 2007), secondary thinking about their primary thoughts and feelings (Jost, Kruglanski, & Nelson, 1998). Of particular importance, people metacognitively appraise the validity of their thoughts and feelings, thereby determining if they should rely on them or not in forming their attitudes (Briñol & Petty, 2009).

The processes we have outlined can be driven by mostly rational cognitive considerations (e.g., seeking knowledge) or antecedents that are highly motivationally and affectively charged (e.g., reactions to threat). As noted earlier, motivational factors might be more likely than purely cognitive ones to result in attitudinal extremism due to the possibility of the person's motivational system becoming imbalanced, and we highlight motivational antecedents to attitude polarization in our discussion. Finally, because we assume that social norms can contribute to polarization through any of these processes, we also briefly discuss normative influence.

Information Exposure and Processing

Repeated exposure to a simple object (e.g., symbol, logo) can affect attitudes toward it (Bornstein, 1989; Zajonc, 1968). For example, repetition can increase liking by making the object easier to process (Winkielman & Cacioppo, 2001). However, mere repetition can also increase the salience of the repeated object (making it stand out versus other objects), and this can polarize one's reaction to it (increasing positivity for positive objects and increasing negativity for negative ones; Mrkva & Van Boven, 2020). Because of the joint operation of these processes, mere exposure is most likely to enhance positivity for objects that are already positively evaluated or neutral.

Repeated exposure to persuasive messages containing substantive information tends to enhance favorability if the arguments are strong, but to decrease favorability if the arguments are weak as repeated exposure gives people greater opportunity to consider and elaborate upon the evidence presented (i.e., generating favorable thoughts to strong arguments and unfavorable thoughts to weak ones; Cacioppo & Petty, 1989). Although many factors besides repetition can affect an individual's opportunity or motivation to think about a message (and therefore result in polarized thoughts and attitudes), people are particularly likely to elaborate upon information when the issue affects them personally (Petty & Cacioppo, 1990) or matches something about them (e.g., their personality; see Teeny, Siev,

Briñol, & Petty, 2021). In contrast, when not motivated or able to think carefully about incoming information, people tend to rely on simple valenced cues (e.g., quick reactions to the attractiveness of the source; Petty & Cacioppo, 1986). When attitudes are based on simple cues rather than extensive thought, those attitudes tend to be less consequential and therefore less likely to result in action (Petty et al., 1995).

People are typically motivated to hold correct attitudes (Festinger, 1950) as this helps them satisfy their needs and pursue their goals (Petty & Cacioppo, 1986). When an accuracy goal dominates, people are open to receiving information on both sides of an unfamiliar issue and process it in a relatively objective way (Petty & Cacioppo, 1990). They are even open to contrary information if they already have an opinion, and the new information is not perceived to be threatening (Hart et al., 2009). However, sometimes accuracy is not at the forefront, and motivational factors such as a desire to avoid cognitive dissonance (Festinger, 1957) or maintain cognitive closure (Kruglanski & Webster, 1996) can produce *selective exposure* (a preference for seeking congenial and/or avoiding uncongenial information; Hart et al., 2009) and motivated reasoning (Kunda, 1990). This biased information seeking and processing can result in attitudes becoming more polarized in their initial direction.

The motivation to defend one's attitudes or bolster them against uncertainty can override accuracy concerns if the perceived threat from the information is sufficiently high. The likelihood of attitude polarization following threat is greatest when the motivation to defend one's attitude overwhelms accuracy and other motives (i.e., motivational imbalance). Among the factors known to increase selective exposure and defensiveness are commitment to and perceived irreversibility of the attitude, the value-inconsistency or otherwise threatening nature of the message, and the individual's dispositional closed-mindedness (Hart et al., 2009; Clark & Wegener, 2013). One of the most studied threats that can affect attitudes is momentary thoughts about one's own death. This *mortality salience* (Greenberg, Solomon, & Pysczynski, 1997) has resulted in polarization of existing beliefs and attitudes (Greenberg et al., 1990; See & Petty, 2006), similar to the impact of other self-threats (e.g., McGregor, Zanna, Holmes, & Spencer, 2001). Research on selective exposure and biased information processing highlights the importance of intervening early in the attitude polarization process. Once people make extremist commitments or irreversible decisions (e.g., supporting an extremist group or perpetrating violence), their motivation to defend or bolster attitudes consistent with those realities will likely strengthen (Gopinath & Nyer, 2009).

Mere Thought

People are active information processors, not mere receptacles of the information to which they are exposed, and thus as just noted, people's positive and negative thoughts in response to the information they receive play an important role in determining what attitudes are formed (Petty, Ostrom, & Brock, 1981). However, thinking about external information is not necessary for attitudes to polarize. Merely thinking about a topic in the absence of any new information can also result in polarization (Tesser, 1978).

Research finds that mere thought polarization effects tend to occur so long as people do not possess conflicting views of the object of thought. While thinking, people generate new reasons for their initial evaluation, and these new attitude-consistent thoughts can polarize attitudes (Clarkson, Tormala, & Leone, 2011). A typical paradigm involves exposing participants to a stimulus (e.g., a person or painting) then randomly assigning them to think about it and rate it immediately, after a short interval, or after a long interval. Generally speaking, attitudes polarize more as the interval increases. However, several factors have been identified that reduce mere thought-based polarization. A process constraint limitation can occur if people have difficulty in rationalizing their current view (Tesser, Leone, & Clary, 1978). A reality constraint limitation can occur if the physical presence of the attitude object renders biased thoughts implausible (Tesser, 1976). For example, it might be easier to become inordinately favorable toward one's country from afar. Polarization can also be stymied if people run out of viewpoint-consistent thoughts with extended thinking and contrary thoughts come to mind (Tormala, Falces, Briñol, & Petty, 2007).

As with information exposure, however, motivation can increase the effect of mere thought on polarization when it makes people resistant to the constraints just mentioned. For example, in line with our discussion of selective exposure and biased processing, when reality challenges core aspects of people's worldviews, they strategically avoid reality testing by framing the reasons underlying their attitudes in unfalsifiable terms (Friesen, Campbell, & Kay, 2015). That is, similar factors, rooted in motivations to defend one's attitudes from threats or bolster them against uncertainty, shape the extent to which external information and mere thought result in attitude polarization.

Metacognitive Validation

When people think extensively about an attitude object, in addition to the positive or negative thoughts they generate, they may also appraise the validity of those thoughts (Petty, Briñol, & Tormala, 2002). The outcome of this metacognitive appraisal process is important because validated thoughts have a greater impact on attitudes, thereby increasing thought-consistent attitude polarization (Briñol & Petty, 2009). Perceptions of thought validity have been implicated in the polarizing impact of numerous variables. For example, thought validity plays a role in mere exposure effects as thinking longer enhances not only the number of attitude-consistent thoughts but also the confidence placed in them (Clarkson et al., 2011). Other variables that have increased the perceived validity of people's thoughts and thereby polarized attitudes include momentary feelings of power (Briñol, Petty, Valle, Rucker, & Becerra, 2007), affirmation of one's values (Briñol, Petty, Gallardo, & DeMarree, 2007), and feeling prepared (Carroll, Briñol, Petty, & Ketchman, 2020). Validation effects are most likely when these variables are salient during or after thinking rather than before. It is also noteworthy in connection to extremism that oppositional and attacking (vs. defensive) mindsets can be associated with confidence (Bizer, Larsen, & Petty, 2011; De Dreu & Gross, 2019) and thereby increase the impact of thoughts on attitudes (e.g., Briñol, Petty, & Requero, 2017; see Briñol, Petty, & DeMarree, 2015, for a review). Overall, individuals likely differ in the sources they tend to look to for validation including radical groups, God, and epistemic authorities (e.g., Evans & Clark, 2012; Huntsinger, 2013).

Motivational factors can also play a role in validation processes. For example, an imbalance in motivation in which the need to maintain certainty (Webster & Kruglanski, 1994) comes to dominate accuracy motivation could enhance perceived thought validity and result in attitude polarization (Hart et al., 2009). In fact, research confirms that ideological polarization is associated with overconfidence (van Prooijen & Krouwel, 2019) and perceiving one's views as superior to others (Toner, Leary, Asher, & Jongman-Sereno, 2013; Brandt et al., 2015) despite basing them on simplistic causal explanatory models (Fernbach, Rogers, Fox, & Sloman, 2013).

Although research indicates that thoughts accompanied by feeling confident typically have a larger impact on attitudes than thoughts accompanied by feeling doubt, some work is consistent with the notion that feeling uncertain can increase thought-attitude correspondence under certain conditions—particularly when the uncertainty is threatening in some way (Briñol, et al., 2015). In these situations, people aim to be confident in order to compensate for and mitigate the anxiety from threat. In one recent study (Horcajo et al., 2020), for example, participants were induced to generate positive or negative thoughts because they evaluated either a strong or a weak resume for a job candidate. Next, participants were asked to think either about the COVID-19 pandemic (inducing threatening uncertainty) or feeling cold (an unpleasant but nonthreatening situation). As predicted, the threat induction enhanced thought confidence more than the nonthreat induction, and this thought confidence produced more polarized attitudes toward the strong and weak job candidates (see also, Horcajo et al., 2008). Thus, the extent to which feeling doubtful increases or decreases attitude polarization might depend on the degree to which the uncertainty is experienced as threatening.

Normative Influence

Influential though they are, the determinants of attitude polarization described so far have focused on individuals without accounting for the social contexts that shape their attitudes and behavior (Visser & Mirabile, 2004). However, normative and group-based processes are widely implicated in the psychology of influence (Cialdini, 2003). Indeed, research on influence from others who are said to hold majority versus minority views is one of the most prolific topics in the domain of attitude change (Martin & Hewstone, 2010). And, normative influence has proven particularly important in the phenomenon of extremism (e.g., Kruglanski et al., 2019; Smith, Blackwood, & Thomas, 2020).

One key process by which norms contribute to attitude polarization is seen in work showing that group discussion can polarize attitudes. In some ways, this research parallels individual thought-based polarization in that each type is facilitated by an initial uniformity of thought. Likewise, both are driven in part by discovering new arguments in support of the initial position, whether by thinking of them oneself or learning them from others (Burnstein & Vinokur, 1977; Levendusky, Druckman, & McLain, 2016). Repetition of known reasons (as when group members reiterate previous points) can also increase their impact on attitudes (Judd & Brauer, 1995). Normative mechanisms have additional impacts in the group discussion context, however. For one thing, attitudes polarize when corroborated (validated) by others (Baron et al., 1996). Moreover, group members might attempt to enhance their image or achieve status by advocating the group's ideal position, typically a more polarized one (Myers, 1978). Notably, those with the most polarized political orientations are most prone to such grandstanding (Grubbs, Warmke, Tosi, Shanti, & Campbell, 2019).

The risks of group polarization are compounded by a related process known as groupthink (Janis, 1972). Proposed partly as an explanation of disasters in US foreign policy decision-making, including the Cuban Missile Crisis, the core idea is that individuals who are motivated to maintain their standing in a group seek consensus at the expense of a realistic analysis of the issues at hand. Although early theoretical treatments of groupthink proposed a variety of antecedent conditions, more recent research has found that most are not required and likeminded groups often fail to consider alternatives to a consensus position even in the absence of additional factors (e.g., cohesion, threat, time pressure; Baron, 2005).

People adhere to social norms and majority positions for both interpersonal reasons (to fit in) and informational reasons (because they perceive them as diagnostic). Thus, norm-driven polarization can be increased when either motivation (relational *or* epistemic) dominates. For example, feelings of ostracism increase openness to even extreme groups because affiliating with them reduces the painful feelings of social exclusion (a relational motive; Hales & Williams, 2018) and feeling humiliated increases attitude polarization in part by increasing the need for cognitive closure (an epistemic motive; Webber et al., 2018). In general, frustration (a reaction to one's goal being thwarted) increases affirmation of social norms, which are seen as offering an alternative means of goal satisfaction (Leander et al., 2020). As such, the frustrated individual's risk of becoming radicalized depends on what exactly—and how polarized and/or unusual—the norms they turn to are.

Determinants of Attitude Certainty

As was the case regarding the determinants of polarized attitudes, there is considerable research on the determinants of attitude certainty. Because there have been several recent reviews of this literature (e.g., Luttrell & Sawicki, 2020; Tormala & Rucker, 2018), our discussion will be brief. In one comprehensive review, Rucker et al. (2014) identified four general sources of attitude certainty. That is, they argued and provided evidence to support four building blocks of attitude confidence.

Specifically, Rucker et al. argued that people infer that they are confident in their attitudes to the extent that they perceive their attitudes are based on (a) accurate information, (b) complete information, (c) relevant/legitimate information, and (d) information that feels right. Each of these general sources of attitude confidence was then tied to several more specific factors. For example, an appraisal regarding accuracy can stem from a perceived social consensus around that information (Visser & Mirabile, 2004) or that the information came from an expert source (Kruglanski et al., 2005). An appraisal of completeness can stem from the quantity of information one has (Smith et al., 2008) or how much thought went into the attitude (Wan, Rucker, Tormala, & Clarkson, 2010). These highlighted determinants of confidence not only apply to attitude confidence but also can contribute to perceptions of confidence in one's thoughts. Finally, it is important to note that this appraisal framework does not require the appraisals to be correct. For example, a person does not have to actually have thought a great deal about an issue to feel certain in the resulting attitude. Rather, it is sufficient to have the mere perception of having thought a lot about one's attitude for confidence in it to be enhanced (e.g., Barden & Petty, 2008; see also, Rucker, Petty, & Briñol, 2008).

Determinants of Attitude Unusualness

Much less research has investigated the determinants of adopting an unusual attitude, but it is nonetheless possible to identify several factors that should increase the likelihood they will develop. We divide these factors into two categories: those that produce attitudes that happen to be unusual and those that specifically produce unusual attitudes. The first category includes attitudes that result from motivational imbalance. As explained, core motivations are usually balanced relative to one another (Kruglanski et al., this volume). As a result, attitudes that result from motivational imbalance are likely to be relatively unusual both with respect to other people's attitudes and with respect to the person's own other attitudes.

Beyond this, people can sometimes be motivated to hold unusual attitudes *because* those attitudes are deviant. For example, although most people prefer to feel moderately distinct from others (Snyder & Fromkin, 1980), those with a strong need for uniqueness (e.g., those high in narcissism; White, Szabo, & Tiliopoulos, 2018) might adopt unusual attitudes in order to satisfy this need. Motivational imbalance could compound this tendency, such as when a uniqueness motive dominates a belonging motive. Similar dynamics occur at the group level, as individuals generally prefer to join groups that are optimally distinct from others (Brewer, 1991), but some prefer especially distinct groups (e.g., Hogg, 2007). Other factors that could increase the tendency to hold unusual attitudes are the goals of drawing attention to and/or communicating something about oneself or the attitude object to others, or simply rejecting normative standards or self-consistency as goals. These questions await investigation.

Properties of Attitudes That Predict Extreme Behavior

Having discussed three key elements involved in possessing and being perceived to possess extremist attitudes, we turn now to the role of attitudes in guiding extremist behavior. We have already noted the difficulties that researchers faced in early efforts to predict when attitudes should produce *any* corresponding behavior (Wicker, 1969) and the prospect that these difficulties might be enhanced when considering extreme behaviors (McCauley & Moskalenko, 2017). Conversely, we also mentioned the possibility that motivational imbalance could increase the propensity to act (in even extreme ways) on one's need-relevant attitudes (Kruglanski et al., 2015). By extreme behavior, we mean behaviors that are polarized and unusual.[3] We also address the role of attitude confidence as a moderator. Prominent measures of extreme behavior used in the literature include willingness to fight and die for one's group or cause (Gómez et al., 2020; Swann, Gómez, Seyle, Morales, & Huici, 2009), to endure intense suffering, or give up all of one's belongings (Bélanger, Caouette, Sharvit, & Dugas, 2014). Although extreme pro-group sacrifice has received the most attention in social psychological literature, the theoretically relevant set of extreme behaviors also includes acts motivated by individual concerns, including those that are largely private such as extreme eating or exercise (see various chapters in this volume). It also includes behaviors that are, in contrast to those just mentioned, widely socially approved but highly unusual, such as extreme prosociality (e.g., giving all of one's money to charity).

Past research has identified *attitude strength* as a key moderator of attitude–behavior consistency (Petty & Krosnick, 1995). Strong attitudes tend to be consequential—persisting over time, resisting persuasive attacks, and most relevant to this chapter, impacting behavior (Krosnick & Petty, 1995). As noted, a key indicator of an attitude's strength is the extent to which it is held with certainty (Rucker et al., 2014), but it also includes the extent to which the attitude is based on direct experience (Fazio & Zanna, 1978), high in importance (Boninger, Krosnick, Berent, & Fabrigar, 1995), based on extensive thinking (Barden & Petty, 2008), composed of uniformly positive or negative versus a mix of evaluations (i.e., ambivalence; e.g., Priester & Petty, 1996), highly accessible (Fazio & Williams, 1986), tied to a strong psychological need (Strahan, Spencer, & Zanna, 2002; Ajzen & Kruglanski, 2019; Kruglanski et al., 2015), imbued with moral significance (Skitka, Bauman, & Sargis, 2005), or based on emotion (Rocklage & Fazio, 2018), among other attributes.

The available research evidence shows that knowing how strong an attitude is can be used to understand when that attitude is likely to predict behavior. For example, there is much research showing that increased certainty in one's attitudes and beliefs enhances their ability to predict various non-extreme (normal) behaviors (e.g., product purchases, voting), and some evidence suggests that certainty can also increase the impact of people's beliefs even for predicting extreme behavior. For example, one recent study (Paredes, Santos, Briñol, Gómez, & Petty, 2020) showed that the more certain people were that their self-concept overlapped with their group (as assessed with the identity fusion scale, Gómez et al., 2020), the more willing they were to engage in high levels of sacrifice for their group. Similarly, enhanced certainty in other self-beliefs (e.g., political ideology; Shoots-Reinhard, Petty, DeMarree, & Rucker, 2015; trait aggressiveness, Santos, Briñol, Petty, Gandarillas, & Mateos, 2019) helps those scales predict relevant behavior better. Moreover, participants in our second attribution of extremism study (described earlier) viewed confident polarized attitudes as more indicative of extremism than uncertain polarized attitudes, suggesting that laypeople expect attitude certainty to increase attitude-consistent extreme behavior.

Although certainty in one's attachment to a group or a position on an issue can increase the impact of that group attachment or position on behavior, as noted earlier, previous research has found that feelings of *un*certainty can motivate behavior under particular circumstances—especially when the uncertainty is experienced as threatening. For instance, research shows that threatening uncertainty can result in a *compensation effect* (e.g., Heine, Proulx, & Vohs, 2006) whereby people attempt to restore confidence by bolstering their attitudinal positions (compensatory conviction; see McGregor, 2003; McGregor et al., 2001), or becoming more confident in their already generated thoughts (Horcajo et al., 2008), which can result in more polarized or confidently held attitudes. Alternatively, people can compensate in a more fluid manner by choosing to act on attitudes irrelevant to the particular threat about which they are already confident. Similar compensatory processes occur whether the object of the uncertainty concerns the self (Baumeister, Bushman, & Campbell, 2000; Bushman & Baumeister, 1998; Hogg, 2007) or one's nonself attitudes and beliefs (Sawicki & Agnew, 2021; Sawicki & Wegener, 2018). At times, people might try to reduce uncertainty symbolically by acting *as if* they were certain (Briñol et al., 2015; Hart, 2014; Jonas et al., 2014;

Landau, Kay, & Whitson, 2015). In such circumstances, greater uncertainty can enhance willingness to act (even extremely) on one's beliefs and attitudes.

As yet, however, it is not entirely clear whether and when feelings of certainty versus uncertainty would be expected to increase the impact of attitudes on extreme behavior. On the one hand, the majority of research findings on attitude certainty suggest that higher certainty usually increases the impact of attitudes on behavior, though the context studied has not tended to be threatening. On the other hand, some research has suggested that a threatening context can enhance individuals' willingness to engage in extreme behaviors that are belief-consistent, although this research did not explicitly consider the moderating impact of certainty. For example, one set of studies (Jasko et al., 2020) examined the extent to which a quest for significance (motivated by beliefs that oneself or one's group was undervalued by others) predicted willingness to support violence in support of one's group. This relationship was studied both in contexts and for people in which radical behavior was salient (high threat context) or not (low threat context). When radical behavior was highly salient (for contextual or personal reasons), the quest for significance showed a stronger relationship with support for violence than in cases in which salience was low. In a conceptually similar study (Paredes, Briñol, Petty, & Gómez, in press), the salience of extreme behavior was manipulated to be high or low by having participants think about making an extreme or a moderate sacrifice for their group. In conditions in which the possibility of radical behavior was made salient, a measure of identity fusion with one's group (Gómez et al., 2020) was more predictive of willingness to self-sacrifice than in conditions where extreme behavior was not made salient. Thus, the salience of threats and extreme behaviors can moderate the relationships among extremism-relevant variables.

Threat as a Moderator of Compensation Effects

Although the research just reviewed shows that people are more willing to engage in self-sacrificial behaviors that are compatible with their beliefs when threats or the possibility of extreme actions are salient, little research has examined the role that attitude certainty plays in compensatory attitude–behavior consistency. One possibility is that threat could enhance compensation effects by producing a motivation to bolster one's low level of attitude certainty in an important domain. If so, threat could increase the likelihood of attitude-consistent behavior especially when people are *uncertain* in their views. In the absence of threat, however, attitude-consistent behavior should be increased as certainty increases.

To examine this interaction hypothesis, for our Study 3, we recruited 299 participants from Amazon's Mechanical Turk platform and asked them to report their attitudes on several nine-point bipolar scales (e.g., −4, *Dislike*; +4, *Like*) along with their attitude certainty (seven-point scale, *not at all certain—very certain*) regarding social distancing during the worldwide COVID-19 pandemic

(May 2020). This was followed by reports of their behavioral proclivities regarding two sets of behaviors (order counterbalanced) related to COVID-19 prevention: one set of behaviors was relatively *moderate* (e.g., avoiding crowds for a month, attempting to persuade others about social distancing) and the other set was rather *extreme* (e.g., completely isolating oneself for a year, fighting others who disagreed). Participants were randomly assigned to respond to questions that referred to the extent of their *willingness*, *desire*, or *intention* to perform each behavior on seven-point scales anchored at *1, not at all* and *7, very much*. Finally, participants reported the amount of threat they perceived from COVID-19 (their concern about themselves or someone they care about becoming seriously ill). The full study design was thus a 2 (behavioral extremity: moderate/extreme; within subjects) × 3 (measure: willingness/desire/intention; between subjects) × attitude toward social distancing (continuous) × attitude certainty (continuous) × perceived threat of COVID-19 (continuous).

First, we expected that higher attitude *certainty* would produce higher attitude–behavior consistency with respect to moderate behaviors when people perceived a low degree of threat from the virus (i.e., when threat was not salient). This is the typical attitude certainty finding—the more certain people are in their attitudes, the more they engage in attitude-consistent behavior. In contrast, we reasoned that higher attitude *uncertainty* would be especially likely to produce attitude–behavior consistency with respect to extreme behaviors when people perceived a high degree of threat from the virus because, as mentioned, uncertainty can be especially motivating when threat is salient. In order to obtain a focused test of our hypotheses, an overall *threat salience* factor was constructed that considered both the behavioral extremity measure and the perceived threat variables. Specifically, the *high threat* condition examined reports of proclivity to engage in extreme behaviors among those participants who were above the median in their reported concern about the virus. The *low threat* condition examined reports of proclivity to engage in moderate behaviors among those participants who were below the median in reported concern about the virus. The threat salience variable was then coded −1 for the low threat condition and +1 for the high threat condition.

We used multiple regression to analyze the data as a 2 (threat salience: low/high) × 3 (measure: willingness/desire/intention) × attitudes (continuous) × attitude certainty (continuous) predicting attitude-consistent behavior. This allows a focused test of the predictions for certainty as a moderator of attitude–behavior consistency in the key high and low threat salience conditions. The analysis returned a highly significant three-way threat salience × attitudes × attitude certainty interaction ($p < .0001$), which is depicted in Figure 2.3.

This three-way interaction resulted from different two-way attitude X attitude certainty interactions in the different threat salience conditions. In the low threat condition (Figure 2.3A), the two-way interaction ($p = .06$) indicated that attitudes impacted action tendencies more when they were held with a relatively high ($p = .0001$) rather than a low degree of certainty ($p = .448$). This

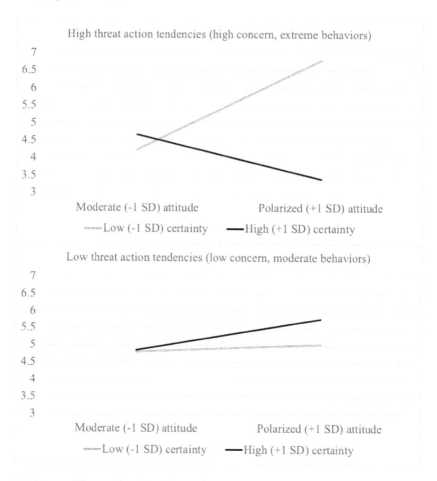

FIGURE 2.3 The moderating effect of certainty on attitude–behavior consistency under conditions of relatively high threat (A: high concern, extreme behaviors) versus low threat (B: low concern, moderate behaviors) in Study 3.

outcome represents the standard attitude strength effect whereby higher certainty attitudes are more predictive of behavior. In contrast, in the high threat condition (Figure 2.3B), the two-way interaction ($p < .0001$) was opposite to the interaction under low threat. Here, attitudes were more predictive of behavior when they were held with relatively low certainty ($p < .001$). For participants who were relatively certain of their attitudes, attitudes were inversely related to extreme action tendencies ($p = .0001$).[4]

In sum, attitude certainty moderated attitude–behavior consistency but did so differently depending on the level of threat salient in the situation. Specifically, when people were relatively unconcerned about COVID-19 and were considering engaging in moderate social distancing behaviors (low threat salience conditions), higher levels of attitude certainty led them to report action tendencies that were more consistent with their attitudes. This pattern is entirely consistent with much previous work on attitude certainty as an attitude strength indicator. In contrast, when people were relatively concerned about the virus and were considering engaging in extreme social distancing behaviors (high threat salience conditions), lower levels of attitude certainty led them to report action tendencies that were more consistent with their attitudes. This pattern is compatible with the notion that when experiencing a threat related to their attitudes in an important domain, people were attempting to compensate for their weak (uncertain) attitudes by expressing a willingness to engage in more attitude-consistent extreme actions.

Conclusions

Extremism is a complex social problem, and multiple levels of analysis must be brought together to understand it fully. We have argued that the study of attitudes can be part of that constellation and offers unique insights to help illuminate the nature and dynamics of extremism. We provided a start on identifying the attitudinal variables that reflect extreme motivations and lead others to see people who hold them as extremists (i.e., polarization, certainty, and unusualness), and we discussed how attitudes with these properties might be fostered. We also addressed the conditions under which attitudinal factors might be useful in predicting engagement in extreme behaviors. In particular, we showed that when people feel threatened by an issue and are considering extreme action, it can be uncertainty rather than certainty in one's attitudes that leads to attitude-consistent actions. More research is needed to better establish the extent and nature of attitudinal processes in relation to extremism, and it is our hope that elaborating these ideas will encourage other researchers to join that effort.

Notes

1. Whether the attitude is considered polarized or not could depend on the method of attitude measurement. For example, if in measuring political attitudes a person was given a −5 (very conservative) to +5 (very liberal) scale, a moderate would score zero and not seem very extreme. However, if the scale asked for attitudes regarding political moderation with −5 representing 'very opposed to moderation' and +5 representing 'strongly favoring moderation', the extremity of the person's attitude toward moderation could more easily be seen.
2. Other attitude strength features can be similarly unusual such as when an attitude is unusually important to the person compared to other people or to the person's other attitudes.
3. Although as noted earlier, considering how people's attitudes toward one object relate to other relevant attitudes could enhance behavioral prediction, we focus on how attitudes toward one particular attitude object relate to behavior as this reflects the most common approach in studies of attitude–behavior consistency.

4. These analyses exclude the mixed or moderate threat conditions (i.e., low concern about the virus with extreme behaviors and high concern about the virus with moderate behaviors). In these cases, certainty did not differentially moderate the impact of attitudes on action tendencies.

References

Abelson, R. P. (1995). Attitude extremity. In R. E. Petty & J. A. Krosnick (Eds.), *Attitude strength: Antecedents and consequences* (pp. 25–41). Hillsdale, NJ: Erlbaum.

Ajzen, I., & Kruglanski, A. W. (2019). Reasoned action in the service of goal pursuit. *Psychological Review, 126*(5), 774–786.

Barden, J., & Petty, R. E. (2008). The mere perception of elaboration creates attitude certainty: Exploring the thoughtfulness heuristic. *Journal of Personality and Social Psychology, 95*(3), 489–509.

Baron, R. S. (2005). So right it's wrong: Groupthink and the ubiquitous nature of polarized group decision making. *Advances in Experimental Social Psychology, 37*, 219–253.

Baron, R. S., Hoppe, S. I., Kao, C. F., Brunsman, B., Linneweh, B., & Rogers, D. (1996). Social corroboration and opinion extremity. *Journal of Experimental Social Psychology, 32*(6), 537–560.

Baumeister, R. F., Bushman, B. J., & Campbell, W. K. (2000). Self-esteem, narcissism, and aggression: Does violence result from low self-esteem or from threatened egotism? *Psychological Science, 9*(1), 26–29.

Bélanger, J. J., Caouette, J., Sharvit, K., & Dugas, M. (2014). The psychology of martyrdom: Making the ultimate sacrifice in the name of a cause. *Journal of Personality and Social Psychology, 107*(3), 494–515.

Bizer, G. Y., Larsen, J. T., & Petty, R. E. (2011). Exploring the valence-framing effect: Negative framing enhances attitude strength. *Political Psychology, 32*(1), 59–80.

Boninger, D. S., Krosnick, J. A., Berent, M. K., & Fabrigar, L. R. (1995). The causes and consequences of attitude importance. In R. E. Petty & J. A. Krosnick (Eds.), *Attitude strength: Antecedents and consequence* (pp. 159–189). Hillsdale, NJ: Erlbaum.

Bornstein, R. F. (1989). Exposure and affect: Overview and meta-analysis of research, 1968–1987. *Psychological Bulletin, 106*(2), 265–289.

Brandt, M. J., Evans, A. M., & Crawford, J. T. (2015). The unthinking or confident extremist? Political extremists are more likely than moderates to reject experimenter-generated anchors. *Psychological Science, 26*(2), 189–202.

Brewer, M. B. (1991). The social self: On being the same and different at the same time. *Personality and Social Psychology Bulletin, 17*(5), 475–482.

Briñol, P., & Petty, R. E. (2005). Individual differences in attitude change. In D. Albarracín, B. T. Johnson, & M. P. Zanna (Eds.), *Handbook of attitudes* (pp. 575–615). Mahwah, NJ: Erlbaum.

Briñol, P., & Petty, R. E. (2009). Persuasion: Insights from the self-validation hypothesis. *Advances in Experimental Social Psychology, 41*, 69–118.

Briñol, P., Petty, R. E., & DeMarree, K. G. (2015). Being threatened and being a threat can increase reliance on thoughts: A self-validation approach. In P. J. Carroll, R. M. Arkin, & A. Wichman (Eds.), *Handbook of Personal Security* (pp. 37–54). New York: Psychology Press.

Briñol, P., Petty, R. E., Gallardo, I., & DeMarree, K. G. (2007). The effects of self-affirmation in non-threatening persuasion domains: Timing affects the process. *Personality and Social Psychology Bulletin, 33*(11), 1533–1546.

Briñol, P., Petty, R. E., & Requero, B. (2017). Aggressive primes can increase reliance on positive and negative thoughts affecting self-attitudes. *Self and Identity, 16*(2), 194–214.

Briñol, P., Petty, R. E., Valle, C., Rucker, D. D., & Becerra, A. (2007). The effects of message recipients' power before and after persuasion: A self-validation analysis. *Journal of Personality and Social Psychology, 93*(6), 1040–1053.

Burnstein, E., & Vinokur, A. (1977). Persuasive argumentation and social comparison as determinants of attitude polarization. *Journal of Experimental Social Psychology, 13*(4), 315–332.

Bushman, B. J., & Baumeister, R. F. (1998). Threatened egotism, narcissism, self-esteem, and direct and displaced aggression: Does self-love or self-hate lead to violence? *Journal of Personality and Social Psychology, 75*(1), 219–229.

Cacioppo, J. T., & Petty, R. E. (1989). Effects of message repetition on argument processing, recall, and persuasion. *Basic and Applied Social Psychology, 10*(1), 3–12.

Carroll, P. J., Briñol, P., Petty, R. E., & Ketchman, J. (2020, July). Feeling prepared increases confidence in any accessible thoughts affecting evaluations unrelated to the original domain of preparation. *Journal of Experimental Social Psychology, 89*, 103962.

Cialdini, R. B. (2003). Crafting normative messages to protect the environment. *Current Directions in Psychological Science, 12*(4), 105–109.

Clark, J. K., & Wegener, D. T. (2013). Message position, information processing, and persuasion: The discrepancy motives model. *Advances in Experimental Social Psychology, 47*, 189–232.

Clarkson, J. J., Tormala, Z. L., & Leone, C. (2011). A self-validation perspective on the mere thought effect. *Journal of Experimental Social Psychology, 47*(2), 449–454.

De Dreu, C. K. W., & Gross, J. (2019). Revising the form and function of conflict: Neurobiological, psychological, and cultural mechanisms for attack and defense within and between groups. *Behavioral and Brain Sciences, 42*, 1–66.

DeMarree, K. G., Wheeler, S. C., Briñol, P., & Petty, R. E. (2014, July). Wanting other attitudes: Actual-desired discrepancies predict feelings of ambivalence and ambivalence consequences. *Journal of Experimental Social Psychology, 53*, 5–18.

Doosje, B., van den Bos, K., Loseman, A., Feddes, A. R., & Mann, L. (2012). 'My in-group is superior!': Susceptibility for radical right-wing attitudes and behaviors in Dutch youth. *Negotiation and Conflict Management Research, 5*(3), 253–268.

Durso, G. R. O., Briñol, P., & Petty, R. E. (2016). From power to inaction: Ambivalence gives pause to the powerful. *Psychological Science, 27*(12), 1660–1666.

Evans, A. T., & Clark, J. K. (2012). Source characteristics and persuasion: The role of self-monitoring in self-validation. *Journal of Experimental Social Psychology, 48*(1), 383–386.

Fazio, R. H. (1990). Multiple processes by which attitudes guide behavior: The MODE model as an integrative framework. *Advances in Experimental Social Psychology, 23*, 75–109.

Fazio, R. H., & Williams, C. J. (1986). Attitude accessibility as a moderator of the attitude—perception and attitude—behavior relations: An investigation of the 1984 presidential election. *Journal of Personality and Social Psychology, 51*(3), 505–514.

Fazio, R. H., & Zanna, M. P. (1978). On the predictive validity of attitudes: The roles of direct experience and confidence. *Journal of Personality, 46*(2), 228–243.

Fernbach, P. M., Rogers, T., Fox, C. R., & Sloman, S. A. (2013). Political extremism is supported by an illusion of understanding. *Psychological Science, 24*(6), 939–946.

Festinger, L. (1950). Informal social communication. *Psychological Review, 57*(5), 271–282.

Festinger, L. (1957). *A theory of cognitive dissonance*. Stanford, CA: Stanford University Press.

Fishbein, M. (1980). A theory of reasoned action: Some applications and implications. In H. Howe & M. Page (Eds.), *Nebraska symposium on motivation 1979* (Vol. 27). Lincoln: University of Nebraska Press.

Fishbein, M., & Ajzen, I. (1975). *Belief, attitude, intention, and behavior: An introduction to theory and research*. Reading, MA: Addison-Wesley.

Friesen, J. P., Campbell, T. H., & Kay, A. C. (2015). The psychological advantage of unfalsifiability: The appeal of untestable religious and political ideologies. *Journal of Personality and Social Psychology, 108*(3), 515–529.

Frimer, J. A., Brandt, M. J., Melton, Z., & Motyl, M. (2019). Extremists on the left and right use angry, negative language. *Personality and Social Psychology Bulletin, 45*(8), 1216–1231.

Gómez, A., Chinchilla, J., Vázquez, A., López-Rodríguez, L., Paredes, B., & Martínez, M. (2020). Recent advances, misconceptions, untested assumptions and future research agenda for identity fusion theory. *Social and Personality Psychology Compass, 14*(6), e12531.

Gopinath, M., & Nyer, P. U. (2009). The effect of public commitment on resistance to persuasion: The influence of attitude certainty, issue importance, susceptibility to normative influence, preference for consistency and source proximity. *International Journal of Research in Marketing, 26*(1), 60–68.

Greenberg, J., Pyszczynski, T. A., Solomon, S., Rosenblatt, A., Veeder, M., Kirkland, S., et al. (1990). Evidence for terror management theory II: The effects of mortality salience on reactions to those who threaten or bolster the cultural worldview. *Journal of Personality and Social Psychology, 58*(2), 308–318.

Greenberg, J., Solomon, S., & Pyszczynski, T. A. (1997). Terror management theory of self-esteem and cultural worldviews: Empirical assessments and conceptual refinements. *Advances in Experimental Social Psychology, 29*, 61–139.

Grubbs, J. B., Warmke, B., Tosi, J., Shanti, J., & Campbell, W. K. (2019). Moral grandstanding in public discourse: Status-seeking motives as a potential explanatory mechanism in predicting conflict. *PLoS One, 14*(10), e0223749.

Hales, A. H., & Williams, K. D. (2018). Marginalized individuals and extremism: The role of ostracism in openness to extreme groups. *Journal of Social Issues, 74*(1), 75–92.

Hart, J. (2014). Toward an integrative theory of psychological defense. *Perspectives on Psychological Science, 9*(1), 19–39.

Hart, W., Albarracín, D., Eagly, A. H., Brechan, I., Lindberg, M. J., & Merrill, L. (2009). Feeling validated versus being correct: A meta-analysis of selective exposure to information. *Psychological Bulletin, 135*(4), 555–588.

Heine, S. J., Proulx, T., & Vohs, K. D. (2006). The meaning maintenance model: On the coherence of social motivations. *Personality and Social Psychology Review, 10*(2), 88–110.

Hogg, M. A. (2007). Uncertainty-identity theory. *Advances in Experimental Social Psychology, 39*, 69–126.

Horcajo, J., Paredes, B., Briñol, P., Petty, R. E., DeMarree, K., & See, Y. H. M (2020). *Threatening uncertainty can enhance thought use: A self-validation analysis*. Unpublished manuscript. Universidad Autonoma de Madrid, Madrid, Spain.

Horcajo, J., See, Y. H. M., Briñol, P., & Petty, R. E. (2008). The role of mortality salience in consumer persuasion. *Advances in Consumer Research, 35*, 782–783.

Huntsinger, J. R. (2013). Incidental experiences of affective coherence and incoherence influence persuasion. *Personality and Social Psychology Bulletin, 39*(6), 792–802.

Janis, I. L. (1972). *Victims of groupthink: A psychological study of foreign-policy decisions and fiascos.* Boston: Houghton Mifflin.

Jasko, K., Webber, D., Kruglanski, A. W., Gelfand, M., Taufiqurrohman, M., Hettiarachchi, M., & Gunaratna, R. (2020). Social context moderates the effects of quest for significance on violent extremism. *Journal of Personality and Social Psychology, 118*(6), 1165–1187.

Jonas, E., McGregor, I., Klackl, J., Agroskin, D., Fritsche, I., Holbrook, C., . . . Quirin, M. (2014). Threat and defense: From anxiety to approach. *Advances in Experimental Social Psychology, 49,* 219–286.

Jost, J. T., Kruglanski, A. W., & Nelson, T. O. (1998). Social metacognition: An expansionist review. *Personality and Social Psychology Review, 2*(2), 137–154.

Judd, C. M., & Brauer, M. (1995). Repetition and evaluative extremity. In R. E. Petty & J. A. Krosnick (Eds.), *Attitude strength: Antecedents and consequences* (pp. 43–71). Hillsdale, NJ: Erlbaum.

Kaiser Family Foundation. (2020). *KFF health tracking poll: Coronavirus, social distancing, and contact tracing, April 15 — April 20, 2020,* [Survey report]. Retrieved from www.kff.org/global-health-policy/issue-brief/kff-health-tracking-poll-late-april-2020/

Kaplan, K. J. (1972). On the ambivalence-indifference problem in attitude theory and measurement: A suggested modification of the semantic differential technique. *Psychological Bulletin, 77*(5), 361–372.

Kelley, H. H. (1967). Attribution theory in social psychology. In D. Levine (Ed.), *Nebraska Symposium on motivation* (Vol. 15, pp. 192–238). Lincoln: University of Nebraska Press.

Krosnick, J. A., & Petty, R. E. (1995). Attitude strength: An overview. In R. E. Petty & J. A. Krosnick (Eds.), *Attitude strength: Antecedents and consequences* (pp. 1–24). Mahwah, NJ: Erlbaum.

Kruglanski, A. W., Bélanger, J. J., & Gunaratna, R. (2019). *The three pillars of radicalization: Needs, narratives, and networks.* New York: Oxford.

Kruglanski, A. W., Jasko, K., Chernikova, M., Dugas, M., & Webber, D. (2017). To the fringe and back: Violent extremism and the psychology of deviance. *American Psychologist, 72*(3), 217–230.

Kruglanski, A. W., Jasko, K., Chernikova, M., Milyavsky, M., Babush, M., Baldner, C., & Pierro, A. (2015). The rocky road from attitudes to behaviors: Charting the goal systemic course of actions. *Psychological Review, 122*(4), 598–620.

Kruglanski, A. W., Raviv, A., Bar-Tal, D., Raviv, A., Sharvit, K., Ellis, S., . . . Mannetti, L. (2005). Says who? Epistemic authority effects in social judgment. *Advances in Experimental Social Psychology, 37,* 345–392.

Kruglanski, A. W., Shah, J. Y., Fishbach, A., Friedman, R., Chun, W. Y., & Sleeth-Keppler, D. (2002). A theory of goal systems. *Advances in Experimental Social Psychology, 34,* 331–378.

Kruglanski, A. W., Szumowska, E., Kopetz, C. H., Vallerand, R. J., & Pierro, A. (2021). On the psychology of extremism: How motivational imbalance breeds intemperance. *Psychological Review, 128*(2), 264–289.

Kruglanski, A. W., & Webster, D. M. (1996). Motivated closing of the mind: "Seizing" and "freezing." *Psychological Review, 103*(2), 263–283.

Kunda, Z. (1990). The case for motivated reasoning. *Psychological Bulletin, 108*(3), 480–498.

Landau, M. J., Kay, A. C., & Whitson, J. A. (2015). Compensatory control and the appeal of a structured world. *Psychological Bulletin, 141*(3), 694–722.

Leander, N. P., Agostini, M., Stroebe, W., Kreienkamp, J., Spears, R., Kuppens, T., . . . Kruglanski, A. W. (2020). Frustration-affirmation? Thwarted goals motivate compliance with social norms for violence and nonviolence. *Journal of Personality and Social Psychology, 119*(2), 249–271.

Levendusky, M. S., Druckman, J. N., & McLain, A. (2016, April–June). How group discussions create strong attitudes and strong partisans. *Research and Politics*, 1–6.

Luttrell, A., Petty, R. E., & Briñol, P. (2016). Ambivalence and certainty can interact to predict attitude stability over time. *Journal of Experimental Social Psychology, 63*, 56–68.

Luttrell, A., & Sawicki, V. (2020). Attitude strength: Predictors versus defining features. *Personality and Social Psychology Compass, 14*(8), e12555.

Martin, R., & Hewstone, M. (Eds.). (2010). *Minority influence and innovation: Antecedents, processes, and consequences*. London: Psychology Press.

McCauley, C., & Moskalenko, S. (2017). Understanding political radicalization: The two-pyramids model. *American Psychologist, 72*(3), 205–216.

McGregor, I. (2003). Defensive zeal: Compensatory conviction about attitudes, values, goals, groups, and self-definition in the face of personal uncertainty. In S. Spencer, S. Fein, & M. Zanna (Eds.), *Motivated social perception: The Ontario symposium* (Vol. 9, pp. 73–92). Mahwah, NJ: Erlbaum.

McGregor, I., Zanna, M. P., Holmes, J. G., & Spencer, S. J. (2001). Compensatory conviction in the face of personal uncertainty: Going to extremes and being oneself. *Journal of Personality and Social Psychology, 80*(3), 472–488.

McGuire, W. J. (1985). Attitudes and attitude change. In G. Lindzey & E. Aronson (Eds.), *Handbook of social psychology* (Vol. 2, pp. 233–346). New York: Random House.

Mrkva, K., & Van Boven, L. (2020). Salience theory of mere exposure: Relative exposure increases liking, extremity, and emotional intensity. *Journal of Personality and Social Psychology, 118*(6), 1118–1145.

Murray, H. A. (1955). Types of human needs. In D. C. McClelland (Ed.), *Studies in motivation* (pp 63–70). New York: Appleton-Century-Crofts.

Myers, D. G. (1978). Polarizing effects of social comparison. *Journal of Experimental Social Psychology, 14*(6), 554–563.

Newby-Clark, I. R., McGregor, I., & Zanna, M. P. (2002). Thinking and caring about cognitive inconsistency: When and for whom does attitudinal ambivalence feel uncomfortable? *Journal of Personality and Social Psychology, 82*(2), 157–166.

Paredes, B., Briñol, P., Petty, R. E., & Gómez, Á. (in press). Increasing the predictive validity of identity fusion in leading to sacrifice by considering the extremity of the situation. *European Journal of Social Psychology*.

Paredes, B., Santos, D., Briñol, P., Gómez, Á., & Petty, R. E. (2020). The role of meta-cognitive certainty on the relationship between identity fusion and endorsement of extreme pro-group behavior. *Self and Identity, 19*(7), 804–824.

Petty, R. E., & Briñol, P. (2012). The elaboration likelihood model. In P. A. M. Van Lange, A. W. Kruglanski, & E. T. Higgins (Eds.), *Handbook of theories of social psychology* (Vol. 1, pp. 224–245). London, England: Sage.

Petty, R. E., Briñol, P., & Tormala, Z. L. (2002). Thought confidence as a determinant of persuasion: The self-validation hypothesis. *Journal of Personality and Social Psychology, 82*(5), 722–741.

Petty, R. E., Briñol, P., Tormala, Z. L., & Wegener, D. T. (2007). The role of meta-cognition in social judgment. In A. W. Kruglanski & E. T. Higgins (Eds), *Social psychology: Handbook of basic principles* (2nd ed., pp. 254–284). New York: Guilford Press.

Petty, R. E., & Cacioppo, J. T. (1981). *Attitudes and persuasion: Classic and contemporary approaches*. Dubuque, IA: Wm. C. Brown.

Petty, R. E., & Cacioppo, J. T. (1986). The elaboration likelihood model of persuasion. *Advances in Experimental Social Psychology, 19*, 123–205.

Petty, R. E., & Cacioppo, J. T. (1990). Involvement and persuasion: Tradition versus integration. *Psychological Bulletin, 107*(3), 367–374.

Petty, R. E., Haugtvedt, C., & Smith, S. M. (1995). Elaboration as a determinant of attitude strength: Creating attitudes that are persistent, resistant, and predictive of behavior. In R. E. Petty & J. A. Krosnick (Eds.), *Attitude strength: Antecedents and consequences* (pp. 93–130). Hillsdale, NJ: Erlbaum.

Petty, R. E., & Krosnick, J. A. (Eds.). (1995). *Attitude strength: Antecedents and consequences*. Hillsdale, NJ: Erlbaum.

Petty, R. E., Ostrom, T. M., & Brock, T. C. (Eds.). (1981). *Cognitive responses in persuasion*. Hillsdale, NJ: Erlbaum.

Petrocelli, J. V., Tormala, Z. L., & Rucker, D. D. (2007). Unpacking attitude certainty: Attitude clarity and attitude correctness. *Journal of Personality and Social Psychology, 92*(1), 30–41.

Priester, J. R., & Petty, R. E. (1996). The gradual threshold model of ambivalence: Relating the positive and negative bases of attitudes to subjective ambivalence. *Journal of Personality and Social Psychology, 71*(3), 431–449.

Rollwage, M. D., Dolan, R. J., & Fleming, S. M. (2018). Metacognitive failure as a feature of those holding radical beliefs. *Current Biology, 28*(24), 4014–4021.

Rocklage, M. D., & Fazio, R. H. (2018), Attitude accessibility as a function of emotionality. *Personality and Social Psychology Bulletin, 44*(4), 508–520.

Rucker, D. D., Petty, R. E., & Briñol, P. (2008). What's in a frame anyway? A meta-cognitive analysis of the impact of one versus two-sided message framing on attitude certainty. *Journal of Consumer Psychology, 18*(2), 137–149.

Rucker, D. D., Tormala, Z. L., Petty, R. E., & Briñol, P. (2014). Consumer conviction and commitment: An appraisal-based framework for attitude certainty. *Journal of Consumer Psychology, 24*(1), 119–136.

Santos, D., Briñol, P., Petty, R. E., Gandarillas, B., & Mateos, R. (2019). Trait aggressiveness predicting aggressive behavior: The moderating role of meta-cognitive certainty. *Aggressive Behavior, 45*(3), 255–264.

Sawicki, V., & Agnew, C. R. (2021). Commitment strength versus commitment bolstering: Uncertainty undermines and promotes relationship success. *Journal of Social Psychology, 161*(1), 47–62.

Sawicki, V., & Wegener, D. T. (2018). Metacognitive reflection as a moderator of attitude strength versus attitude bolstering: Implications for attitude similarity and attraction. *Personality and Social Psychology Bulletin, 44*(5), 638–652.

See, Y. H. M., & Petty, R. E. (2006). Effects of mortality salience on evaluation of ingroup and outgroup sources: The impact of pro- versus counterattitudinal positions. *Personality and Social Psychology Bulletin, 32*(3), 405–416.

Shoots-Reinhard, B., Petty, R. E., DeMarree, K. G., & Rucker, D. D. (2015). Personality certainty and politics: Increasing the predictive utility of individual difference inventories. *Political Psychology, 36*(4), 415–430.

Skitka, L. J., Bauman, C. W., & Sargis, E. G. (2005). Moral conviction: Another contributor to attitude strength or something more? *Journal of Personality and Social Psychology, 88*(6), 895–917.

Skowronski, J. J., & Carlston, D. E. (1989). Negativity and extremity biases in impression formation: A review of explanations. *Psychological Bulletin, 105*(1), 131–142.

Smith, L., Blackwood, L., & Thomas, E. (2020). The need to refocus on the group as the site of radicalization. *Perspectives on Psychological Science, 15*(2), 327–352.

Smith, S. M., Fabrigar, L. R., MacDougall, B. L., & Wiesenthal, N. L. (2008). The role of amount, cognitive elaboration, and structural consistency of attitude-relevant knowledge in the formation of attitude certainty. *European Journal of Social Psychology, 38*(2), 280–295.

Snyder, C. R., & Fromkin, H. L. (1980). *Uniqueness, the human pursuit of difference.* New York: Plenum Press.

Sterling, J., Jost, J. T., & Bonneau, R. (2020). Political psycholinguistics: A comprehensive analysis of the language habits of liberal and conservative social media users. *Journal of Personality and Social Psychology, 118*(4), 805–834.

Strahan, E. J., Spencer, S. J., & Zanna, M. P. (2002). Subliminal priming and persuasion: Striking while the iron is hot. *Journal of Experimental Social Psychology, 38*(6), 556–568.

Swann, W. B., Jr., Gómez, Á., Seyle, D. C., Morales, J. F., & Huici, C. (2009). Identity fusion: The interplay of personal and social identities in extreme group behavior. *Journal of Personality and Social Psychology, 96*(5), 995–1011.

Teeny, J. T., Siev, J. J., Briñol, P., & Petty, R. E. (2021). A review and conceptual framework for understanding personalized matching effects in persuasion. *Journal of Consumer Psychology, 31*(2), 382–414.

Tesser, A. (1976). Attitude polarization as a function of thought and reality constraints. *Journal of Research in Personality, 10*(2), 183–194.

Tesser, A. (1978). Self-generated attitude change. *Advances in Experimental Social Psychology, 11,* 289–338.

Tesser, A., Leone, C., & Clary, E. G. (1978). Affect control: Process constraints versus catharsis. *Cognitive Therapy and Research, 2*(3), 265–274.

Toner, K., Leary, M. R., Asher, M. W., & Jongman-Sereno, K. P. (2013). Feeling superior is a bipartisan issue. Extremity (not direction) of political views predicts perceived belief superiority. *Psychological Science, 24*(12), 2454–2462.

Tormala, Z. L., Falces, C., Briñol, P., & Petty, R. E. (2007). Ease of retrieval effects in social judgment: The role of unrequested cognitions. *Journal of Personality and Social Psychology, 93*(2), 143–157.

Tormala, Z. L., & Rucker, D. D. (2018). Attitude certainty: Antecedents, consequences, and new directions. *Consumer Psychology Review, 1*(1), 72–89.

van Prooijen, J.-W., & Krouwel, A. P. M. (2019). Psychological features of extreme political ideologies. *Current Directions in Psychological Science, 28*(2), 159–163.

Visser, P. S., & Mirabile, R. R. (2004). Attitudes in the social context: The impact of social network composition on individual-level attitude strength. *Journal of Personality and Social Psychology, 87*(6), 779–795.

Wan, E. W., Rucker, D. D., Tormala, Z. L., & Clarkson, J. J. (2010). The effect of regulatory depletion on attitude certainty. *Journal of Marketing Research, 47*(3), 531–541.

Webber, D., Babush, M., Schori-Eyal, N., Moyano, M., Hettiarachchi, M., Bélanger, J. J., … Gelfand, M. J. (2018). The road to extremism: Field and experimental evidence that significance loss-induced need for closure fosters radicalization. *Journal of Personality and Social Psychology, 114*(2), 270–285.

Webster, D. M., & Kruglanski, A. W. (1994). Individual differences in need for cognitive closure. *Journal of Personality and Social Psychology, 67*(6), 1049–1062.

Wegener, D. T., & Petty, R. E. (1997). The flexible correction model: The role of naive theories of bias in bias correction. *Advances in Experimental Social Psychology, 29*, 141–208.

White, D., Szabo, M., & Tiliopoulos, N. (2018). Exploring the relationship between narcissism and extreme altruism. *The American Journal of Psychology, 131*(1), 65–80.

Wicker, A. W. (1969). Attitudes versus actions: The relationship of verbal and overt behavioral responses to attitude objects. *Journal of Social Issues, 25*(4), 41–78.

Winkielman, P., & Cacioppo, J. T. (2001). Mind at ease puts a smile on the face: Psychophysiological evidence that processing facilitation elicits positive affect. *Journal of Personality and Social Psychology, 81*(6), 989–1000.

Zajonc, R. B. (1968). Attitudinal effects of mere exposure. *Journal of Personality and Social Psychology, 9*(2, Pt.2), 1–27.

Zmigrod, L., Rentfrow, P. J., & Robbins, T. W. (2020). The partisan mind: Is extreme political partisanship related to cognitive inflexibility? *Journal of Experimental Psychology: General, 149*(3), 407–418.

3

ON EXTREME BEHAVIOR AND OUTCOMES

The Role of Harmonious and Obsessive Passion

Robert J. Vallerand and Virginie Paquette

ON THE ROLE OF PASSION IN EXTREME BEHAVIOR A LOOK AT OUTCOMES AND DETERMINANTS

At first glance, extremism seems fairly easy to identify. Consider the following. A total of 19 terrorists were trained for months to obtain their flying license and then flew planes in the Twin Towers and the Pentagon killing themselves and over 3,000 US residents in the process. Environmentalist Paul Watson and his Sea Shepherd Society have sunk dozens of ships and their sailors that killed protected endangered sea creatures. This is clearly extremism, right? But what about video gamers who played for days without stopping and ended up dying in the process, is this extremism as well? Or what about Mr. Fauja Singh who started running seriously at age 89 and then successfully completed his first marathon at age 90 and ran a marathon at the age 100? Or former NBA Referee Dick Bavetta who officiated 2,600 consecutive games in the National Basketball Association (that's over 30 seasons straight at the highest level), sometimes injured and ill, until the age of 74? So, are these examples all representative of extremism?

We suggest that they are. Indeed, according to Kruglanski, Szumowska, Kopetz, Vallerand, and Pierro (2021), extremism refers to 'infrequent phenomena whose rarity issues from a pronounced intensity or magnitude of their underlying motivation'. Thus, based on this definition, all of the aforementioned examples pertain to extreme behavior as extremism refers to intense behavior or outcomes as well as results from an extreme motivational force. In line with the aforementioned definition, we wish to make three further points. First, the behavior can be extreme irrespective of its nature. Thus, the activity may be targeted at someone (like the 9/11 terrorist attack) or performed alone (like the marathon achieved

DOI: 10.4324/9781003030898-5

by Mr. Singh at age 100). Second, the activity may lead to negative outcomes (such as sinking ships and onboard sailors) or positive consequences (like Dick Bavetta's unbelievable run of 2,600 consecutive games refereed). Finally, the third issue deals with the underlying motivational force that is hypothesized to be of important magnitude. It may be reasoned that such extreme behavior must result from very high forms of motivation. Indeed, much research reveals that individuals who display extreme behavior are typically *passionate* about the activity they participate in (Vallerand, 2015). However, research also shows that not all passions are equivalent and while some may lead to extreme behavior and others do not, at least not to the same extent. Thus, it is crucial to explore the role of different types of passion in extreme behavior and outcomes.

The purpose of this chapter is to address the role of passion in extreme behavior. We first present the construct of passion and our underlying theoretical model, the Dualistic Model of Passion (Vallerand, 2015). Then, second, we review research on the role of passion in extreme behavior. It will be seen that passion plays a role in a variety of extreme behavior at both the interpersonal and intrapersonal levels. Third, we address the role of the determinants of passion. Finally, we highlight key conclusions.

A Dualistic Model of Passion

On the Concept of Passion

Passion has a long history but a short past. Indeed, although much discourse on passion dates back to the Greek philosophers, psychological research on the topic of passion for activity engagement basically started less than 20 years ago (see Vallerand, 2015, Chapter 2). In 2003, Vallerand and his colleagues then proposed and adduced evidence for the Dualistic Model of Passion (Vallerand et al., 2003) to fill the void in the psychological literature on the determinants and consequences of passion. Since then, hundreds of studies have been conducted, and the DMP has become the main theoretical model on passion (see Vallerand, 2008, 2010, 2015; Vallerand & Houlfort, 2019).

The foundation of the DMP rests on the basic premise that activities people engage in are vital to personal growth, as they provide the necessary ingredients for psychological needs, socialization, and human development. Such a premise is in line with the organismic perspective (e.g., Deci & Ryan, 2000; White, 1959) that posits that needs for autonomy, competence, and relatedness represent key nutrients for psychological well-being. The DMP further proposes that people engage in various activities throughout life to grow as individuals. After a period of trial and error that would appear to start in early adolescence (Erikson, 1968), most people eventually start to show a preference for some activities, especially those that are perceived as particularly enjoyable and important, provide support for psychological needs, and have some resonance with how they see themselves.

These activities become passionate activities. Although a passion for certain activities typically starts to develop in adolescence at a time when they engage in identity construction (Erikson, 1968), people can develop a passion at any point in life. Two examples are Tiger Woods at a very young age (he started golf at 2!) and Mr. Singh's passion for running at 90 years of age.

The DMP defines passion as a strong inclination toward a specific object, activity, concept, or person that one loves (or at least strongly likes), highly values, invests time and energy on a regular basis, and that is part of identity (see Vallerand, 2015, Chapter 2 for an elaborated discussion). Because passion provides an important means to reach self-growth, most people (around 80%) are passionate toward at least one activity in their life (Vallerand, 2015). Thus, passion is not a trait but is oriented toward a specific activity (see Vallerand, 2015). For instance, one may be passionate for one activity (e.g., politics) but not another (playing basketball). And even within politics, some may be passionate for a given ideology (e.g., promoting a strong and unified Canada) but not another (e.g., the independence of Quebec from Canada). Finally, the last dimension of passion is that it takes a *dualistic* form and can lead to adaptive or maladaptive outcomes, including some extreme ones. Thus, as we shall see later, there are two types of passion, one that is more adaptive (harmonious) and another that is less so (obsessive).

On Harmonious and Obsessive Passion

One of the key postulates from the DMP is that activities that people love eventually get internalized in the person's identity to the extent that these are highly valued and meaningful for the person (Aron, Aron, & Smollan, 1992; Csikszentmihalyi, Rathunde, & Whalen, 1993). Furthermore, in line with research from self-determination theory (Ryan & Deci, 2017) that posits that values and regulations concerning *noninteresting* activities can be internalized in either a controlled or an autonomous fashion, the DMP posits that enjoyable activities can also be internalized. In fact, two types of passion, obsessive and harmonious, can be distinguished depending on whether the passionate activity has been internalized in a controlled or autonomous way.

OP results from a controlled internalization of the activity into one's identity. Such an internalization process leads the activity representation and values and regulations associated with the activity to be at best partially internalized in identity, and at worse to be internalized in complete isolation of other self-elements. In other words, the activity representation is compartmentalized in identity (Bélanger, Lafreniere, Vallerand, & Kruglanski, 2013; Philippe, Vallerand, Bernard-Desrosiers, Guilbault, & Rajotte, 2017) and as such is more difficult to regulate. A controlled internalization is controlled in the sense that it originates from intrapersonal and interpersonal pressure that forces you to behave in a certain way. It typically occurs because certain contingencies are attached to the activity such as feelings of social acceptance or self-esteem (Lafrenière, Bélanger, Sedikides, &

Vallerand, 2011; Mageau, Carpentier, & Vallerand, 2011), or a lack of emotional regulation (St-Louis, Rapaport, Poirier, Vallerand, & Dandeneau, 2021) where one's sense of excitement derived from activity engagement becomes uncontrollable. Thus, with OP, people experience an uncontrollable urge to partake in the activity they view as important and enjoyable. They cannot help but to engage in the passionate activity that they love, sometimes at the expense of other activities in their life. Such rigid persistence may lead to some important adaptive outcomes such as achieving high performance. However, at the same time, such rigid persistence may also lead to life conflicts and other negative affective, cognitive, and behavioral consequences (Vallerand, Chichekian, Verner-Filion, & Rapaport, 2021). For instance, if a university professor is working on an interesting paper and it is time to go home, she might not be able to resist the urge to keep on working on the paper. If she does, she might feel guilty and upset with herself while writing for having stayed at work and may have difficulties focusing on the task at hand, preventing her from experiencing positive affect and flow while writing. It is thus proposed that with OP individuals come to display a rigid persistence toward the activity and extreme behavior as oftentimes they can't help but to engage in the passionate activity even when they should not. While such rigid persistence may lead to some benefits (e.g., a very impressive number of publications), it may also come at a cost for the individual, potentially leading to less than optimal functioning within the confines of the passionate activity (not having as much fun or even creativity while writing) as well as outside of it (e.g., relationship problems; see Vallerand, 2015; Vallerand et al., 2021).

Conversely, HP results from an autonomous internalization of the activity into the person's identity and self. Such internalization occurs when individuals have freely accepted the activity as important for them without any contingencies attached to it. This type of internalization facilitates the integration of the activity in a coherent fashion with other self-elements. It then leads to a motivational force to engage in the activity willingly with a sense of volition and personal endorsement about pursuing the activity. When HP is at play, individuals freely choose to engage in the beloved activity. With this type of passion, the activity occupies a significant but not overpowering space in the person's identity and is internalized in harmony with other self-elements and aspects of the person's life. With HP, the person can fully partake in the passionate activity with an open, nondefensive, and mindful engagement (St-Louis, Verner-Filion, Bergeron, & Vallerand, 2018) that is conducive to positive experiences. Consequently, people with a HP should be able to fully focus on the task at hand and experience positive outcomes both during (e.g., positive affect, concentration, flow) and after engaging the task (e.g., general positive affect, satisfaction). Thus, there should be little or no conflict between the person's passionate activity and his/her other life activities. Furthermore, when prevented from engaging in their passionate activity, people with a HP should be able to adapt well to the situation and focus their attention and energy on other tasks that need to be completed. Thus, when confronted with the possibility of staying at work to continue writing the ever-important article or going home, our university professor with a HP for work can readily jot down ideas about the paper in order to be ready to start writing the next day and proceed to look forward to the evening's activities with the family without thinking about the missed opportunity to write. Thus, behavioral engagement in the passionate activity can be seen as flexible, and the person can let go and engage in other activities when it is important to do so.

Thus, two types of passion exist, HP and OP. They are, respectively, internalized either in harmony or in isolation from other self-elements, entail two different types of engagement in the passionate activity with different implications for behavior and outcomes. With OP, such a self-organization leads to extreme engagement very often leading to extreme behavior and outcomes. This is because OP creates a goal-shielding mechanism (Bélanger et al., 2013) where engagement in the passionate activity is often pursued at the expense of other activities in the absence of other countervailing forces. Thus, extreme engagement, behavior, and outcomes are then free to take place. Conversely, with HP, other important self-elements remain activated as HP is part of a coherent and integrated self-structure (Bélanger et al., 2013; Vallerand, 2015). Thus, people can then fully engage in the passionate activity while also having access to other self-elements. They can then stop engagement if other needs are more pressing or other life activities need attention. Thus, in that sense, with HP, activity engagement remains under the control of the person leading to more moderate, and less extreme, behavior and outcomes, although if necessary extreme engagement remains possible in the short term (Vallerand et al., 2021). This analysis leads to the prediction that although both HP and OP lead to full engagement in the passionate activity, it is an OP that should typically yield the most extreme behavior and outcomes.

Research Methods on Passion

In our passion research, two methodological tools are used. First, we use the Passion Scale (Marsh et al., 2013; Vallerand et al., 2003) to measure HP and OP. Results of exploratory and confirmatory factor analyses in our first study (Vallerand et al., 2003, Study 1) provided support for the validity of the two subscales assessing OP and HP. Sample items for OP include 'I almost have an obsessive feeling toward this activity' and 'I have the impression that my activity controls me', and sample items for HP include 'This activity is in harmony with other activities in my life' and 'This activity is well integrated into my life'. The factor validity of the Passion Scale has been replicated in well over 20 studies and in at least 12 different languages (see Vallerand, 2015, Chapter 4; Vallerand & Rahimi, in press). In addition, in a subsequent study with over 3,000 participants, Marsh et al. (2013) have shown that both subscales were positively related to the passion criterion subscale (love and valuation of the activity, time commitment, and the activity being part of identity and perceived as a passion) as well as differently related to life satisfaction (positive relationship for HP and no relationship for OP) and conflict and rumination (positive relationship for OP and no relationship for HP) as posited to by the DMP. Marsh et al. (2013) also showed that the scale is invariant as a function of language (French and English), gender, and five different types of activities. Furthermore, internal consistency analyses have shown that both subscales are reliable (typically .75 and above), and test–retest correlations over periods ranging from four to 6 weeks revealed moderately high stability values (in the .80s; Rousseau, Vallerand, Ratelle, Mageau, & Provencher, 2002). Finally, hundreds of studies have been conducted with the Passion Scale showing that HP leads to more adaptive outcomes than OP, thereby providing support for the construct validity of the Passion Scale and for the DMP as well.

In addition to measuring passion using the Passion Scale, a second methodological tool pertains to experimentally inducing either OP or HP. The situational induction of passion that we have developed is in line with theory and research to the effect that psychological constructs are dynamic and can be operationalized in terms of both individual differences and short-lived situations (for review and discussion, see Avnet & Higgins, 2003). Thus, passion can be seen as a psychological construct that is relatively stable toward a specific activity (one can generally display a HP or OP toward a specific activity) while being also able to fluctuate at the state level (one can be made to behave like someone who has a predominant harmonious or OP toward a given activity at a given point in time). Thus, to the extent that someone has a passion for a given activity, it is possible to trigger one type of passion or the other because both types of passion have been internalized in identity, albeit to different degrees and are thus available and accessible (Vallerand, 2015). Similar manipulations have proven to be effective in activating specific psychological constructs, such as different regulatory focus modes (Avnet & Higgins, 2003; Pierro et al., 2008).

The manipulation entails the following procedures. Typically, participants are asked to 'identify an activity that you love, that is important to you, and in which you invest a significant amount of time on a regular basis'. (just like participants who are asked to complete the Passion Scale). Subsequently, participants engage in a writing task for a period of 5 minutes where they are randomly assigned to one of the three writing tasks (the HP, OP, or control condition inductions).

In the *HP condition*, participants are instructed to:

> Write about a time when your favorite activity was in harmony with other things that are part of you and you felt that your favorite activity allowed you to live a variety of experiences. Recall this event vividly and include as many details as you can to relive the experience.

Participants in the *OP condition* are assigned to a similar writing task but are instructed to:

> Write about a time where you had difficulties controlling your urge to do your favorite activity and you felt that your activity was the only thing that

really turned you on. Recall this event vividly and include as many details as you can to relive the experience.

Research has shown that these manipulations produce the same effects as the OP and HP subscales (e.g., Bélanger et al., 2013; Lafrenière, Vallerand, & Sedikides, 2013; Schellenberg & Bailis, 2015). Furthermore, because participants have been randomly assigned to these different conditions, researchers can make causal claims regarding the role of passion in outcomes.

In sum, initial research (Vallerand et al., 2003, Study 1) provided support for the psychological constructs of OP and HP as well as for the validation of the Passion Scale and the experimental inductions of HP and OP. Overall, several hundreds of studies have been conducted on the construct of passion using the DMP perspective (see Curran, Hill, Appleton, Vallerand, & Standage, 2015; Vallerand, 2010, 2015, for reviews). We now turn to our review of passion and extreme behavior.

Passion and Extreme Behavior

As mentioned previously, several studies have focused on extreme behavior. It is impossible to review all of these outcomes in this chapter. In the following, we review some of these studies mainly focusing on our own research while mentioning key studies conducted by other scientists. Such research is presented as a function of the intrapersonal and interpersonal nature of the activity. In line with the DMP, it is expected that OP should be associated with more extreme outcomes, whereas it will be less likely for HP. This is especially the case because with OP, the passionate activity has been internalized in identity in isolation (instead of harmony) with other self-elements leading OP to focus on the activity at the expense of other concerns (Vallerand, 2015). Thus, there are no (or limited) internal constraints (Kruglanski et al., 2021), and OP is then allowed to lead to extreme behavior. In addition, such lack of control over the activity and rigid persistence may also lead to negative affective and cognitive task experiences that in and of themselves may further contribute to extreme behavior. Such is not the case for HP where the activity is integrated with other aspects of the self, thereby allowing the person to pay attention to other self and life elements and to display flexibility allowing the person to physically and mentally disengage from the passionate activity when it is necessary to attend to other goals (Bélanger et al., 2013). In addition, such flexibility allows the person to experience a number of positive affective and cognitive task experiences that will prevent extreme behavior and lead to adaptive life outcomes (Vallerand, 2015).

Passion and Extreme Interpersonal Behavior

A number of studies have been conducted to test the role of passion in extreme interpersonal behavior. A first study (Rip, Vallerand, & Lafrenière, 2012, Study 1) examined the role of passion toward a political ideology and the intention to use radical and mainstream behavior to promote a cause. Participants were members of political parties that promoted the independence of Quebec from Canada. They were asked to complete the Passion Scale toward their ideology and their intention to engage in democratic and radical political activism to reach the cause. Results revealed, as expected, that OP was related to the endorsement of radical and aggressive political tactics (e.g., subversion, sabotage, readiness to use arms, and to give one's life), while HP was associated with the intention to engage in peaceful and democratic political acts (e.g., making financial contributions, door-to-door discussion to convince others peacefully). These results on the role of OP in violent and extreme behavior have been replicated with other political activists (Bélanger et al., 2019, Study 1b) as well as with environmental activists and in this case while using experimental inductions of HP and OP (Bélanger et al., 2019, Study 3). Inducing OP with the activists led to higher levels of violent behavior (e.g., intentions of bombing a polluting company) toward those who do not respect the environment, whereas inducing HP led to more adaptive behavior (e.g., talking and exchanging ideas with people).

If OP influences a number of extreme interpersonal behaviors, then what are the psychological processes mediating these effects? The DMP posits that cognitions and states experienced during task engagement should represent important mediators of such effects. Indeed, much research has shown that OP typically leads to negative affect and HP to positive affect during task engagement (Curran et al., 2015; Vallerand, 2010, 2015). These two types of affective states have been found to lead to different types of outcomes with negative affect such as anger leading to extreme negative behavior. Conversely, positive emotions have mainly been found to lead to more adaptive and typically less extreme outcomes (Fredrickson, 2001). This is because positive emotions open up the mind to new behavior and build up an adaptive repertoire that is likely to open up to several new avenues and reducing extreme behavior (Fredrickson, 2013). In addition, the DMP posits that rumination about the activity that one is passionate about when engaged in something else is likely to trigger some extreme negative outcomes as well (Carpentier, Mageau, & Vallerand, 2012).

Research has provided support for the aforementioned hypothesis regarding the mediating role of negative emotions between OP and extreme interpersonal behavior (Gousse-Lessard, Vallerand, Carbonneau, & Lafrenière, 2013; Philippe, Vallerand, Richer, Vallieres, & Bergeron, 2009, Study 3; Rip et al., 2012, Study 2; Vallerand, Ntoumanis, et al., 2008, Study 2). For instance, in a study with devout Muslims contacted at Montreal mosques (Rip et al., 2012, Study 2), the authors measured passion, hatred against others, as well as intentions of engaging in violent or peaceful behavior. Results from structural equation modeling (SEM) analyses revealed that OP predicted hatred that, in turn, led to the endorsement of religious extremism and intentions of violence toward others (e.g., engaging in a Fatwah or a Holy war). The effects were even stronger when participants with OP read a threatening excerpt against Islam (actual statement from Pope Benedict against Islam). Conversely, HP positively and directly predicted peaceful religious activism (e.g., promotion of reconciliation and rebuilding of relationships).

The findings of the study by Rip et al. (2012, Study 2) were largely replicated in a study on environmental activism (Gousse-Lessard et al., 2013, Study 3) using a broader measure of negative emotions. Results from SEM analyses revealed that OP for the environmental cause led to negative emotions that, in turn, predicted extreme radical behaviors. Conversely, HP was related to positive emotions that, in turn, predicted peaceful activism. It should be noted that OP was also positively related to positive emotions, although less so than HP. This last finding underscores the complex nature of OP and the fact that it may sometimes yield conflicted psychological experiences during task engagement that may be both positive and negative in nature. Such conflicted emotional experiences and outcomes were replicated in the study with soccer fans during the 2006 World Cup of Soccer (Vallerand, Ntoumanis et al., 2008, Study 2). OP led to hatred toward fans of other teams that led to mocking them in the street. Clearly, doing so can lead to violent exchanges and fights between gangs. OP may thus be involved in hooligan behavior. Conversely, HP only predicted pride that led to peaceful celebrations. In these various studies, HP was not involved in violent behavior. It should be mentioned that in this study OP also predicted pride and thus indirectly contributed to positively celebrating peacefully. These findings suggest that the behavior that is fueled by OP is complex and conflicted and includes extreme negative interpersonal behavior and at times more moderate reactions (peaceful celebrations). Conversely, more moderate and adaptive behavior always results from HP. In addition, affective experiences during task engagement seem to be responsible for both extreme and moderate behavioral effects.

Other studies have also looked at aggression and immoral behavior. For instance, in two studies in sports, basketball players with OP displayed more aggression toward opponents, especially when winning was on the line (Donahue, Rip, & Vallerand, 2009, Studies 1 and 2). Research has also shown that OP leads to immoral behaviors such as cheating in paintball games (Bureau, Vallerand, Ntoumanis, & Lafrenière, 2013, Study 1) as well as in sports (Bureau et al., 2013, Study 2). As pertains to immoral behavior, Bélanger et al. (2019) conducted a series of studies involving participants passionate for the environmental cause (Bélanger et al., 2019, Studies 1a, 2, 3, and 4) and politics (Bélanger et al., 2019, Study 1b). The results of this series of studies revealed that OP and HP were positively and negatively, respectively, associated with moral disengagement that, in turn, led to the support of violent behaviors against those who don't respect the cause they are passionate about. Conversely, HP directly predicted a more moderate response, namely, the promotion of mainstream activism (e.g., peaceful protest).

Of interest, in one of these studies (Study 2), Bélanger et al. used a particularly ingenious procedure to examine how passion was related to dehumanization. Participants were instructed to watch computer-created images where a human face is morphed into a doll-like face and to indicate at which point the on-screen image no longer has 'a mind' (doll-like face). They were also told that these faces represented oil company executives. Thus, the quicker one responds, the more one perceives a human face as nonhuman. This is a measure of situational dehumanization, one specific dimension of moral disengagement. Then participants were asked to indicate to what extent they would be ready to engage in violent or peaceful actions toward people who engage in actions against the environment. The results of a SEM analysis indicated that OP was related to quicker dehumanization that, in turn, was associated with the promotion of violent behaviors. HP was not related to dehumanization and predicted intention to use peaceful behaviors.

In another series of studies (Philippe et al., 2009, Studies 1, 2, and 3), we ascertained the role of passion in another type of extreme behavior, road rage driving. In the first two studies (including Study 2 that used a representative sample of 450 drivers from the Province of Quebec), we showed that OP for driving positively correlated with measures of verbal and physical aggression while generally driving, as well as aggressive use of the car, whereas HP did not. Study 3 went one step further and took place with passionate drivers in a driving simulator, under controlled laboratory conditions. In this controlled situation, participants sat in a real car and could see other cars were displayed on the screen, including a yellow car that constantly slowed the participant. Passion for driving, anger toward the driver of the car (who was supposedly in the laboratory next door), as well as self-reported aggression toward the yellow car were all measured. In addition, participants were videotaped and assessed by judges for the level of displayed verbal and physical aggression. Results from a path analysis revealed that OP predicted anger that led to both self-reported and objective aggression (as assessed by judges on the video). Thus, OP seems to be involved in road rage and anger mediates the effects. Such was not the case with HP. Of interest, one of the participants with a high level of OP at some point had enough and wanted to go next door and let the driver of the yellow car have it. Of course, we stopped him in his tracks . . . but it makes for an interesting story nevertheless!

Interestingly, research has revealed that passion is also related to extreme behavior in romantic relationships. In an important series of four studies, Bélanger et al. (under review) have looked at the role of romantic passion in stalking toward a previous lover. Participants completed the romantic passion scale (Ratelle, Carbonneau, Vallerand, & Mageau, 2013) as well as scales assessing stalking (i.e., trespassing private and legal boundaries, harassing and intimidating, coercion and threat, and violence) toward a past lover. Research provided support for the role of OP in stalking behavior. Specifically, OP was related to being aggressive and even engaging in illegal action to rekindle the relationship. Such was not the case for HP where more subdued and moderate behavior was emitted such as engaging in discussion with one's former lover. Of importance are the findings of Study 3 in which these findings were replicated while using *experimental* inductions of OP and HP.

Carbonneau and Vallerand (2013) also looked at conflict behavior between romantic partners in ongoing relationships. They found that OP predicted engaging in destructive conflict behavior, whereas HP predicted engaging in repair behavior after conflict. In another series of three studies, Carbonneau, Vallerand, Lavigne, & Paquet (2016) assessed how passion affects the personal outcomes of personal development experienced within the purview of the relationship. The authors reasoned that with OP, individuals are willing to go to the extreme length to maintain their romantic relationship, including letting go of friendships and personal activities to limit themselves fully to the loved one. Conversely, with HP, it was expected that individuals could grow and internalize dimensions of the loved one within themselves, as well as maintain their own identity and other past friendships and activities. That is exactly what the findings showed, including changes over time using a prospective design (Study 2). Of major importance were the results of Study 3, which replicated these findings with the assessment of the participant's best friend. It would seem that with OP, individuals are ready to go to extreme levels in order to maintain the romantic relationship by letting go of who they were and stall self-growth within the purview of the relationship. However, the exact opposite takes place with HP where one maintains his or her sense of identity and adds new dimensions from the loved one. Thus, with OP, one ends up losing one's sense of identity, which is rather extreme as an outcome, whereas with HP, one grows even more and expands one's sense of identity.

Finally, Philippe, Vallerand, Guilbault, and their colleagues have also applied the DMP to sexual behavior, including some extreme sexual behavior and ramifications. Philippe and colleagues reasoned that sexual passion is distinct from romantic passion as sexual activities can be performed alone or with someone else, and in this latter case, within or outside the relationship. These authors have conducted a total of ten studies that show that the distinction between HP and OP matters greatly with respect to sexual behavior (Guilbault, Bouizegarene, Philippe, & Vallerand, 2019, Studies 1 and 2; Philippe et al., 2017, Studies 1–5; 2019, Studies 1–3). For instance, some of these studies show that OP for sexuality leads one to experience intrusive cognitive sexual thoughts (Philippe et al., 2017, Study 1), attention to alternative partners, and biased processing of stimuli, such as seeing sex-related words where there are none (Philippe et al., 2017, Study 4). In addition, obsessive sexual passion leads to violent actions toward the loved one, especially under threat (Philippe et al., 2017, Study 2; 2019, Study 1), as well as engagement in extradyadic sex (Guilbault et al., 2019, Studies 1 and 2; Philippe et al., 2019, Study 3), and the greater likelihood of breaking up over time (Philippe et al., 2017, Study 3). Such was not the case for HP that was unrelated to these extreme outcomes while being positively related to sexual satisfaction (Philippe et al., 2017, Study 1; 2019, Studies 1 and 3) and relationship quality (Philippe et al., 2017, Studies 2 and 3; 2019, Studies 1, 2, 3). In sum, OP (but not HP) matters with respect to a number of extreme interpersonal sexual behaviors.

Overall, the aforementioned review reveals that OP leads to a number of extreme interpersonal behaviors that include violent religious, political, and cause-related activism, stalking, romantic, sexual behavior, sport fanatic behavior, and road rage. The fact that one construct, namely, OP, can predict a variety of diverse range of extreme behaviors provides support for the DMP on the role of OP in maladaptive outcomes, relative to HP. These findings are also in line with Kruglanski et al.'s (2021) claim that extreme behavior is fueled by a powerful motivational force. In addition, affective and cognitive on-task experiences such as anger, negative emotions, and dehumanization mediate the effects of OP. This last finding provides additional support to the DMP that posits that what is experienced during task engagement plays a crucial role in immediate as well as subsequent behavior and outcomes for both the passionate person and other individuals (Vallerand, 2015). Conversely, HP is related to positive emotions and more adaptive outcomes such as mainstream and peaceful behaviors and positive relationship quality.

Passion and Extreme Intrapersonal Behavior

A first area where the role of passion in extreme intrapersonal behavior has been documented is gambling addiction. It is important here to underscore that passion refers to the quality of one's engagement in the activity and addiction to one of the types of consequences that can result from such engagement. Furthermore, although OP entails loving the activity, addiction may not, as sometimes one wishes that he or she could stop doing the activity such as gambling. Finally, OP is seen as a predictor of addiction. Thus, because with OP engagement in gambling gets out of control, then one likely consequence is addiction and pathological gambling. However, addiction would not take place with HP as engagement remains under the person's control. In an early study (Rousseau et al., 2002), results showed that OP was positively related to some problematic gambling behaviors and self-perceptions such as being a heavier gambler, as gambling more money and for longer periods of time than intended. Conversely, HP was unrelated to such extreme behavior. Other research also clearly shows that involvement fueled by OP leads to other undesired outcomes such as experiencing severe gambling problems (e.g., Back, Lee, & Stinchfield, 2011; Morvannou, Dufour, Brunelle, Berbiche, & Roy, 2017a, 2017b; Philippe & Vallerand, 2007; Ratelle, Vallerand, Mageau, Rousseau, & Provencher, 2004; Skitch & Hodgins, 2005) as well as negative emotional consequences both while gambling and following a gambling episode (e.g., anxiety, feelings of guilt; Lee, Chung, & Bernhard, 2013; Mageau, Vallerand, Rousseau, Ratelle, & Provencher, 2005; Ratelle et al., 2004). Such was not the case with HP.

Subsequent research went one step further and looked at pathological gambling. In a study with Montreal residents who were representative of the general population of adults aged 55 years and over (Philippe & Vallerand, 2007), participants completed the Passion Scale for gambling and a measure of pathological gambling: the South Oaks Gambling Scale-Revised (SOGS-R; Lesieur & Blume, 1993). The SOGS-R categorizes participants into three categories: nonpathological gamblers, at-risk gamblers, and pathological gamblers. Because only a small percentage of the population displays pathological gambling (less than 3% of the general population; Ladouceur, 1996), it represents an extreme form of behavior. The results revealed that for participants in the first two categories (nonpathological gamblers and at-risk gamblers), their score on the HP subscale was significantly higher than that on the OP subscale. However, for participants who were diagnosed as pathological gamblers, there was a complete reversal as OP was significantly higher than their HP. In fact, their OP scores were twice as high as their HP scores. Finally, in a 1-year prospective study, Morvannou et al. (2017a) showed that having an OP for gambling *doubled* the risk of experiencing serious gambling problems 1 year later! HP did not relate to gambling problems. In fact, it even appears that gambling in games such as poker out of HP can even lead to psychological well-being (Oikonomidis, Palomäki, & Laakasuo, 2019).

In another study (Vallerand et al., 2003, Study 4), we looked at passion in a group of gamblers who have very serious problems: those who have opted for self-exclusion from casinos. Because they have lost control over their gambling, such gamblers need the authorities of the casino to control it for them and bar them from entry. When a gambler asks for self-exclusion, this is because he or she has hit rock bottom! In this study, Vallerand et al. compared self-excluded gamblers' passion with a group of regular casino players. The results showed that while neither group differed on HP, self-excluded gamblers had significantly higher OP scores than regular gamblers. Of interest, using OP and HP to predict group membership (self-excluded vs. regular gamblers) allowed us to correctly predict 79% of the self-excluded participants and 81% of the regular casino players in their respective group. Thus, the results supported our hypotheses on the opposite role of OP and HP in extreme gambling behavior.

Other research has looked at the role of OP in other types of addiction. One is substance addiction. Substance addiction is typically a rare event that characterizes only a very small portion of the population (around 1%). Studies have revealed that, while both types of passion were related to alcohol and marijuana consumption, OP was more strongly related to it (Davis, Arterberry, Bonar, Bohnert, & Walton, 2018; Davis & Arterberry, 2019; Steers et al., 2015). One should expect such findings as both OP and HP represent a passion for consuming alcohol or marijuana and thus should predict regularly engaging in such behavior. As with gambling, however, OP was also positively associated with substance consumption-related problems (e.g., frequent intoxication, neglecting responsibilities), whereas HP was not and was even positively related to adaptive outcomes such as life satisfaction (Davis, 2016; Davis et al., 2018; Davis & Arterberry, 2019; Dolan, Arterberry, & Davis, 2020; Steers et al., 2015). These findings are important as they suggest that although OP for drinking and smoking marijuana leads to addiction problems, HP for such behavior does not and may even lead to positive psychological outcomes such as enjoyment while drinking. Future research is necessary to determine whether HP and OP may lead to different physiological effects that may translate into various drinking and smoking experiences and consequences.

Other research has also shown that OP, but not HP, is positively related to other types of addictive behaviors such as addiction to Internet and video games (Burnay, Billieux, Blairy, & Larøi, 2015; Lafrenière, Vallerand, Donahue, & Lavigne, 2009; Wang & Chu, 2007) and exercise dependence (Kovacsik, Soós, de la Vega, Ruíz-Barquín, & Szabo, 2018; Paradis, Cooke, Martin, & Hall, 2013; Parastatidou, Doganis, Theodorakis, & Vlachopoulos, 2014; Sicilia, Alcaraz-Ibáñez, Lirola, & Burgueño, 2017; Stenseng, Haugen, Torstveit, & Høigaard, 2015, Study 2). Regarding exercise dependence, it should be noted that when controlling for the volume of exercise, only OP predicts exercise addiction and HP does not (see Szabo & Kovacsik, 2019).

The aforementioned research on the role of OP in exercise addition opens the door to the possibility that OP also leads to physical injuries. Indeed, obsessively passionate individuals, because of their rigid persistence, should pursue their favorite activity even when the conditions indicate that they should stop for their own safety, thus jeopardizing their health. Conversely, harmoniously passionate people, because they have a flexible engagement in their activity, should be able to momentarily disengage from their activity when the conditions are too risky. This is exactly what research reveals (Akehurst & Oliver, 2014).

Other research has looked at-risk behavior for injuries. In a study conducted by Vallerand and his colleagues (2003, Study 3), it was shown that assessing OP for cycling during the summer in Canadian cyclists allowed us to predict who would engage in winter cycling under dangerous conditions later that winter. HP was not related to such risky behavior. Furthermore, a study with dancers (Rip, Fortin, & Vallerand, 2006) revealed that, while both types of passion protected against acute injuries because passionate individuals train regularly, OP was related to chronic injuries. Indeed, when they were injured, obsessively passionate dancers kept dancing which gave them no opportunity to totally heal from their injuries, thereby leading to chronic injuries. Such was not the case with HP as dancers were then able to momentarily stop when they were injured and take some time to heal, thus preventing chronic injuries. Other research in France with regular runners has shown that OP (but not HP) is a significant predictor of injuries (Stephan, Deroche, Brewer, Caudroit, & Le Scanff, 2009).

Other studies have looked at health issues. For instance, in a series of three studies with individuals involved in humanitarian causes, including some who went on an international mission, St-Louis and her colleagues (2016) found that OP was positively associated with health problems related to their humanitarian work. Of interest is Study 3 that used a 3-month cross-lagged panel design to assess the interplay between passion and health outcomes before and after the return from a humanitarian mission abroad. The results from SEM analyses showed that the overall direction of effects was from passion to changes in health outcomes. Specifically, OP led to increases in negative physical symptoms (e.g., rashes, colds) and decreases in a validated measure of self-rated health. Conversely, HP led to increases in positive health ratings and decreases in negative physical symptoms. In addition, OP was found to predict self-neglect while away during the mission that, in turn, predicted health problems upon return of a humanitarian mission. OP also positively predicted rumination about one's humanitarian work that, over time, led to posttraumatic stress disorder following the international mission. HP was unrelated to these negative outcomes and was positively related to satisfaction with one's involvement in the cause. Overall, these findings underscore the fact that OP fosters a mindset leading to self-neglect and constant rumination that may have both intrapersonal health costs as well as interpersonal diminished effectiveness to help others. Future research on this issue is necessary to replicate these findings over a longer period of time.

As seen earlier, OP leads to extreme behavior as the person just can't let go and continues until something breaks or gives. We have seen earlier that one negative outcome is posttraumatic stress disorder. Another health variable of importance that we've looked at is burnout. Because OP triggers rigid persistence, it should promote burnout. Such should not be the case with HP as it promotes a flexible persistence, taking time off, thereby preventing burnout and actually promoting mental health (Vallerand et al., 2021). Research has supported these hypotheses. For instance, in a study conducted by Carbonneau and her colleagues (2008), more than 490 teachers completed scales assessing their passion for teaching, work satisfaction, and burnout at two points in time over a 3-month period. Results showed that, as expected, OP and HP were positively and negatively, respectively, related to burnout at both Time 1 and Time 2. In addition to protecting one against burnout, HP also predicted increases in work satisfaction over time. These findings provide support for the role of OP in extreme psychological behavior at work, namely, burnout, while HP seems to protect one against it.

Several other studies have replicated these findings on the role of OP in burnout with workers in different types of organizations (Amarnani, Lajom, Restubog, & Capezio, 2019, Sample 1; Birkeland & Buch, 2015, Studies 1 and 2), as well as teachers (Fernet, Lavigne, & Vallerand, 2014, Studies 1 and 2; Trépanier, Fernet, Austin, Forest, & Vallerand, 2014, Study 2), students (Saville, Bureau, Eckenrode, & Maley, 2018), and nurses from both Canada and France (Amarnani et al., 2019, Sample 2; Trépanier et al., 2014, Study 1; Vallerand, Paquet, Philippe, & Charest, 2010, Studies 1 and 2) and Italy (Bélanger, Pierro, Kruglanski, Vallerand, De Carlo, & Falco, 2015, Study 2). Such findings have also been found with athletes of different levels of expertise (Curran, Appleton, Hill, & Hall, 2011, 2013; Demirci & Çepikkurt, 2018; Gustafsson, Hassmén, & Hassmén, 2011; Jafari, 2019; Kent, Kingston, & Paradis, 2018; Lopes & Vallerand, 2020, Studies 1 and 2; Martin & Horn, 2013; Moen, Myhre, & Sandbakk, 2016a,

2016b; Moen, Bentzen, & Myhre, 2018; Schellenberg, Gaudreau, & Crocker, 2013). Just like research in work organizations, HP for the sport has been associated with a more moderate and healthier way of engaging in sport, thereby leading to psychological well-being. Of interest, Bélanger, Raafat, Nisa, and Schumpe (in press) found that OP leads to sleeping problems, thereby underscoring its role in extreme persistence and the health problems it may cause in people's lives. Thus, the persistence in an obsessively passionate activity may translate into additional problems such lack of sleep that may contribute to burnout.

If OP leads to burnout and HP protects against it, then what are the processes involved in these effects? Different variables have been found to mediate these relationships. In two studies conducted on nurses from two cultures (France and from the Province of Quebec), Vallerand and his colleagues (2010) found that OP for work was positively related to the conflict between the passionate activity and other spheres in the individual's life that, in turn, predicted burnout. Conversely, HP was positively associated with work satisfaction (a positive psychological work experience) that, in turn, protected against burnout. These results remained true when predicting *changes* in burnout over time. Of importance is the research by Donahue et al. (2012, Study 2) on nurses that showed that while OP positively predicts burnout at work, HP leads to taking time off from work to engage in recovery activities that, in turn, led to the prevention of burnout.

Lopes and Vallerand (2020, Study 2) have replicated these basic findings on the role of conflict as a mediator of the effects of OP for sport on burnout in athletes. Moreover, they pursued the work by Donahue et al. (2012) by looking at the type of passion that people have for their recovery activities. Finally, these authors have also looked at need satisfaction (of competence, autonomy, and relatedness) while engaging in their sport and in recovery activities as another mediator of the passion–burnout relationship. Overall, Lopes and Vallerand found that need satisfaction acts a mediator between passion and burnout, preventing it from taking place. Of greater interest is the fact that only HP for the leisure (recovery) activities prevented burnout from taking place. In fact, engaging in such activities out of OP led to further increases in burnout. Thus, it is not the recovery activity per se that matters, but how one engages in it. Engaging in a recovery activity out of OP does not help and in fact can even hurt even more the person, leading to additional increases in burnout. Clearly, extreme engagement in a second activity does not save one from extreme engagement in a first one; it simply compounds the problem! In line with the study by Vallerand (1997), these results also underscore the fact that assessing passion for more than one life domain is necessary in order to fully understand the nature of motivational processes at play in one specific area, such as burnout.

Before leaving this section, we would like to underscore that although much of the review on extreme behavior has focused on negative outcomes, there is one area where extreme behavior is positive and it is performance. Because expert performance in various fields is something achieved only by a limited few, then one can see high performance also as 'extreme' in nature. Much research has shown that passion matters greatly with respect to performance (Vallerand, 2015). Research on expert performance reveals that high-level performers spend several years of considerable engagement in deliberate practice (i.e., engagement in the activity with clear goals of improving on certain task components), in order to reach excellence in their chosen field of expertise (see Ericsson & Charness, 1994). For instance, to train to reach the Olympics and/or the professional levels takes roughly 10,000 hours of deliberate practice defined as activities specifically oriented toward mastering difficult cognitive or motor sequences (Ericsson & Charness, 1994). This amounts to 4 hours a day, 6 days a week for 10 years! The DMP proposes that such as training regimen is so demanding that passion is necessary to go through it fully. In fact, the DMP hypothesizes that the two types of passion (harmonious and obsessive) lead to engagement in deliberate practice that, in turn, leads to improved performance.

A first study to test this basic model was conducted with some of the best student comedians in the Province of Quebec (Vallerand, Salvy et al., 2007, Study 1). Male and female student comedians completed scales assessing their passion for dramatic arts as well as deliberate practice in arts (based on Ericsson & Charness, 1994). Then, later, their teachers assessed their performance individually. Thus, performance was not a self-report measure but an 'objective' measure from the students' teachers. Results from a SEM analysis provided support for the basic model. Both types of passion led to engagement in deliberate practice that, in turn, led to objective performance. Also of interest is the finding that in this study psychological well-being was also assessed. Results revealed that only HP was positively and significantly related to life satisfaction, whereas OP was unrelated to it. This is in line with research discussed previously on the positive role of HP in psychological well-being (Vallerand, 2012).

These findings on the role of both OP and HP in predicting engagement in deliberate practice that led to performance have been replicated and expanded upon in a number of studies (see Vallerand & Verner-Filion, 2020 on this issue). Such research has also shown that achievement goals (Elliot & Church, 1997) also play a role in the passion–performance relationship. Specifically, OP leads to performance approach (e.g., beating others) and especially performance avoidance goals (e.g., avoiding doing worse than others) and to a minimal extent to mastery goals (e.g., trying to improve), whereas HP only predicts mastery goals. In turn, mastery goals predict deliberate practice that leads to performance, while performance avoidance and (sometimes) performance approach goals undermine performance. These findings have been obtained in sports including with members of the Canadian Junior Waterpolo teams (Vallerand, Mageau et al., 2008, Study 2) as well as with world-class classical musicians (Bonneville-Roussy, Lavigne, & Vallerand, 2011). In fact, that model has also been found to predict which hockey players were likely to play professional hockey 15 years later (Verner-Filion, Vallerand, Amiot, & Mocanu, 2017)! Once more, in several of these studies, HP predicted psychological well-being while engaging in high-performance activities, whereas OP did not.

Two points would appear in order here. First, both HP and OP are predictors of high-level performance. Thus, both HP and OP (and not simply OP) about a given activity will lead the person to do what it takes to work hard and develop into an international (Vallerand, Mageau et al., 2008, Study 2) or even professional athlete (Verner-Filion et al., 2017) or a world-class classical musician (Bonneville-Roussy et al., 2011). Thus, both HP and OP are key drivers in achieving excellence in a given field (see also Vallerand, 2015, Chapter 10 for a review). However, the processes triggered by the two types of passion are not exactly the same, and there seem to be two roads to high performance: the harmonious and the obsessive roads. The harmonious road allows one to engage in highly demanding activities specifically aimed at skill improvement (deliberate practice) that leads to high levels of performance, while allowing one to experience psychological well-being in the process. The second, obsessive, road also leads to engaging in deliberate practice and to high performance. However, the obsessive road seems to come at the expense of one's psychological well-being. Thus, contrary to what is typically believed, one can be a high-level performer and lead a relatively balanced, happy life at the same time, to the extent that HP underlies one's engagement in the field of excellence. These findings should be replicated with other activities and outcomes in other fields.

In sum, OP leads to extreme intrapersonal outcomes such as addiction, burnout, and physical and mental health. With OP, the individual is unable to quit or to momentarily stop engaging his or her favorite activity even when his or her physical and mental health is in jeopardy, which leads to ill-being. With HP, however, individuals engage in the activity with a more moderate and flexible perspective, which allows them to disengage from the activity when necessary, thereby preventing extreme engagement and outcomes and even leading to various positive activity and life outcomes such as positive emotions and life satisfaction.

Determinants of Harmonious and Obsessive Passion

We have seen so far in this chapter that OP leads to extreme intrapersonal and interpersonal outcomes. How does this rather extreme form of motivational force develop? Similarly, how does HP that leads to more moderate and adaptive outcomes come about? As mentioned previously, the DMP posits that people will gravitate around activities that will satisfy their psychological needs of competence, autonomy, and relatedness and their quest for self-growth. Thus, activities that will provide people with the experience of such needs and growth when engaging in the activity are likely to become passionate for these individuals. Thus, need satisfaction experienced when engaging in the activity should be involved in both types of passion. However, it is important to underscore that the DMP posits that the development of HP and OP also depends on whether psychological needs are satisfied for the rest of one's life. Because one is seeking need satisfaction, the level of need satisfaction in one's life matters greatly. Thus, if one cannot satisfy his or her need in life in general, one may overcompensate in areas where some need satisfaction is possible. In sum, the DMP proposes that the functionality of passion depends on two types of need satisfaction: the one that takes place when engaging in the activity and the type that typically exists in the person's life *outside* the passionate activity. It is these two types of need satisfaction that jointly determine which type of passion will develop.

In light of this context, the DMP posits that while need satisfaction within the activity is an important determinant for both HP and OP, OP for a given activity is likely to develop when there is a *lack* of need satisfaction in the rest of the person's life. Thus, with OP, the need satisfaction provided by the activity serves to compensate for the relative lack of need satisfaction in the person's life. Thus, the functionality with OP is in part compensatory in nature, and as such one can predict that extreme forms of engagement and outcomes will take place to make up for what is missing in the person's life. Conversely, with HP, there is already need satisfaction in other areas of his or her life. The need satisfaction experienced within the purview of the activity does not serve a compensatory role but rather adds to the person's life. This is like a bonus effect! As such, the person does not rely solely on this specific activity for need satisfaction and growth as there are other sources of need satisfaction in the person's life. Thus, the person can let go of this activity more readily when necessary. The involvement is not as extreme as there is no need imbalance as proposed by Kruglanski et al. (2021). However, there is need imbalance with OP and such imbalance largely takes place between the lack of need satisfaction in one's life and need satisfaction within the purview of the activity.

These hypotheses were investigated by Lalande et al. (2017) in a series of four studies using various designs, including some that were prospective and experimental in nature. For instance, in Study 3, we asked basketball players to complete the Passion Scale with respect to basketball, as well as the extent to which their psychological needs of competence, autonomy, and relatedness (Ryan & Deci, 2017) were satisfied both when playing basketball as well as in their life in general outside of basketball. Overall, the results provided support for the hypotheses. Specifically, need satisfaction *within* the passionate activity was conducive to both HP and OP (albeit more strongly for HP). However, lack of need satisfaction *outside* of the purview of the passionate activity (i.e., in one's life in general) predicted OP, but not HP. Thus, as hypothesized, OP resulted from the joint contribution of lack of need satisfaction in one's life *and* the presence of need satisfaction experienced while playing basketball. One can therefore see the compensatory nature of OP, where basketball may represent the only context wherein the psychological needs may be experienced. Such is not the case for HP, where other sources of need satisfaction exist besides the one where one is passionate about.

These findings were further replicated with another sample of workers while using a 6-month longitudinal study (Study 4) where changes in HP and OP for work were predicted as hypothesized. Specifically, an increase in OP resulted from both lack of need satisfaction in one's life and the presence of need satisfaction at work and an increase in HP over time resulted only from the latter. Of importance, another study (Lalande et al., 2017, Study 2) used an experimental design in which lack of need satisfaction in one's life was experimentally manipulated and need satisfaction in one's favorite activity and passion were assessed. Results from a SEM analyses revealed that lack of need satisfaction in one's life and need satisfaction in the activity jointly predicted OP whereas only need satisfaction positively predicted HP. Overall, it thus appears that OP results at least in part from a compensatory engagement in a need-satisfying activity when other life activities fail to provide such opportunities, whereas HP simply results from need satisfaction experienced within the passionate activity.

Finally, of additional interest for our discussion is that in Studies 2 to 4, psychological outcomes were also assessed. For example, in Study 3, we assessed an extreme outcome, namely, burnout, as well as a more positive one, namely, life satisfaction. Results showed that only OP positively predicted burnout, whereas only HP positively predicted life satisfaction. These predictions with respect to outcomes are important because they show that although the two types of passion result in part from the positive contribution of need satisfaction when engaged in the activity, they nevertheless lead to diametrically opposed outcomes. Thus, the type of passion matters greatly!

In line with Lavigne, Vallerand, and Crevier-Braud (2011), it would appear that the motivational functionality of OP reflects a mix of growth-oriented and need deficit reduction perspectives as one seeks to reduce need deficit in one's life by experiencing some growth in a specific activity. Such a perspective is far from ideal as it leads to the extreme form of motivational force that triggers an unbridled activity engagement and less than ideal outcomes. Conversely, HP seems to display simply a more moderate motivational force that leads to a more adaptive form of activity engagement and growth-oriented outcomes. In addition, in line with the study by Vallerand (1997), this series of findings underscores the importance of taking into account other life contexts in one's life when making predictions about motivational processes in one specific context such as work or sports as these may interact. Thus, considering various sources of need satisfaction appears to be important when examining the determinants of HP and OP, as well as the type of outcomes that they generate, both in terms of quality and extremism.

Conclusions

The purpose of this chapter is to address the role of passion in extreme behavior. The present review leads to a number of conclusions. First, in line with the study by Kruglanski et al.'s (2021), it was seen that extremism is not limited to certain activities like terrorism, but pertains to a number of activities whose nature is of intrapersonal and interpersonal dimensions that can have positive or negative outcomes for both the perpetrator and others. We started this chapter with a number of seemingly disparate examples of extreme behavior that pertained to terrorism, activism, sport involvement, and performance. In our review, we went beyond these and also addressed additional outcomes that include stalking, romantic behavior, road rage, various forms of pathological gambling, video game, exercise, and substance addiction, and performance.

Because of such diversity of extreme behavior, it becomes important and useful to have a broad concept that provides a comprehensive analysis of the processes at play. The concept of passion provides such a broad understanding of the effects and ramifications of the underlying motivational force behind 'extreme behavior'. However, as presented in our discussion of the DMP, not all passions are equal, and it was empirically shown that extreme behavior is mostly the lot of OP. This represents a second conclusion from this chapter. OP represents a very powerful motivational force where the activity that one is passionate about has been internalized in identity in isolation of other self-elements. This leads to goal shielding (Bélanger et al., 2013) and compartmentalization of the passionate activity representation in identity (Philippe et al., 2017). One of the consequences that follow is that the activity can take up a life of its own leading to a lack of self-regulation over the activity and extreme forms of activity engagement and corresponding outcomes. Of additional interest, because one may be passionate for more than one activity (e.g., Schellenberg & Bailis, 2015), it would be intriguing in future research to assess whether those OP self-elements are grouped together in identity or if they all remain individually compartmentalized. What would be the ensuing consequences of such internal structures?

A third and less obvious conclusion is that another less extreme but nevertheless intense form of activity involvement is HP. Although HP is not as extreme as OP, it still leads to a number of adaptive outcomes as pertains to a variety of life domains. Thus, one does not need to be obsessive about an outcome, and to suffer while engaging in its pursuit, to reach it. With HP, you can typically have your cake and eat it too! Of course, this is a big claim. However, the fact that our research has been conducted in a variety of real-life settings while using state-of-the-art research methods supports it. Such methodology includes experimental designs where HP and OP are experimentally induced as well as longitudinal designs where changes in outcomes take place sometimes over several years. Thus, we are confident in the causal role of OP in extreme behavior and that of HP in more adaptive outcomes. An interesting question for future research is how one can be so involved in the activity as is the case with HP and still not fall prey to imbalance as with OP? One answer may lie in the positive states that one experiences with HP. Much research has shown that HP leads to positive emotions, flow, deep concentration (see Curran et al., 2015, for a meta-analysis), and even mindfulness (St-Louis et al., 2018). Such states are known to lead to adaptive functioning and well-being and may protect one against imbalance (Vallerand, 2015). Future research on this issue would appear in order.

A fourth conclusion pertains to one type of extreme behavior, namely, high-level performance, where HP was also found to contribute to it to the same extent as OP. In fact, there was even a bonus effect as HP (but not OP) also led to psychological well-being while expert performance was pursued and achieved. We suggested that there were two roads to expert performance: the harmonious and obsessive roads. Future research is necessary to further probe these two roads in order to better delineate the nuances in the processes characterized by each one. One promising avenue pertains to the persistence necessary to reaching success in a given activity. In line with the DMP, recent research of ours reveals that OP and HP lead to different types of persistence with OP leading to rigid persistence and HP to flexible persistence (Vallerand et al., 2021). However, both types of persistence lead to overcoming obstacles on the way to success, flexible (but not rigid) persistence also leads one to engaging in other satisfying activities, thus contributing to psychological well-being. Future research on the mediating role of the two types of persistence in high-level performance would appear important.

A fifth conclusion is that passion results from the joint forces of two types of need satisfaction: one that takes place within the activity that one is passionate about and the other that takes place in the rest of one's life. To the extent that both types of need satisfaction take place, HP will develop, whereas OP results from the *absence* of need satisfaction in one's life in combination with the presence of need satisfaction in the passionate activity. This lack of need satisfaction in the person's life besides the passionate activity renders the latter even more crucial in the person's life and leads to some dependence on it for need gratification. This is in line with the study by Kruglanski et al. (2021) who posit that extreme behavior results from a motivational imbalance. These authors suggested that the imbalance lies among goals where some are neglected at the expense of one specific (passionate) goal. The findings of studies reviewed in this chapter clearly support this analysis as pertains to OP. Furthermore, the research by Lalande et al. (2017) presented herein suggests that such a *motivational* imbalance may very well result from the *need* imbalance that takes place between a passionate activity and the rest of one's life. Specifically, we showed that with OP, people compensate for the need satisfaction that's missing in one's life by over engaging in the passionate activity that provides some satisfaction. This analysis leads to the surprising suggestion that one way to reduce OP is to include need-satisfying activities in one's life in addition to the passionate activity. Future research on this issue would appear important for both applied and theoretical reasons.

Moreover, additional research is necessary as pertains to the determinants of passion. To this end, we underscore that much research exists on the role of three key factors as determinants of need satisfaction: personal, social, and task factors

(Vallerand, 1997). Future research on the role of these three factors in both HP and OP has started and shows the importance of all three (Vallerand, 2015; Vallerand & Houlfort, 2019). However, we have not considered how these factors should be taken simultaneously into account in both the passionate activity and the person's life. Such research would appear important in light of the contribution of task, person, and contextual factors in need satisfaction in life and the activity and how these translate to passion. For instance, we have not distinguished in our research between specific psychological needs and have rather considered them together. It may prove important in future research to determine whether motivational imbalance between specific needs takes place such that people are willing to let certain needs become unsatisfied (e.g., relatedness) in life in general in order to satisfy others in one context (e.g., competence). The Olympic athlete who focuses only on excelling (competence) in sport at the expense of other needs (relatedness) in all other life contexts comes to mind. Future research on this issue, bridging social, personal, and task factors with the need satisfaction perspective, would appear important to better understand the processes involved in the strategy adopted by the Olympic athlete and its costs.

A final issue not addressed so far in this chapter pertains to individuals at the end of the passion continuum, namely, nonpassionate individuals. First, it is important to underscore that it is possible, even likely, that one can be nonpassionate for a given activity such as work and be passionate for another such as playing basketball. This is why in our research on nonpassionate people (Philippe et al., 2009), we have asked them to identify their favorite activity and to complete the Passion Scale for this activity. Thus, if participants don't meet the passion criteria and have low levels of HP and OP for this activity, then they truly are nonpassionate individuals. The little research conducted on this issue (Philippe et al., 2009, Studies 1 and 2) has shown that nonpassionate individuals display lower levels of psychological well-being than those with HP but are similar to those with OP. In addition, they experience a significant *decrease* in psychological well-being over time (1 year) just like people with OP, whereas those with HP experience an increase in well-being. These findings show that nonpassionate people are less happy than those with an HP but are not different than those with an OP. This is intriguing and raises several questions. For example, would nonpassionate individuals engage in extreme behavior like those with OP? This would seem unlikely, as they don't have a meaningful activity to engage in. One possibility, however, is that their search for such a meaningful activity to engage in and to identify with (and thus to become passionate about) may lead them to overcompensate like those participants in the study by Lalande et al. (2017) once they find one and to develop an OP for this activity. If they did, then, they would be at risk of eventually engaging in extreme behavior in this new passionate activity. Such behavior is often observed with those who have found a new cause to promote with zeal and even frenzy. Research on this hypothesis is interesting, from both conceptual and applied perspectives.

In sum, as proposed by Kruglanski and colleagues (2021) and as exemplified in this chapter, extreme behavior encompasses a variety of relative rare behavior that ranges from the individual to the interpersonal with outcomes that can be adaptive or maladaptive. We have shown that passion, and especially OP, plays an intricate role in such behavior and have posited that another powerful and more adaptive motivational force, namely, HP, deserves consideration. Future research on this issue would appear important to provide a better understanding of extremism and its ramifications.

References

Akehurst, S., & Oliver, E. J. (2014). Obsessive passion: A dependency associated with injury-related risky behaviour in dancers. *Journal of Sports Sciences*, *32*(3), 259–267. doi:10.1080/02640414.2013.823223

Amarnani, R. K., Lajom, J. A. L., Restubog, S. L. D., & Capezio, A. (2019). Consumed by obsession: Career adaptability resources and the performance consequences of obsessive passion and harmonious passion for work. *Human Relations*, *73*(6), 811–836. doi:10.1177/0018726719844812

Aron, A., Aron, E. N., & Smollan, D. (1992). Inclusion of other in the self scale and the structure of interpersonal closeness. *Journal of Personality and Social Psychology*, *63*, 596–612. doi:10.1037/0022-3514.63.4.596

Avnet, T., & Higgins, E. T. (2003). Locomotion, assessment, and regulatory fit: Value transfer from "how" to "what". *Journal of Experimental Social Psychology*, *39*(5), 525–530. doi:10.1016/s0022-1031(03)00027-1

Back, K. J., Lee, C. K., & Stinchfield, R. (2011). Gambling motivation and passion: A comparison study of recreational and pathological gamblers. *Journal of Gambling Studies*, *27*(3), 355–370. doi:10.1007/s10899-010-9212-2

Bélanger, J. J., Collier, K., Nisa, C., & Schumpe, B. (under review). Crimes of passion: When romantic obsession leads to abusive relationships. *Journal of Personality*.

Bélanger, J. J., Lafreniere, M. A. K., Vallerand, R. J., & Kruglanski, A. W. (2013). When passion makes the heart grow colder: The role of passion in alternative goal suppression. *Journal of Personality and Social Psychology*, *104*(1), 126–147. doi:10.1037/a0029679

Bélanger, J. J., Pierro, A., Kruglanski, A. W., Vallerand, R. J., De Carlo, N., & Falco, A. (2015). On feeling good at work: The role of regulatory mode and passion in psychological adjustment. *Journal of Applied Social Psychology*, *45*(6), 319–329. doi:10.1111/jasp.12298

Bélanger, J. J., Raafat, K. A., Nisa, C. F., & Schumpe, B. M. (in press). Passion for an activity: A new predictor of the quality of sleep. *Sleep*.

Bélanger, J. J., Schumpe, B. M., Nociti, N., Moyano, M., Dandeneau, S., Chamberland, P. E., & Vallerand, R. J. (2019). Passion and moral disengagement: Different pathways to political activism. *Journal of Personality*, *87*(6), 1234–1249. doi:10.1111/jopy.12470

Birkeland, I. K., & Buch, R. (2015). The dualistic model of passion for work: Discriminate and predictive validity with work engagement and workaholism. *Motivation and Emotion*, *39*(3), 392–408. doi:10.1007/s11031-014-9462-x

Bonneville-Roussy, A., Lavigne, G. L., & Vallerand, R. J. (2011). When passion leads to excellence: The case of musicians. *Psychology of Music*, *39*(1), 123–138. doi:10.1177/0305735609352441

Bureau, J. S., Vallerand, R. J., Ntoumanis, N., & Lafreniere, M. A. K. (2013). On passion and moral behavior in achievement settings: The mediating role of pride. *Motivation and Emotion, 37*(1), 121–133. doi:10.1007/s11031-012-9292-7

Burnay, J., Billieux, J., Blairy, S., & Larøi, F. (2015). Which psychological factors influence Internet addiction? Evidence through an integrative model. *Computers in Human Behavior, 43*, 28–34. doi:10.1016/j.chb.2014.10.039

Carbonneau, N., & Vallerand, R. J. (2013). On the role of harmonious and obsessive passion in conflict behavior. *Motivation and Emotion, 37*, 743–757. doi:10.1007/s11031-013-9354-5

Carbonneau, N., Vallerand, R. J., Fernet, C., & Guay, F. (2008). The role of passion for teaching in intrapersonal and interpersonal outcomes. *Journal of Educational Psychology, 100*(4), 977–987. doi:10.1037/a0012545

Carbonneau, N., Vallerand, R. J., Lavigne, G. L., & Paquet, Y. (2016). "I'm not the same person since I met you": The role of romantic passion in how people change when they get involved in a romantic relationship. *Motivation and Emotion, 40*(1), 101–117. doi:10.1007/s11031-015-9512-z

Carpentier, J., Mageau, G. A., & Vallerand, R. J. (2012). Ruminations and flow: Why do people with a more harmonious passion experience higher well-being? *Journal of Happiness Studies, 13*(3), 501–518. doi:10.1007/s10902-011-9276-4

Csikszentmihalyi, M., Rathunde, K., & Whalen, S. (1993). *Talented teenagers: The roots of success and failure.* New York: Cambridge University Press.

Curran, T., Appleton, P. R., Hill, A. P., & Hall, H. K. (2011). Passion and burnout in elite junior soccer players: The mediating role of self-determined motivation. *Psychology of Sport and Exercise, 12*(6), 655–661. doi:10.1016/j.psychsport.2011.06.004

Curran, T., Appleton, P. R., Hill, A. P., & Hall, H. K. (2013). The mediating role of psychological need satisfaction in relationships between types of passion for sport and athlete burnout. *Journal of Sports Sciences, 31*(6), 597–606. doi:10.1080/02640414.2012.742956

Curran, T., Hill, A. P., Appleton, P. R., Vallerand, R. J., & Standage, M. (2015). The psychology of passion: A meta-analytical review of a decade of research on intrapersonal outcomes. *Motivation and Emotion, 39*(5), 631–655. doi:10.1007/s11031-015-9503-0

Davis, A. K. (2016). The dualistic model of passion applied to recreational marijuana consumption. *Addiction Research & Theory, 25*(3), 188–194. doi:10.1080/16066359.2016.1242722

Davis, A. K., & Arterberry, B. J. (2019). Passion for marijuana use mediates the relations between refusal self-efficacy and marijuana use and associated consequences. *Journal of Psychoactive Drugs, 51*(4), 343–350. doi:10.1080/02791072.2019.1596334

Davis, A. K., Arterberry, B. J., Bonar, E. E., Bohnert, K. M., & Walton, M. A. (2018). Why do young people consume marijuana? Extending motivational theory via the dualistic model of passion. *Translational Issues in Psychological Science, 4*(1), 54–64. doi:10.1037/tps0000141

Deci, E. L., & Ryan, R. M. (2000). The "what" and "why" of goal pursuits: Human needs and the self-determination of behavior. *Psychological Inquiry, 11*, 227–268. doi:10.1207/s15327965pli1104_01

Demirci, E., & Çepikkurt, F. (2018). Examination of the relationship between passion, perfectionism and burnout in athletes. *Universal Journal of Educational Research, 6*(6), 1252–1259. doi:10.13189/ujer.2018.060616

Dolan, S., Arterberry, B., & Davis, A. (2020). A quadripartite model of passion for marijuana use: Associations with consumption, consequences, craving, and satisfaction with life. *Addiction Research & Theory*, 1–6. doi:10.1080/16066359.2020.1718117

Donahue, E. G., Forest, J., Vallerand, R. J., Lemyre, P. N., Crevier-Braud, L., & Bergeron, É. (2012). Passion for work and emotional exhaustion: The mediating role of rumination and recovery. *Applied Psychology: Health and Well-Being, 4*(3), 341–368. doi:10.1111/j.1758-0854.2012.01078.x

Donahue, E. G., Rip, B., & Vallerand, R. J. (2009). When winning is everything: On passion, identity, and aggression in sport. *Psychology of Sport and Exercise, 10*(5), 526–534. doi:10.1016/j.psychsport.2009.02.002

Elliot, A. J., & Church, M. A. (1997). A hierarchical model of approach and avoidance achievement motivation. *Journal of Personality and Social Psychology, 72*(1), 218–232. doi:10.1037/0022-3514.72.1.218

Ericsson, K. A., & Charness, N. (1994). Expert performance. *American Psychologist, 49*, 725–747. doi:10.1037/0003-066x.49.8.725

Erikson, E. H. (1968). *Identity: Youth and crisis*. New York: W.W. Norton. doi:10.1126/science.161.3838.257

Fernet, C., Lavigne, G., & Vallerand, R. J. (2014). Fired up with passion: The role of harmonious and obsessive passion in burnout in novice teachers. *Work and Stress, 28*, 270–288. doi:10.1080/02678373.2014.935524

Fredrickson, B. L. (2001). The role of positive emotions in positive psychology: The broaden-and build theory of positive emotions. *American Psychologist, 56*, 218–226. doi:10.1037//0003-066x.56.3.218

Fredrickson, B. L. (2013). Positive emotions broaden an build. *Advances in Experimental Social Psychology, 47*, 1–53.

Gousse-Lessard, A. S., Vallerand, R. J., Carbonneau, N., & Lafrenière, M. A. K. (2013). The role of passion in mainstream and radical behaviors: A look at environmental activism. *Journal of Environmental Psychology, 35*, 18–29. doi:10.1016/j.jenvp.2013.03.003

Guilbault, V., Bouizegarene, N., Philippe, F. L., & Vallerand, R. J. (2019). Understanding extradyadic sex and its underlying motives through a dualistic model of sexual passion. *Journal of Social and Personal Relationships, 37*(1), 281–301. doi:10.1177/0265407519864446

Gustafsson, H., Hassmén, P., & Hassmén, N. (2011). Are athletes burning out with passion? *European Journal of Sport Science, 11*(6), 387–395. doi:10.1080/17461391.2010.536573

Jafari, Z. M. (2019). Effect of passion for physical activity on (physical) burnout in student athletes. *Sport TK: Revista Euroamericana de Ciencias del Deporte, 8*(2), 125–132. doi:10.6018/sportk.391851

Kent, S., Kingston, K., & Paradis, K. F. (2018). The relationship between passion, basic psychological needs satisfaction and athlete burnout: Examining direct and indirect effects. *Journal of Clinical Sport Psychology, 12*(1), 75–96. doi:10.1123/jcsp.2017-0030

Kovacsik, R., Soós, I., de la Vega, R., Ruíz-Barquín, R., & Szabo, A. (2020). Passion and exercise addiction: Healthier profiles in team than in individual sports. *International Journal of Sport and Exercise Psychology, 18*(2), 176–186. doi:10.1080/1612197x.2018.1486773

Kruglanski, A. W., Szumowska, E., Kopetz, C. H., Vallerand, R. J., & Pierro, A. (2021). On the psychology of extremism: How imbalance breeds intemperance. *Psychological Review, 128*, 264–289. doi:10.1037/rev0000260

Ladouceur, R. (1996). The prevalence of pathological gambling in Canada. *Journal of Gambling Studies, 12*, 129–142. doi:10.1007/bf01539170

Lafrenière, M.-A. K., Bélanger, J. J., Sedikides, C., & Vallerand, R. J. (2011). Self-esteem and passion for activities. *Personality and Individual Differences, 51*, 541–544. doi:10.1016/j.paid.2011.04.017

Lafrenière, M-A. K., Vallerand, R. J., Donahue, R., & Lavigne, G. L. (2009). On the costs and benefits of gaming: The role of passion. *CyberPsychology and Behavior, 12*, 285–290. doi:10.1089/cpb.2008.0234

Lafrenière, M.-A. K., Vallerand, R. J., & Sedikides, C. (2013). On the relation between self-enhancement and life satisfaction: The moderating role of passion. *Self and Identity, 12*(6), 597–609. doi:10.1080/15298868.2012.713558

Lalande, D., Vallerand, R. J., Lafrenière, M. A. K., Verner-Filion, J., Laurent, F. A., Forest, J., & Paquet, Y. (2017). Obsessive passion: A compensatory response to unsatisfied needs. *Journal of Personality, 85*(2), 163–178. doi:10.1111/jopy.12229

Lavigne, G. L., Vallerand, R. J., & Crevier-Braud, L. (2011). The fundamental need to belong: On the distinction between growth and deficit-reduction orientations. *Personality and Social Psychology Bulletin, 37*(9), 1185–1201. doi:10.1177/0146167211405995

Lee, C. K., Chung, N., & Bernhard, B. J. (2013). Examining the structural relationships among gambling motivation, passion, and consequences of internet sports betting. *Journal of Gambling Studies, 30*(4), 845–858. doi:10.1007/s10899-013-9400-y

Lesieur, H., & Blume, S. (1993). Revising the South Oaks Gambling Screen in different settings. *Journal of Gambling Studies, 9*, 213–223. doi:10.1007/bf01015919

Lopes, M., & Vallerand, R. J. (2020). The role of passion, need satisfaction, and conflict in athletes' perceptions of burnout. *Psychology of Sport and Exercise*, 1–47. doi:10.1016/j.psychsport.2020.101674

Mageau, G. A., Carpentier, J., & Vallerand, R. J. (2011). The role of self-esteem contingencies in the distinction between obsessive and harmonious passion. *European Journal of Social Psychology, 41*(6), 720–729. doi:10.1002/ejsp.798

Mageau, G. A., Vallerand, R. J., Rousseau, F. L., Ratelle, C. F., & Provencher, P. J. (2005). Passion and gambling: Investigating the divergent affective and cognitive consequences of gambling. *Journal of Applied Social Psychology, 35*(1), 100–118. doi:10.1111/j.1559-1816.2005.tb02095.x

Marsh, H. W., Vallerand, R. J., Lafreniere, M. A. K., Parker, P., Morin, A. J. S., Carbonneau, N., . . . Paquet, Y. (2013). Passion: Does one scale fit all? Construct validity of two-factor passion scale and psychometric invariance over different activities and languages. *Psychological Assessment, 25*, 796–809. doi:10.1037/a0032573

Martin, E. M., & Horn, T. S. (2013). The role of athletic identity and passion in predicting burnout in adolescent female athletes. *The Sport Psychologist, 27*(4), 338–348. doi:10.1123/tsp.27.4.338

Moen, F., Bentzen, M., & Myhre, K. (2018). The role of passion and affect in enhancing the understanding of coach burnout. International *Journal of Coaching Science, 12*(1), 3–34.

Moen, F., Myhre, K., & Sandbakk, Ø. (2016a). Psychological determinants of burnout, illness and injury among elite junior athletes. *The Sport Journal, 19*, 1–14.

Moen, F., Myhre, K., & Stiles, T. C. (2016b). An exploration about how passion, perceived performance, stress and worries uniquely influence athlete burnout. *Journal of Physical Education and Sports Management, 3*(1), 88–107. doi:10.15640/jpesm.v3n1a7

Morvannou, A., Dufour, M., Brunelle, N., Berbiche, D., & Roy, É. (2017a). One-year prospective study on passion and gambling problems in poker players. *Journal of gambling studies*, *34*(2), 379–391. doi:10.1007/s10899-017-9706-2

Morvannou, A., Dufour, M., Brunelle, N., Berbiche, D., & Roy, É. (2017b). Passion for poker and the relationship with gambling problems: A cross-sectional study. *International Gambling Studies*, *17*(2), 176–191. doi:10.1080/14459795.2017.1311354

Oikonomidis, A., Palomäki, J., & Laakasuo, M. (2019). Experience and passion in poker: Are there well-being implications? *Journal of Gambling Studies*, *35*(2), 731–742. doi:10.1007/s10899-018-9795-6

Paradis, K. F., Cooke, L. M., Martin, L. J., & Hall, C. R. (2013). Too much of a good thing? Examining the relationship between passion for exercise and exercise dependence. *Psychology of Sport and Exercise*, *14*(4), 493–500. doi:10.1016/j.psychsport.2013.02.003

Parastatidou, I. S., Doganis, G., Theodorakis, Y., & Vlachopoulos, S. P. (2014). The mediating role of passion in the relationship of exercise motivational regulations with exercise dependence symptoms. *International Journal of Mental Health and Addiction*, *12*(4), 406–419. doi:10.1007/s11469-013-9466-x

Philippe, F. L., & Vallerand, R. J. (2007). Prevalence rates of gambling problems in Montreal, Canada: A look at old adults and the role of passion. *Journal of Gambling Studies*, *23*(3), 275–283. doi:10.1007/s10899-006-9038-0

Philippe, F. L., Vallerand, R. J., Beaulieu-Pelletier, G., Maliha, G., Laventure, S., & Ricard-St-Aubin, J. S. (2019). Development of a dualistic model of sexual passion: Investigating determinants and consequences. *Archives of Sexual Behavior*, *48*(8), 2537–2552. doi:10.1007/s10508-019-01524-w

Philippe, F. L., Vallerand, R. J., Bernard-Desrosiers, L., Guilbault, V., & Rajotte, G. (2017). Understanding the cognitive and motivational underpinnings of sexual passion from a dualistic model. *Journal of Personality and Social Psychology*, *113*(5), 769–785. doi:10.1037/pspp0000116

Philippe, F. L., Vallerand, R. J., Richer, I., Vallieres, É., & Bergeron, J. (2009). Passion for driving and aggressive driving behavior: A look at their relationship. *Journal of Applied Social Psychology*, *39*(12), 3020–3043. doi:10.1111/j.1559-1816.2009.00559.x

Pierro, A., Leder, S., Mannetti, L., Higgins, E. T., Kruglanski, A. W., & Aiello, A. (2008). Regulatory mode effects on counterfactual thinking and regret. *Journal of Experimental Social Psychology*, *44*(2), 321–329. doi:10.1016/j.jesp.2007.06.002

Ratelle, C. F., Carbonneau, N., Vallerand, R. J., & Mageau, G. (2013). Passion in the romantic sphere: A look at relational outcomes. *Motivation and Emotion*, *37*(1), 106–120. doi:10.1007/s11031-012-9286-5

Ratelle, C. F., Vallerand, R. J., Mageau, G. A., Rousseau, F. L., & Provencher, P. (2004). When passion leads to problematic outcomes: A look at gambling. *Journal of Gambling Studies*, *20*(2), 105–119. doi:10.1023/b:jogs.0000022304.96042.e6

Rip, B., Fortin, S., & Vallerand, R. J. (2006). The relationship between passion and injury in dance students. *Journal of Dance Medicine & Science*, *10*(1–2), 14–20.

Rip, B., Vallerand, R. J., & Lafrenière, M. A. K. (2012). Passion for a cause, passion for a creed: On ideological passion, identity threat, and extremism. *Journal of Personality*, *80*(3), 573–602. doi:10.1111/j.1467-6494.2011.00743.x

Rousseau, F. L., Vallerand, R. J., Ratelle, C. F., Mageau, G. A., & Provencher, P. J. (2002). Passion and gambling: On the validation of the Gambling Passion Scale (GPS). *Journal of Gambling Studies*, *18*(1), 45–66. doi:10.1023/a:1014532229487

Ryan, R. M., & Deci, E. L. (2017). *Self-determination theory: Basic psychological needs in motivation, development, and wellness*. New York, NY: Guilford Publications. doi:10.7202/1041847ar

Saville, B. K., Bureau, A., Eckenrode, C., & Maley, M. (2018). Passion and burnout in college students. *College Student Journal, 52*(1), 105–117.

Schellenberg, B. J., & Bailis, D. S. (2015). Can passion be polyamorous? The impact of having multiple passions on subjective well-being and momentary emotions. *Journal of Happiness Studies, 16*(6), 1365–1381. doi:10.1007/s10902-014-9564-x

Schellenberg, B. J., Gaudreau, P., & Crocker, P. R. (2013). Passion and coping: Relationships with changes in burnout and goal attainment in collegiate volleyball players. *Journal of Sport and Exercise Psychology, 35*(3), 270–280. doi:10.1123/jsep.35.3.270

Sicilia, Á., Alcaraz-Ibáñez, M., Lirola, M. J., & Burgueño, R. (2017). Influence of goal contents on exercise addiction: Analysing the mediating effect of passion for exercise. *Journal of Human Kinetics, 59*(1), 143–153. doi:10.1515/hukin-2017-0154

Skitch, S. A., & Hodgins, D. C. (2005). A passion for the game: Problem gambling and passion among university sstudents. *Canadian Journal of Behavioural Science/Revue canadienne des sciences du comportement, 37*(3), 193–197. doi:10.1037/h0087256

Steers, M. L. N., Neighbors, C., Christina Hove, M., Olson, N., & Lee, C. M. (2015). How harmonious and obsessive passion for alcohol and marijuana relate to consumption and negative consequences. *Journal of Studies on Alcohol and Drugs, 76*(5), 749–757. doi:10.15288/jsad.2015.76.749

Stenseng, F., Haugen, T., Torstveit, M. K., & Høigaard, R. (2015). When it's "all about the bike"—Intrapersonal conflict in light of passion for cycling and exercise dependence. *Sport, Exercise, and Performance Psychology, 4*(2), 127–139. doi:10.1037/spy0000028

Stephan, Y., Deroche, T., Brewer, B. W., Caudroit, J., & Le Scanff, C. (2009). Predictors of perceived susceptibility to sport-related injury among competitive runners: The role of previous experience, neuroticism, and passion for running. *Applied Psychology: An International Review, 58*, 672–687. doi:10.1111/j.1464-0597.2008.00373.x

St-Louis, A. C., Carbonneau, N., & Vallerand, R. J. (2016). Passion for a cause: How it affects health and subjective well-being. *Journal of Personality, 84*(3), 263–276. doi:10.1111/jopy.12157

St-Louis, A. C., Rapaport, M., Poirier, L. C., Vallerand, R. J., & Dandeneau, S. (2021). On emotion regulation strategies and well-being: The role of passion. *Journal of Happiness Studies, 22*(4), 1791–1818. doi:10.1007/s10902-020-00296-8

St-Louis, A. C., Verner-Filion, J., Bergeron, C., & Vallerand, R. J. (2018). Passion and mindfulness: Accessing adaptive self-processes. *The Journal of Positive Psychology, 13*, 155–164. doi:10.1080/17439760.2016.1245771

Szabo, A., & Kovacsik, R. (2019). When passion appears, exercise addiction disappears. *Swiss Journal of Psychology, 78*(3–4), 137–142. doi:10.1024/1421-0185/a000228

Trépanier, S.-G., Fernet, C., Austin, S., Forest, J., & Vallerand, R. J. (2014). Linking job demands and resources to burnout and work engagement: Does passion underlie these differential relationships? *Motivation and Emotion, 38*, 353–366. doi:10.1007/s11031-013-9384-z

Vallerand, R. J. (1997). Toward a hierarchical model of intrinsic and extrinsic motivation. *Advances in Experimental and Social Psychology, 29*, 271–360. doi:10.1016/s0065-2601(08)60019-2

Vallerand, R. J. (2008). On the psychology of passion: In search of what makes people's lives most worth living. *Canadian Psychology, 49*, 1–13. doi:10.1037/0708-5591.49.1.1

Vallerand, R. J. (2010). On passion for life activities: The dualistic model of passion. In M. P. Zanna (Ed.), *Advances in experimental social psychology* (Vol. 42, pp. 97–193). New York: Academic Press. doi:10.1016/s0065-2601(10)42003-1

Vallerand, R. J. (2012). The role of passion in sustainable psychological well-being. *Psychological Well-Being: Theory, Research, and Practice, 2*, 1–21. doi:10.1186/2211-1522-2-1

Vallerand, R. J. (2015). *The psychology of passion: A dualistic model*. New York: Oxford. doi:10.1093/acprof:oso/9780199777600.003.0003

Vallerand, R. J., Blanchard, C., Mageau, G. A., Koestner, R., Ratelle, C., Léonard, M., . . . Marsolais, J. (2003). Les passions de l'ame: On obsessive and harmonious passion. *Journal of Personality and Social Psychology, 85*(4), 756–767. doi:10.1037/0022-3514.85.4.756

Vallerand, R. J., Chichekian, T., Verner-Filion, J., & Rapaport, M. (2021). *A look at the persistence of passionate individuals: On the role of rigid and flexible persistence in activity and life outcomes*. Manuscript submitted for publication.

Vallerand, R. J., & Houlfort, N. (Eds.). (2019). *Passion for work: Theory, research and applications*. New York: Oxford. doi:10.1093/oso/9780190648626.001.0001

Vallerand, R. J., Mageau, G. A., Elliot, A. J., Dumais, A., Demers, M. A., & Rousseau, F. (2008). Passion and performance attainment in sport. *Psychology of Sport and Exercise, 9*(3), 373–392. doi:10.1016/j.psychsport.2007.05.003

Vallerand, R. J., Ntoumanis, N., Philippe, F. L., Lavigne, G. L., Carbonneau, N., Bonneville, A., Lagacé-Labonté, C., & Maliha, G. (2008). On passion and sports fans: A look at football. *Journal of Sports Sciences, 26*(12), 1279–1293. doi:10.1080/02640410802123185

Vallerand, R. J., Paquet, Y., Philippe, F. L., & Charest, J. (2010). On the role of passion in burnout: A process model. *Journal of Personality, 78*, 289–312. doi:10.1111/j.1467-6494.2009.00616.x

Vallerand, R. J., & Rahimi, S. (in press). On the psychometric properties of the passion scale. In Anastasia Efklides, Itziar Alonso-Arbiol, Tuulia Ortner, Willibald Ruch, & F. J. R. van de Vijver (Eds.), *Psychological Assessment in Positive Psychology*. New York: Hogrefe.

Vallerand, R. J., Salvy, S. J., Mageau, G. A., Elliot, A. J., Denis, P. L., Grouzet, F. M., & Blanchard, C. (2007). On the role of passion in performance. *Journal of Personality, 75*(3), 505–534. doi:10.1111/j.1467-6494.2007.00447.x

Verner-Filion, J., Vallerand, R. J., Amiot, C. E., & Mocanu, I. (2017). The two roads from passion to sport performance and psychological well-being: The mediating role of need satisfaction, deliberate practice, and achievement goals. *Psychology of Sport & Exercise, 30*, 19–29. doi:10.1016/j.psychsport.2017.01.009

Wang, C. C., & Chu, Y. S. (2007). Harmonious passion and obsessive passion in playing online games. *Social Behavior and Personality: An International Journal, 35*(7), 997–1006. doi:10.2224/sbp.2007.35.7.997

White, R. W. (1959). Motivation reconsidered: The concept of competence. *Psychological Review, 66*(5), 297–333. doi:10.1037/h0040934

4

THE EXTREME GROUP

John M. Levine and Arie W. Kruglanski

Extremism, an intense and infrequent form of motivational imbalance in which a single need dominates all others, can be observed in a variety of domains (Kruglanski, Szumowska, Kopetz, Vallerand, & Pierro, 2021). Other chapters in this volume provide compelling analyses of the causes and consequences of extremism at the level of the individual. Examples include substance abuse, eating disorders, extreme greed, political activism, exceptional humanism, and engagement in extreme sports, among others. Important as they are, these forms of extremism do not provide an exhaustive picture of the role that motivational imbalance plays in human affairs. In particular, they do not shed light on how extremism operates at the level of the group, which is the focus of the present chapter.

Group extremism has long fascinated scholars (e.g., historians, political scientists, sociologists) and social commentators alike, and the reason is not hard to understand. Groups that eschew moderation (balanced efforts to pursue multiple goals) in favor of extremism (a single-minded focus on one goal) have the potential to produce substantial benefits *and* costs for their members and, in many cases, for outsiders as well. Examples include violent political groups (e.g., al Qaeda, neo-Nazis), ideologically oriented military formations (e.g., Waffen SS, Iranian Revolutionary Guards), cults (e.g., Heaven's Gate, Unification Church), religious orders (e.g., Benedictine Monks, Poor Clare Nuns), single-issue political groups (e.g., Operation Rescue, Climate Direct Action), and health-related groups (e.g., pro-anorexia [Pro-Ana], Vaccine Choice Canada).

The Present Chapter

The goal of this chapter is to analyze the psychological and social processes that underlie group extremism, defined as willful collective behavior that substantially violates the norms of expected conduct in a given context (cf. Kruglanski, Jasko, Chernikova, Dugas, & Webber, 2017). Two features of this definition are worth pointing out. By 'collective behavior', we do not refer to individuals who independently exhibit the same behavior to a common stimulus. Instead, we refer to groups of people who share a common identity, pursue collective goals, exert influence over one another, interact under the constraints of norms, roles, and status systems, and are interdependent regarding actions and outcomes. By 'behavior that substantially violates the norms of expected conduct', we refer to deviation from descriptive norms, which specify how groups *do* behave in a particular context, without regard to whether this deviation also violates injunctive norms, which specify how groups *should* behave (Cialdini & Trost, 1998). This distinction is important to outsiders, however, who typically evaluate a group's descriptive deviance in light of injunctive norms. Consider, for example, the approbation showered on unusually altruistic groups versus the opprobrium heaped on unusually aggressive groups. In fact, many definitions of extremism combine the two forms of deviation, as in the following: 'Extremism connotes views and practices on the right or left of the political spectrum that *lie far outside mainstream societal attitude*s and usually *draw broadly negative reactions*' [italics ours] (Futrell, Simi, & Tan, 2019, p. 619).

Our analysis is informed by the model of extremism recently advanced by Kruglanski et al. (2021), which views extremism as a motivational imbalance in which one need dominates and effectively elbows aside other needs. This dominance reduces the constraints associated with simultaneously satisfying multiple needs, thereby justifying extreme actions designed to satisfy the focal need alone. Stated differently, conventional prohibitions against a single-minded pursuit of the focal need are eliminated, and hence 'anything goes'. The model stipulates that extremism, although rare, has wide-ranging cognitive, affective, and behavioral consequences. These include heightened rumination about the focal need; the assumption that others share this need; willingness to endure hardship (i.e., make sacrifices) to satisfy the need; reduced motivation to satisfy alternative needs; preference for means, including extreme means, that serve the focal need; strong affective reactions to need satisfaction/frustration; and emotional volatility. Although the model is based on research dealing with individual behavior, it is potentially applicable to groups as well.

One type of extreme group, namely, terrorist groups that use violent means to achieve their goals, has received a great deal of attention in recent years. Much is known about the types of people who join these groups, their internal dynamics, and the endogenous and exogenous factors that influence their success (for overviews and bibliographies, see Cronin, 2009; Gupta, 2008; Hoffman, 2006; McCauley, 2002; Sageman, 2004, 2008, 2016; Silke, 2018; Tinnes, 2017a, 2017b, 2018).

A useful perspective on violent extremism is provided by SQT (Kruglanski, Belanger, & Gunaratna, 2019; Kruglanski, Gelfand et al., 2014; Kruglanski, Jasko,

Webber, Chernikova, & Molinario, 2018), which is a special case of the model of extremism outlined earlier (Kruglanski et al., 2021). According to SQT, the dominant need underlying violent extremism is a desire for personal significance, which can be stimulated either by events that reduce, or promise to reduce, significance (e.g., failure in an important life domain) or by aspirations for extreme significance (e.g., hero or martyr status). In most cases, however, a desire for personal significance only produces violence if two additional conditions are present. One is a salient ideological narrative that helps group members make sense of their situation and justifies violence as an appropriate means for improving it. The second is a network of like-minded people who provide social validation for the narrative (via informational influence) and rewards for engaging in narrative-consistent behavior (via normative influence) (Kruglanski et al., 2017).

In addition to violent groups, SQT is applicable to nonviolent groups that have a prepotent desire for significance and use extreme means for attaining this goal (e.g., cults, religious orders, single-issue political groups). Not only do such groups eschew violence, they often engage in decidedly prosocial behavior. Examples include religious groups providing medical care to the poor and political groups using peaceful protest to address societal inequities. What extreme violent and nonviolent groups have in common is the willingness to endure hardship and suffering in the service of their goals. This makes sense from the perspective of SQT, because extreme self-sacrifice ' . . . unambiguously communicates one's commitment to a cause from which personal significance can be derived' (Kruglanski et al., 2018, p. 113). The crucial point is that the quest for significance can elicit a wide variety of extreme (i.e., intense and infrequent) behaviors, depending on the salient narrative specifying how the need should be met.

In applying SQT to extreme groups, we pay particular attention to the network component, which has received less emphasis than the motivational (quest for significance) and narrative components. Rather than being accorded co-equal status in determining group extremism, networks are typically portrayed as little more than vehicles for enhancing the salience and validity of narratives, as in the statement, 'networks exert their influence via narratives. . . . It is the narratives, in turn, that ultimately serve as the platforms for actions' (Kruglanski et al., 2018, p. 113). We would argue that networks are in fact the core component of the model. Besides creating and sustaining narratives, they play another critical role that SQT does not explicitly address, namely, providing the *means* by which extreme groups engage in narrative-specified behavior designed to satisfy their quest for significance.

Quest for Significance

A large number of motives have been proposed to explain group extremism. In our view, all of these derive, in one way or another, from the universal human need for significance, which subsumes such specific needs as achievement, esteem, meaning, competence, and control (Kruglanski et al., 2019, in press).

Activating the Quest for Significance: Significance Loss

The most frequently mentioned motive for group extremism is shared grievance based on the perception that the group is unfairly deprived of valued material and/or symbolic resources (cf. Smith, Pettigrew, Pippin, & Bialosiewicz, 2012). Numerous analyses have suggested that this perception and the emotional responses it engenders (e.g., anger, contempt) predispose groups to engage in collective action in general (e.g., Becker & Tausch, 2015; van Stekelenburg, Klandermans, & Walgrave, 2019; van Zomeren, 2015) and extreme action in particular (e.g., Futrell et al., 2019; McCauley & Moskalenko, 2011; Obaidi, Bergh, Akrami, & Anjum, 2019). Related evidence indicates that perceiving one's group to be disrespected, dehumanized, discriminated against, or marginalized increases endorsement of violent action (e.g., Golec de Zavala, Cichocka, Eidelson, & Jayawickreme, 2009; Kteily & Bruneau, 2017; Lyons-Padilla, Gelfand, Mirahmadi, Farooq, & van Egmond, 2015; Victoroff, Adelman, & Matthews, 2012; see also Betts & Hinsz, 2013). It is important to note, however, that shared grievance is not sufficient to produce collective action but rather operates in conjunction with other factors, most notably perceived group efficacy, moral beliefs, and social identity (e.g., van Stekelenburg et al., 2019; van Zomeren, Kutlaca, & Turner-Zwinkels, 2018).

Why is shared grievance an important motivator of group extremism? The answer, according to SQT, is that it threatens the need to feel recognized, respected, and valued—in other words, the need to feel significant. This assumption is consistent with evidence that social identity increases the likelihood that shared grievance will produce collective action (e.g., van Stekelenburg & Klandermans, 2017; van Zomeren et al., 2018). According to social identity theory, people strive for positive self-esteem, which is heavily dependent on their group memberships. More specifically, people experience positive social identity when their ingroup is superior to outgroups and negative social identity in the opposite case (Tajfel & Turner, 1979). Social identity theory postulates that one way groups respond to threatened social identity is by engaging in competition with outgroups. Self-categorization theory (Turner, Hogg, Oakes, Reicher, & Wetherell, 1987) elaborates the cognitive bases of social identity, positing that people who identify strongly with their group depersonalize, or self-stereotype, themselves in terms of ingroup attributes. If so, high identifiers, compared to low identifiers, should experience greater threat to collective significance when they perceive their group is treated unfairly (cf. Ellemers, Spears, & Doosje, 2002), which in turn should lead them to engage in more extreme forms of group competition, including aggression. Consistent with this line of reasoning, social identity has been identified as a major determinant of violent extremism (e.g., McCauley &

Moskalenko, 2011; Sageman, 2016, 2017; for an exception, see Jimenez-Moya, Spears, Rodriguez-Bailon, & de Lemus, 2015).

Activating the Quest for Significance: Significance Gain

Although the loss of significance has been commonly identified as a highly motivating factor responsible for political activism (e.g., Jasko, Szastok, Szastok, Grzymala-Moszczynska, Maj, & Kruglanski, 2019; Molinario, Jaśko et al., 2021), violent extremism (Kruglanski et al., 2019; Kruglanski, Webber & Koehler, 2019), and suicidal terrorism (Webber, Klein, Kruglanski, Brizi, & Merari, 2017), the quest for significance may also be activated by an opportunity for significance gain, the earning of admiration and respect as a hero or a martyr. In this vein, a recent Arabic-language video by Boko Haram, the Nigerian Jihadi group associated with ISIS, seeks to inspire viewers by the prospect of tremendous significance gain through victory or martyrdom:

> We are people whom Allah honored with Islam, so you O' worshipers should accept the call of Allah and pursue jihad, and join the group of Allah, under the one banner, to please Your creator. . . . My beloved, know that we do have two options and not a third: it is either a victory that we attain, or martyrdom, which is our highest goal.
>
> *(in Weimann, this volume, p. 10)*

The incentive of significance may be conveyed by exposure to successful role models, and/or generally admired figures, as their example may inspire individuals to pursue lofty goals. Several studies illustrate this phenomenon. Lockwood and Kunda (1997, 1999) demonstrated how recalling superstars in one's domain of endeavor encouraged research participants to reach high levels of achievement in that domain by making these levels appear more attainable. Hasbrouck (2020) asked participants to name a personal hero and to write a short paragraph describing how that person inspired them and influenced their hopes, goals, and aspirations. This led participants to evince a higher desire for significance as compared to controls who wrote on a neutral topic. Willer (2009, Study 4) found that participants who received status for their contributions to the group subsequently increased their contributions and formed a more positive impression of the group. Molinario, Elster et al. (2021) reported that participants who read a bogus newspaper article, which described how the need for doing something important and meaningful leads to success in several life domains, expressed an elevated desire for significance as compared to controls. Skitka, Hanson, and Wisneski (2017) showed that feelings of pride stemming from higher levels of moral convictions were associated with greater intentions to engage in activism for the relevant moral causes. In other words, the likely prospect of significance gain apparently motivated individuals to embark on activism. Finally, Jasko, Szastok et al. (2019)

found that the more important the values embodied in a political movement are to individuals, the more personally significant they feel when actively engaged in supporting its cause, and consequently the greater their willingness to self-sacrifice and act for the cause in the future.

Findings reviewed earlier attest that the prospect of significance gain can be highly motivating, akin in its stirring powers to a significance loss. Consistent with the view of behavioral learning theorists (e.g., Hull, 1951; Spence, 1956), the motivational influence of drive, based on (e.g., food) *deprivation*, is separate from that of *incentivization* (for discussion, see Kruglanski, Chernikova, Rosenzweig, & Kopetz, 2014). In the present context, both significance deprivation and the incentive of significance gain may induce the quest for significance and motivate joining groups that promise significance attainment.

Desire to Reduce Uncertainty

Self-Uncertainty

Another motive often proposed to explain group extremism is people's desire for clear-cut knowledge (certainty) regarding who they are and how they should feel, think, and act. This motive is highly relevant to the quest for significance, in that the absence of such knowledge poses a fundamental threat to people's views of themselves as competent actors who understand the demands of their social world and their capacity to meet these demands. Moreover, as in the case of collective grievance, social identity plays an important role in how the desire for certainty affects extremism. In contrast to Tajfel and Turner's (1979) emphasis on group members' motivation to gain positive self-esteem, identity-uncertainty theory focuses on members' motivation to reduce psychological uncertainty about themselves (Hogg, 2007a). According to this theory, group identification satisfies the need for certainty via depersonalization, in which a consensual group prototype specifies how members should feel, think, and act. Moreover, highly entitative groups, characterized by clear boundaries and internal structure, homogeneous membership, a high degree of social interaction, and common goals and fate, are most effective in reducing self-uncertainty. In applying uncertainty-identity theory to extreme groups, Hogg (2012, 2014; Hogg & Adelman, 2013) argues that such groups, besides being highly entitative, exhibit other characteristics (e.g., inflexible customs, rigid authority structures, ethnocentrism, intolerance of dissent, suspicion of outsiders, rigid ideological systems) that make them attractive to people who have a strong need to reduce uncertainty and, we would argue, thereby satisfy their quest for significance. In other words, the certainty that group identification affords informs individuals what they need to do, think, and feel in order to be accepted and deemed worthy by their fellow members, in this sense contributing to individuals' self-esteem after all.

Desire for self-certainty is also highlighted in other analyses of extremism (e.g., Doosje, Loseman, & van den Bos, 2013; Hogg, Kruglanski, & van den Bos, 2013; Jost & Napier, 2012; Moghaddam & Love, 2012). As an example, McGregor and Marigold (2003) measured participants' self-esteem and then in one condition ('uncertainty threat') asked them to think of an unresolved personal dilemma, which made them feel very uncertain. In a control condition, participants engaged in other activities that did not affect their sense of self (e.g., they thought about a friend's dilemma). It was found that in the uncertainty threat, but not in the control, condition participants with high self-esteem responded more quickly to questions like 'I seldom experience conflict between the different aspects of my personality' and 'In general, I have a clear sense of who I am and what I am'. They also showed greater certainty with respect to such societally important issues as capital punishment and abortion. Ironically then, making individuals anxiously uncertain produced an overcompensation and an inflated sense of (defensive) certainty. These findings illuminate the relation between self-uncertainty and self-esteem (or self-worth as McGregor and Marigold refer to it). Specifically, persons with high (vs. low) self-esteem seem to be particularly bothered by uncertainty that questions their worth and particularly defensive in response.

Need for Cognitive Closure

Although self-uncertainty parallels the need for specific closure (a positive view of the self), the need for nonspecific closure—discomfort with uncertainty of any kind (even one irrelevant to one's self concept)—can also motivate attraction to extreme groups. In this vein, Kruglanski (2004) posits that a heightened desire for clear-cut and unambiguous answers motivates people to accept and then refuse to relinquish ideas that provide certainty. This perspective suggests that people who (for personality or situational reasons) are high in need for nonspecific cognitive closure should be attracted to the simple, clear-cut ideologies propounded by extreme groups (Kruglanski, Pierro, Mannetti, & De Grada, 2006; Kruglanski et al., 2018; Kruglanski & Orehek, 2012). In other words, people may be attracted to clear-cut ideologies not only because they offer significance but also because they offer certainty as such.

The desire for certainty as such could stem from prior or present association of uncertainty with the potential for highly negative outcomes, which might arise, for instance, from growing up in times of great physical and economic scarcity. In this vein, Webber et al. (2017), using samples of suspected militants and former terrorists, found that participants' feelings of humiliation were positively associated with their need for closure. Kruglanski's analysis also helps explain why intense religious beliefs, which satisfy the need for certainty, are often associated with violent extremism (e.g., Bromley & Melton, 2004; Hoffman, 2006; McGregor, Nash, & Prentice, 2012).

Sacrifice and Identity Fusion

As discussed earlier, we define extremism as a form of motivational imbalance in which one goal displaces all others. As a consequence of this prioritization, the behavioral constraints associated with the renounced goals are lifted, and pursuit of the dominant goal becomes easier. However, this benefit comes with an important cost, namely, the rewards forgone by not pursuing the renounced goals. In this sense, then, sacrifice is an inherent component of extremism (e.g., Bélanger, Caouette, Sharvit, & Dugas, 2014; Dugas et al., 2016; Olivola & Shafir, 2013).

The role of sacrifice in extreme behavior is a key feature of identity fusion theory (for reviews, see Gomez et al., 2020; Swann & Buhrmester, 2015; Swann, Jetten, Gomez, Whitehouse, & Bastian, 2012). This formulation owes an intellectual debt to social identity theory (Tajfel & Turner, 1979) but differs from it in several ways (e.g., by focusing on intragroup rather than intergroup relations and assuming that personal and social identities are independent rather than antagonistic). According to Swann and colleagues, highly fused members experience a visceral feeling of oneness with their group, which reflects a union of their personal and social selves. As a result, they feel strongly connected to one another and the group as a whole, which motivates them to engage in self-sacrificial forms of pro-group behavior, including dying for the group (e.g., Swann, Buhrmester et al., 2014; Swann, Gómez et al., 2014; Swann, Gómez, Dovidio, Hart, & Jetten, 2010). When intergroup conflict is involved, extreme behavior that helps the ingroup simultaneously harms the outgroup, as in the cases of suicide terrorists (e.g., Bélanger et al., 2014; Hoffman, 2006; Whitehouse, 2018) and Japanese Kamikaze pilots (Inoguchi, Nakajima, & Pineau, 1958). However, identity fusion can also produce self-sacrificial behavior that helps the ingroup without harming the outgroup (e.g., Buhrmester, Fraser, Lanman, Whitehouse, & Swann, 2015; Gómez, Morales, Hart, Vázquez, & Swann, 2011).

Swann and colleagues' analysis of the psychological underpinnings of identity fusion is quite consistent with a quest for significance interpretation of group members' motivation. They argue, for example:

> because group members who are highly fused with the group channel their feelings of personal agency into the priorities of the group, the effectively bolster the *collective* efficacy of the group. . . . Individuals may also benefit from being fused. That is, by channeling their feelings of agency into the agendas that they share with the group, highly fused persons are able to act in accordance with a meaning system that extends beyond their own needs and desires. . . . Fusion may therefore offer individuals a pathway to a meaningful existence.
>
> *(Swann et al., 2012, p. 12)*

Several lines of evidence are consistent with a quest for significance interpretation of identity fusion. For example, shared negative experiences (grievances), which we argue previously threaten feelings of significance, produce identity fusion (Jong, Whitehouse, Kavanagh, & Lane, 2015). Moreover, the impact of fusion on extreme pro-group behavior is mediated by perceptions of invulnerability and agency (Gómez, Brooks, Buhrmester, Vázquez, Jetten, & Swann, 2011, Study 9). And a threat directly linked to a loss of significance, namely, ostracism (Williams, 2001), increases extreme pro-group actions among fused, but not non-fused, group members (Gómez, Morales et al., 2011).

Charismatic and Other Extreme Groups

Identity fusion is typically operationalized as willingness to sacrifice one's life on behalf of the ingroup. Although the theory has not emphasized one particular form of sacrifice, namely, committing suicide on the orders of group leaders, this extreme behavior sometimes occurs in religious cults (e.g., Branch Davidians, Heaven's Gate, People's Temple). Cults are a subcategory of charismatic groups, which are characterized by a shared belief system, a high level of social cohesiveness, strong adherence to the group's behavioral norms, and the imputation of special (charismatic or divine) powers to the group or its leaders (Galanter, 1999; Galanter & Forest, 2006; see also McCauley & Segal, 1987). As in the case of conventional religions, a powerful motivation for joining and remaining in religious cults is a desire for significance and certainty.

Although the majority of religious cults do not demand suicide, they all require members to prioritize the group's welfare over all other needs and to engage self-sacrificial behaviors. These behaviors, which often parallel those demanded by utopian communities (Kanter, 1968), include abstinence of various kinds (e.g., food, personal adornment, sex), assignment of money or property to the group, acceptance of restrictions on privacy and personal possessions, participation in group confessions, adherence to rigid rules of conduct, and renunciation of family ties inside and outside the group. A striking example of family renunciation is contained in a Shaker hymn reported by Kanter (1968, p. 508):

> Of all the relations that ever I see
> My old fleshly kindred are furthest from me
> So bad and so ugly, so hateful they feel
> To see them and hate them increases my zeal
>
> O how ugly they look!
> How ugly they look!
> How nasty they feel!

In addition to religious cults, charismatic groups also include 'spiritual recovery movements' (Galanter, 1999), which claim to ameliorate physical or psychological problems, operate outside conventional medicine, and attribute their effectiveness to metaphysical factors (e.g., Alcoholics Anonymous). Such movements fit our definition of extremism because their members prioritize one motive over all others and engage in unconventional behaviors (e.g., confessing self-control failures to strangers). Because undesirable conditions such as alcoholism are socially stigmatizing, they reduce members' feelings of significance, which in turn elicits extreme behaviors designed to eliminate the condition.

A different process occurs in groups that encourage and help members to engage in maladaptive and socially stigmatized behaviors. In such groups, members' desire for significance elicits extreme behaviors designed to maintain and thereby affirm the value of the stigmatized condition. One example is Pro-Ana groups, which treat anorexia as a positive behavior rather than a dangerous medical condition and provide 'thinspiration' for losing weight via extreme dieting (Borzekowski, Schenk, Wilson, & Peebles, 2010). Another example is 'Deaf community' groups, which oppose cochlear implants for deaf children as well as 'oralism' (the use of spoken language) (Tucker, 1998). This opposition can take extreme forms, as in the following example:

> When the America's Got Talent finalist Mandy Harvey first appeared on the show, she caused a social media storm. Harvey became deaf due to an illness, but decided to pursue her love of music, feeling the beat of the music through her feet. However, her singing caused a backlash among a very small minority, who sent Harvey death threats for promoting a 'hearing' activity.
> (The Guardian. *'When a deaf singer gets deaf threats from other deaf people, something's wrong'*. Josh Salisbury. November 27, 2017)

To this point, we have focused on extremism designed to benefit ingroup members, including the self. But extremism can have a different, albeit less common, goal, namely, benefitting outgroup members (cf. Louis et al., 2020). Examples include Catholic nuns who minister to the poor; Northern whites who participated in Southern civil rights marches in the 1960s; and Christians who aided Jews in Nazi-occupied Europe during World War II (Righteous Among the Nations). In these and similar cases, people prioritize one goal over all others and use extreme (unconventional) means to attain this goal, often incurring substantial costs in the process. As noted earlier, this altruistic behavior often earns the praise of third parties as well as recipients. Like pro-ingroup extremism, pro-outgroup extremism can be interpreted in terms of the quest for significance, which is often intimately tied to religious values. For example, in explaining why he helped Jews during the German occupation of the Netherlands, Seine Otten stated:

> I was teaching in the Christian school and we were living in Nieuwlande. . . . My wife and I believed very strongly in the Bible, and we tried to live by it. In 1941, when we began to realize that Jews would be persecuted, we told them, 'When you are in need, you may come to Nieuwlande'
>
> *(Block & Drucker, 1992, p. 62)*

It is worth noting that the definition of 'outgroup' can be quite elastic, even stretching to members of other species (e.g., the Animal Liberation Front).

Narratives

While the quest for significance is the driving force behind group extremism, it is not sufficient to produce extreme behavior. Another critical component is a salient narrative that (a) reinforces group members' feelings of significance threat, (b) provides a simple and compelling explanation for this threat, and (c) specifies and legitimizes nonnormative means for reducing the threat. Thus, narratives are shared belief systems, or ideologies, that give members 'good' reasons to engage in extreme behavior.[1] It is important to emphasize that although all extreme groups desire significance, they subscribe to many different narratives, which explains why extremism can take such diverse forms (e.g., extraordinary altruism, extraordinary violence).

A vivid example is a composite narrative reflecting the perspective of militant extremists in Europe, the Middle East, Africa, South Asia, East Asia, Latin America, and North America (Saucier, Akers, Shen-Miller, Knezevic, & Stankov, 2009, p. 265):

> We (i.e., our group, however defined) have a glorious past, but modernity has been disastrous, bringing on a great catastrophe in which we are tragically obstructed from reaching our rightful place, obstructed by an illegitimate civil government and/or by an enemy so evil that is does not deserve to be called human. This intolerable situation calls for vengeance. *Extreme measures are required* [italics ours]; indeed, any means will be justified for reaching our sacred end. We must think in military terms to annihilate this evil and purify the world of it. It is a duty to kill the perpetrators of evil, and we cannot be blamed for carrying out this violence. Those who sacrifice themselves in our cause will attain glory, and supernatural powers should come to our aid in this struggle. In the end, we will bring our people to a new world that is paradise.

Whether ideology should be considered a major *cause* of extremism is controversial (Holbrook & Horgan, 2019). Nonetheless, even skeptics agree that ideology plays a critical role in rationalizing and justifying extreme forms of collective action (e.g., McCauley & Moskalenko, 2011). A prime example is religion. As discussed earlier, religious beliefs are a common source of legitimacy for extreme acts, which can profoundly affect the welfare of both ingroup and outgroup members (e.g., Block & Drucker, 1992; Bromley & Melton, 2004; Galanter, 1999; Hoffman, 2006).

Religion is one source of sacred values, which can be defined as:

> any preferences regarding objects, beliefs, or practices that people treat as incompatible or non-fungible with profane issues or economic goods, such as belief in God or country. They tend to be highly stable and difficult to influence socially. . . . They are also insensitive to spatial and temporal discounting . . . matters linked to sacred values—however far removed in time or space—are more important than mundane concerns.
>
> *(Atran, 2021, pp. 6.8–6.9)*

Sacred values, in conjunction with identity fusion, are the major building blocks of the Devoted Actor model of extremism (e.g., Atran, 2016; Atran & Sheikh, 2015). Consistent with the model, people who view group values as sacred, believe these values are threatened, and identify strongly with the group are particularly likely to engage in extreme behavior on behalf of the group (Sheikh, Gómez, & Atran, 2016).

Narratives are important because they provide a bridge between the motive underlying extremism (quest for significance) and the behavior employed to satisfy this motive. For example, group narratives have been shown to strengthen the link between quest for significance and support for both prosocial and antisocial behavior toward outgroups (e.g., Jasko et al., 2020; Leander et al., 2020). However, not all narratives perform this bridging function equally well, which raises the question of why some narratives are more persuasive than others. Narratives are persuasive to the extent that their source has epistemic authority, defined as the 'right' to define how things are and how they should be (Kruglanski et al., 2005). Reference groups, which people use as frames of reference for evaluating their attitudes and beliefs, possess such authority. Although reference groups are not always membership groups, we focus here on those that serve both functions. Because such groups satisfy members' social (e.g., acceptance, status) as well as epistemic needs, they are likely to be seen as highly trustworthy sources of information, which in turn should increase members' willingness to accept their narratives. Thus, rather than being subjected to critical examination, these narratives are likely to be 'swallowed hook, line, and sinker'.

Group narratives reflect a consensus, or shared reality, regarding the group's collective goal and the appropriate means for achieving this goal. Shared reality can be defined as 'the product of the motivated process of experiencing with others a personal connection and commonality of inner states (judgements, beliefs, feelings, attitudes) about some target' (Echterhoff & Higgins, 2017, p. 176). It is important because it satisfies two fundamental human motives—the epistemic need for a confident and accurate understanding of the world and the affiliative/relational need for a connection with other people (see Echterhoff, Higgins, & Levine, 2009; Higgins, 2019). Although most of the research on shared reality has focused on how it affects individual cognition, it also plays a central role in the formation, maintenance, and functioning of groups (Levine, 2018; Levine & Higgins, 2001; Levine, Resnick, & Higgins, 1993).

Narratives and Moral Disengagement

To the extent that extreme groups engage in violent acts against outgroups, their members often suffer pangs of conscience for violating conventional norms of conduct. For this reason, the narratives of such groups typically help members morally disengage from their actions. One facilitator of moral disengagement is the perception that the outgroup is less than human. In fact, dehumanization is a common feature of extremist group narratives, as illustrated by a statement by a member of the Red Army Faction, a left-wing German terrorist group:

> We say the guy in uniform is a pig, he is not a human being, and we have to tackle him from this point of view.... It is wrong to talk to these people at all and of course the use of guns is allowed.
> *(Meinhof, 1970, p. 75; quoted in Wasmund, 1983, p. 234)*

According to Haslam (2006), dehumanization can involve viewing people as either animal-like (i.e., lacking human uniqueness attributes, such as self-control or rationality) or object-like (i.e., lacking human nature attributes, such as warmth or emotion). Dehumanization has negative cognitive, affective, and behavioral consequences for the people so designated, including reduced moral status, contempt, and aggression (Haslam, 2015). Two implications of dehumanization, namely, that outgroup characteristics are shared by all group members and are immutable, are particularly relevant to extremist group narratives. To the extent outgroups are perceived in this essentialist fashion, it is easier to view them as implacable enemies, to feel hatred toward them, and to use violent means in fighting them (cf. Giner-Sorolla, Leidner, & Castano, 2012; McCauley & Moskalenko, 2011).

Dehumanization, though quite common, is only one way group narratives can help members avoid self-censure for aggressive acts (Bandura, 1999; Waller, 2007). Other forms of moral disengagement include euphemistic labeling of harmful acts; advantageous comparisons that frame group actions as moderate responses to extreme provocations; displacement of responsibility for distasteful actions to specific figures (e.g., authorities); diffusion of responsibility to the group as a whole; distortion or mimimization of the negative consequences of group actions; and a sense of moral justification. Both moral justification and dehumanization are evident in the comments of Heinrich Himmler, Reichsfuhrer of the German SS, in a speech to senior SS officers:

> Most of you know what it means when one hundred corpses are lying side by side, or five hundred, or one thousand. . . . This is a page of glory in our history which has never to be written and is never to be written. . . . Because we have exterminated a germ, we do not want in the end to be infected by the germ and die of it. . . . Wherever it may form, we will cauterize it. Altogether, however, we can say that we have fulfilled this most difficult duty for the love of our people. And our spirit, our soul, our character has not suffered injury from it.
>
> *(Himmler speech, October 4, 1943; quoted in Gilbert, 1985, pp. 615–616)*

Several features of these comments—that this 'page of glory' is never to be written, that it involved a most difficult duty, and that it did not injure the spirit, soul, or character—reveal the potential psychological costs of extreme violence and hence perpetrators' embrace of narratives that allow them to morally disengage from their actions.

Narratives and Group Socialization

Narratives are often transmitted to potential group members by charismatic communicators who stress the humiliation and loss they suffer by identifying with a social category, such as their religion, ethnicity, or race. Thus, in addition to articulating the group's ideology, such narratives are likely to increase potential members' quest for significance. Consider the following message from Abu Yehia Al Libi, Al Qaeda's major propagandist (killed by US drone strike in 2012):

> Jihad in Algeria today is YOUR hope with permission from Allah in redemption from the hell of the unjust ruling regimes whose prisons are congested with YOUR youths and children if not with YOUR women; [regime] which thrust its armies, police, and intelligence to oppress YOU, for which they opened the doors to punish YOU. . . . So join YOUR efforts to theirs, add YOUR energies to theirs . . . and know that their victory is YOUR victory. . . . Their salvation is YOUR salvation.

Al Libi's fiery message to young Muslims is an example of a recruiter reaching out to prospective group members. In other instances, they may seek out the narrative (e.g., on the Internet) or encounter it in everyday situations. At any rate, prior to joining a group, prospective members typically want to learn what the group stands for, how members are expected to behave, and how these behaviors correspond to their needs (Levine, Bogart, & Zdaniuk, 1996). Prospective members' efforts to acquire this information are not always successful, because some groups are unwilling to reveal their core beliefs to outsiders. However, as the Al Libi example shows, many extreme groups are highly motivated to acquaint prospective members with their narrative and assess their susceptibility to it. This is not surprising, as these narratives require members to accept unconventional beliefs and engage in costly behaviors. An example is a workshop for prospective members run by the Unification Church (Galanter, 1999, p. 129):

> The program for the weekend ran from eight in the morning to eleven at night. The main component was a series of ninety-minute lectures, each concentrating on major points from the Divine Principle, followed by a half hour of small group discussion that explored the meaning of these religious principles relative to the individual participants' lives.
>
> Communication was regulated in subtle but nonetheless effective ways by workshop leaders. For example, recreational activities such as singing, sports, and skits were interspersed with the lectures and discussions, but these seemingly casual functions were well structured and provided a context in which information-sharing and communication about the church took place. Conversations and ideas that did not bear on the themes under discussion were discouraged in a congenial but firm manner.
>
> The balance between active members and nonmember recruits during the small group discussions also assured control by the leaders over the context of communication . . . Thus, the majority were committed to supporting the church's philosophy through either overt or covert management of communication. Under these circumstances, it was possible to suppress deviant points of view, often before they were expressed.

In some cases, prospective members are allowed to join a group only after demonstrating that they fully understand and accept the group narrative. In other cases, they need only express a hazy understanding of, and general sympathy with, the narrative. When this occurs, new members enter the group on probation and undergo a period of socialization, during which they are expected to learn and internalize the group narrative (Levine & Moreland, 1994; see Networks section for a description of the socialization process). Extreme groups take socialization very seriously and hence make substantial demands on new members. These demands are remarkably similar across various types of extreme groups, including religious cults, utopian communities, and terrorist cells (e.g., Galanter & Forest, 2006; Horgan, Taylor, Bloom, & Winter, 2017; Kanter, 1968; McCauley & Segal, 1987; Wasmund, 1983). They include isolating new members from outside contacts; exposing them to intensive indoctrination about the value of the group and its goals; forcing them to endure physical and psychological discomfort;

creating opportunities for them to participate in activities (e.g., shared hardship, rituals) that foster group identification and cohesion; and monitoring and testing their adherence to group norms and commitment to group goals.

The effectiveness of group socialization varies as a function of the characteristics and behaviors of both current and new members. For example, socialization is smoother and faster if current members are motivated to integrate new members into the group and provide feedback regarding their behavior. New members are more receptive to socialization if they want to fit into the group and make efforts to learn its norms and values. Thus, the probability that new members will accept the group narrative is heavily dependent on factors that affect minority members' conformity to pressure from majority members (see Levine & Tindale, 2015, for a review).

Creating a Narrative

So far, we have focused on groups that possess an extreme narrative. In such cases, new (and prospective) members are expected to wholeheartedly embrace the narrative *in toto*. In other words, the narrative is treated as sacred and immutable, and any efforts to shape or alter it are punished, often harshly.[2] The situation is quite different in groups that do *not* possess an extreme narrative. This might occur, for example, when formerly unacquainted people first come together or when long-time acquaintances (e.g., friends) pursue prosaic goals. In such cases, in order for an extreme narrative to exist, the group must either import or create it. We focus here on the latter mechanism.

How does a group create an extreme narrative? At first glance, Sherif's (1935, 1936) classic studies of norm formation might seem relevant to this question (Abrams & Levine, 2017). By using the autokinetic effect, Sherif had small groups of participants make private and then public judgments about the movement of a light in a dark room. He found evidence of social influence in the latter setting, a necessary condition for the development of extreme narratives. However, because participants' public responses converged toward the central tendency of their private responses, Sherif's findings demonstrated response moderation, not extremitization.

Another form of reciprocal influence, namely, group polarization, is more relevant to the creation of extreme narratives. This well-documented phenomenon occurs when, as a function of discussion, the group's modal position becomes more extreme in the direction that members initially favored. Several explanations for polarization have been proposed. These include persuasive arguments theory, based on informational influence; social comparison theory, based on normative influence; and self-categorization theory, based on ingroup identification and movement toward the prototypic ingroup position (for reviews, see Isenberg, 1986; Levine & Moreland, 1998). All of these theories have empirical support, and their underlying mechanisms are likely to operate together to produce polarization.[3]

Polarization is a ubiquitous feature of extreme groups (McCauley & Segal, 1987; McCauley & Moskalenko, 2011). Sageman's (2004, 2008) description of self-radicalization by informal groups of young Muslims ('bunches of guys') provides a vivid example:

> With the gradual intensity of interaction within the group and the progressive distance from former ties, they changed their values. From secular people, they became more religious. From material rewards, they began to value spiritual rewards, including eventually otherworldly rewards. From the pursuit of short-term opportunities, they turned to a long-term vision of the world. They abandoned their individual concerns for community concerns and became ready to sacrifice for comrades and the cause. Martyrdom, the ultimate sacrifice for the group and the cause, became their ultimate goal and the true path to glory and fame with respect to their friends. . . . This belief encourages active engagement. The mutual group reinforcement also allows them to leave behind traditional societal morality for a more local morality, preached by the group.
>
> *(Sageman, 2008, pp. 86–87)*

Polarization is often discussed in the context of 'groupthink', a form of mutual influence often blamed for poor group decisions. According to Janis (1982), several antecedent conditions (e.g., strong group cohesion, insulation from outside influences, homogeneity of member attitudes, perception of external threat) stimulate concurrence seeking among group members, which is manifested in overestimation of the ingroup vis-a-vis outgroups, closed-mindedness, and pressures for uniformity. Concurrence seeking, in turn, produces defective decision-making processes (e.g., inadequate information search, biased risk assessment), which finally yield flawed (and often polarized) collective decisions. Given the weak and inconsistent evidence for the antecedent conditions of Janis's model, one might ask why it is so popular. The answer, according to Baron (2005), is that many of the phenomena in the model, such as polarization, are so common that they do not require special antecedent conditions. This view is consistent with the ubiquity of polarization in extreme groups.

The role of group discussion in producing polarization has been highlighted in analyses of collective action. For example, van Stekelenburg and Klandermans (2017, p. 127) argue that:

> The effect of interaction in networks on the propensity to participate in politics is contingent on the amount of political discussion that occurs in social networks and the information that people are able to gather about politics as a result.

Similarly, McGarty, Thomas, and their colleagues assert that, in order for supporters of a cause to mobilize on its behalf, they must belong to an 'opinion-based group' (Smith, Thomas, & McGarty, 2015; Thomas, McGarty, & Mavor, 2009). Such groups emerge as members come to agreement (i.e., develop a shared reality) regarding (a) their preferred strategies for furthering the shared cause, (b) the efficacy of these strategies, and (c) the value of publicly expressing their views (Bongiorno, McGarty, Kurz, Haslam, & Sibley, 2016; see also Smith, Blackwood, & Thomas, 2020). In applying this framework to extremism, Thomas, McGarty, and Louis (2014) found that group discussion led to endorsement of radical means for attaining a group goal only when members believed a prestigious outside source advocated such means. These findings indicate that, in order for group discussion to produce extremism, radical solutions must be seen as legitimate. This legitimization can come from inside, as well as outside, the group, and leaders, because of their epistemic authority (Kruglanski et al., 2005), often perform this function (cf. Hogg & Adelman, 2013).

Networks

According to SQT, networks (or groups) play a critical role in extremism by increasing the salience and perceived validity of narratives linking the quest for significance to behaviors for achieving this goal. In the previous section, we provided several examples of how networks, via their impact on narratives, can increase group members' *readiness* to engage in extreme behaviors. Moreover, as van Stekelenburg et al. (2019) suggest, networks have the same impact on 'moderate' behaviors, such as participating in street demonstrations: 'It is within these networks that processes such as grievance formation, empowerment, identification, and group-based emotions all synthesize into a motivational constellation preparing people for action and building mobilization potential' (p. 378).

However, as we argued earlier, networks do much more than prime particular behaviors—they also provide the *means* for enacting these behaviors. In the present section, we focus on factors that affect the ability of extreme groups to achieve their proximal and distal goals. We discuss these factors in relation to two temporal processes that occur in all groups—socialization and development (Moreland & Levine, 1988). The former refers to changes over time in the relationship between a group and its members; the latter refers to changes over time in the group as a whole.

Group Socialization

A group's ability to achieve its goals is critically affected by its composition, defined as the number and characteristics of the people who belong to the group. This is particularly true in extreme groups, which require their members to prioritize one goal above all others, to accept unconventional (and what, to outsiders, may seem outlandish) belief systems, and to engage in behaviors that require great self-sacrifice. Such groups offer their members a sense of significance and meaning in exchange for their devotion, which far exceeds what they could expect from participation in more moderate groups. In order for extreme groups to be successful, they must attract and retain a sufficient number of members who already possess, or can be taught, the beliefs, values, and skills they need to function effectively. For many (perhaps most) extreme groups, the right set of members can produce triumph, whereas the wrong set can lead to tragedy. Groups manage their composition through the process of socialization, which we briefly discussed in the section on narratives.

Socialization involves three psychological processes: evaluation, commitment, and role transition (Levine & Moreland, 1994; Levine, Moreland, & Hausmann, 2005; Moreland & Levine, 1982). Evaluation involves efforts by the group and the individual to assess the past, present, and probable future rewardingness of their own and alternative relationships. Evaluation produces feelings of commitment on the part of the group and the individual, which can rise or fall over time depending on changes in the other party's perceived rewardingness. Commitment in turn affects the likelihood that the individual will undergo a role transition and move from one phase of group membership to another. These transitions occur when the group's and the individual's commitment rise or fall to particular levels (decision criteria) that the parties believe warrant a major change in the individual's relationship with the group. Transitions are typically marked by ceremonies, some quite elaborate, designed to clarify for both parties that the individual's responsibilities and rights are now different. Following a role transition, the group's and the individual's expectations for one another change, which initiates a new round of evaluations that in turn may produce changes in commitment and (if decision criteria are reached) additional role transitions. In this way, an individual's passage through a group can be conceptualized as a series of role transitions between different phases of group membership. Figure 4.1 illustrates a typical passage involving four role transitions (entry, acceptance, divergence, and exit) and five membership phases (investigation, socialization, maintenance, resocialization, and remembrance). An important feature of the model is that the individual, as well as the group, can influence the timing and course of this process.[4] In the following, we discuss group socialization in terms of the first four membership phases, with special emphasis on how socialization operates in extreme groups.

Investigation Phase

During investigation, the group engages in recruitment, looking for prospective members who can contribute to achieving its goals, and prospective members engage in reconnaissance, looking for groups that can contribute to satisfying their needs. More attention has been devoted to the recruiting activities of

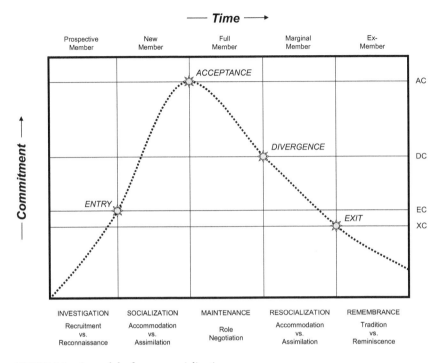

FIGURE 4.1 A model of group socialization

extreme groups than to the reconnaissance activities of prospective members, although both are important to the success of investigation.

The first step in recruiting members to an extreme group is identifying people with a high need for significance because they feel disrespected and discriminated against. An example is the Kouachi brothers, disenfranchised youths with no steady jobs and dim prospects for the future, who perpetrated the 2015 Charlie Hebdo attack in Paris. In addition, past criminal activity and the marginal status it confers renders individuals vulnerable to the glory and significance promised by extreme groups. An example is Ali La Pointe, a thief and pickpocket immortalized in Gilo Pontecorvo's film the 'Battle of Algiers', who died a hero and martyr of the Front de Liberation Nationale (FLN). As Basra and Neumann (2016, p. 25) note, not only does the lofty cause of the extreme group offer criminals significance, but also their background and expertise are highly useful to the group:

> The prevalence of criminal backgrounds amongst European jihadists is remarkable. Whether amongst 'foreign fighters' that have travelled to Syria and Iraq, or amongst those involved in terrorism in Europe, criminal pasts are common . . . jihadism can . . . offer redemption from past sins. . . (and) criminals develop skills that can be useful for them as extremists, such as

access to weapons and forged documents, as well as the psychological 'skill' of familiarity with violence.

In order to recruit effectively, then, extreme groups must first identify the basis of potential members' need for significance and then tailor their recruiting strategies accordingly, as in this statement by a member of a right-wing terrorist group in the United States:

> Really the way that I see it is I think that the recruiting is all about preying on kids that are susceptible. That's never admitted. We would never admit that. It's kids that don't have families, 'We'll be your family'. Kids that need protection or whatever, for different reasons. 'We'll be whatever you need'.
> *(Kevin, Blood and Honour, July 7, 2014, quoted in Simi, Windisch, & Sporer, 2016, p. 60)*

A good deal is known about how extreme groups go about trying to recruit new members (e.g., Bloom, 2017; Galanter, 1999; Hunter, Shortland, Crayne, & Ligon, 2017; McCauley & Segal, 1987; Simi et al., 2016; Wasmund, 1983; Windisch, Logan, & Ligon, 2018). Given that such groups demand substantial commitment and loyalty from their members, it would not be surprising if they devoted considerable time and energy to recruiting, evaluated prospective members carefully, and used rigorous standards (high entry criteria [EC]) in deciding whether these people should be allowed to join the group. Although these generalizations are often true, extreme groups vary in the intensity of their recruiting efforts, as well as the strategies they employ to identify and attract new members. For example, changes in staffing requirements, produced by member attrition or shifts in needed skills, can affect how groups recruit new members (e.g., Bloom, 2017; Hunter et al., 2017). Also important are the number of potential members, their presumed motives, and their perceived attraction to the group. For example, when a sufficient number of people are eager to join, active recruiting is unnecessary:

> One of the surprising aspects of the global Salafi movement, given its notoriety and ubiquity, is the relative lack of resources invested in any recruitment drive. I did not detect any active top-down organizational push to increase al Qaeda's membership. The pressure came from the bottom up. Prospective mujahedin were eager to join the movement.
> *(Sageman, 2004, p. 117)*

However, extreme groups are not always so fortunate, as is true of many right-wing terrorist groups (Simi et al., 2016). In such cases, the group may not even know where to target its recruiting activities. One mechanism for identifying and making contact with potential recruits is distributing information in public places:

> We tried to get a hold of folks that showed scientific proof of differences between whites and blacks. I remember a flyer that I made about the difference between the whites and the blacks brains . . . comparing the brain of the African-Americans to apes and chimpanzees and stuff, so I was distributing that at the Mall.
> *(Harry, Las Vegas Skinheads, June 8, 2013; quoted in Simi et al., 2016, p. 63)*

In recent years, rather than passing out information in malls, many extreme groups have turned to the Internet (e.g., Kruglanski et al., 2019; Sageman, 2008). By utilizing existing social media, creating websites, and developing online magazines, such groups can reach very large audiences.

Another mechanism is face-to-face interaction with potential members, which allows recruiters to interact with several people at once, to deliver detailed pro-group messages, to provide incentives for listening to these messages, and to evaluate interest in the group:

> We'd start passing flyers for parties, saying come join us, come check us out. We're not saying you have to join us now, but come check us out, see what it's like. We had a keg every single night, no matter what. For about two summers, we had a keg every single day where at least kids could come in. There's minors, 15, 16 years old, we didn't care. You come in, join our ranks and see what it's like, and we'd do the same thing. Pushing women and girls out there. You'd get to know which girls were the easy ones, so you'd push them on to the guys. They'd just turn heads. These guys were like,' Oh man, we'd get this all the time. It's just one big party, life's great.
> *(Richard, White Aryan Resistance, September 21, 2013; quoted in Simi et al., 2016, p. 65)*

Finally, a third, and particularly effective, mechanism is introducing potential recruits to the 'white power' music scene, where they are exposed to negative depictions of outgroups (e.g., Blacks, Jews, Muslims) together with positive portrayals of violence:

> Music was the propaganda. The music was the most powerful tool to recruit anybody. You could teach through music, you could make things make sense with lyrics. You could inspire, motivate, insight, that's what I was good at.
> *(Chester, Volksfront, October 22, 2013; quoted in Simi et al., 2016; p. 68)*

> It's the message that [is] in it. Hate rock is one of the best recruitment tools . . . It's very aggressive. You throw hate rock on top of the aggression or the aggression on top of the [hate rock] music and you're going to spit out some good Nazis.
>
> *(Denis, Golden State Skinheads, July 27, 2014; quoted in Simi et al., 2016, p. 69)*

As these examples reveal, violent extreme groups often keep their narratives under wraps early in the recruiting process, revealing them only gradually to those prospective members who seem amenable to them. This covert recruiting strategy is also used by nonviolent extreme groups (e.g., religious cults), as Galanter (1999, p. 126) found in studying the Unification Church:

> In the San Francisco Bay area . . . the [recruitment] process was generally surreptitious. Potential members were approached under the guise of introducing them to an innocuous socially oriented group and only when the potential recruits became fully involved in the group were they made aware of its association with the church.

As noted above, less attention has been devoted to the reconnaissance activities of prospective members of extreme groups than to the recruiting activities of these groups. Prospective members express a wide range of specific motives for wanting to join extreme groups, which can be subsumed under the quest for significance discussed earlier. Reconnaissance activities involve identifying potentially interesting groups, evaluating the likelihood that membership will satisfy one's needs, and finally persuading chosen groups that one is a good candidate for membership (Moreland & Levine, 1982). Each of these steps can be more or less difficult, depending on the group's desire to recruit new members.

Finally, it is important to note that potential members of extreme groups (both violent and nonviolent) are not always strangers. In many cases, they are acquaintances, friends, or relatives of current members (e.g., della Porta, 1988; Kruglanski et al., 2019; Sageman, 2004), which is consistent with the 'social network' analysis of recruitment to social movements (e.g., Lofland & Stark, 1965; Snow, Zurcher, & Ekland-Olson, 1980). For obvious reasons, prior (positive) relationships between prospective and current members greatly facilitate the recruitment process.

In the investigation phase, the group and the individual engage in reciprocal evaluation designed to provide information about the desirability of forming a relationship. If, as a function of this evaluation, both parties' commitment levels rise to their EC, as depicted in Figure 4.1, the individual undergoes the role transition of entry and enters the socialization phase as a new member. Entry is an 'inclusive' transition, in that it signals promotion to a higher status in the group.

Socialization Phase

During socialization, the group attempts to change the individual in ways that will enhance his or her contribution to group goals, and the individual attempts to change the group in ways that will enhance its contribution to personal needs. Because in most cases the group has more leverage than the individual, the latter type of influence is relatively rare (e.g., when a new member brings uniquely valuable skills; see Levine & Choi, 2011; Levine, Choi, & Moreland, 2019). Thus, socialization is more likely to involve assimilation on the part of new members, rather than accommodation on the part of the group.

Earlier we mentioned the rigorous demands that extreme groups make on new members in an effort to transmit the group narrative. To recapitulate, these include isolating members from outside contacts; exposing them to intensive indoctrination about the value of the group and its goals; forcing them to endure physical and psychological discomfort; creating opportunities for them to participate in activities (e.g., shared hardship, rituals) that foster group identification and cohesion; and monitoring and testing their adherence to group norms and commitment to group goals.

The purpose of socialization, however, is much broader than consolidating new members' cognitions regarding the group narrative. It is also designed to increase their loyalty to the group and their motivation to engage in extreme, self-sacrificial behavior on its behalf, as in the following example from a German Neo-Nazi:

> The older members would just punch one of the younger members for no reason, or head-butt someone and break their nose or things like that. And then it was often justified with arguments like 'we have to toughen you guys up for the fight, so that you know what its like to be in a fight and you can show here and now that you are man enough to take it'.
>
> (Kruglanski et al., 2019, p. 144)

In addition, socialization in extreme groups has yet a third goal, namely, to ensure that new members acquire the competencies they will later need to carry out their assigned tasks. The major strategies used to achieve this goal are formal and informal training and practice. Training is heavily emphasized in terrorist groups, as indicated by the training philosophy of the al-Qaida veteran Mustafa bin Abd al-Qadir Sethmariam Nasar (aka Abu Musab al-Suri):

> The last building block in al-Suri's training doctrine, namely training through action and fighting, is derived directly from his experience of the Syrian Islamist uprising in 1980–1982. Al-Suri did not intend to allow untrained recruits to undertake complicated operations, which would contradict his principle of sequence 'will . . . preparation . . . launch.' Rather,

he describes how a gradual introduction of untrained recruits into an operative role can take place to allow 'expertise [to] develop through battle.' Recruits should first participate in action only as bystanders or witnesses. Later, they will serve in a minor auxiliary function without directly intervening. Finally, when deemed qualified, they will operate directly in main operations under the command of senior members.

(Lia, 2008, p. 536)

Extreme groups are not unique in desiring particular competencies, as indicated by the effort work teams expend to ensure that their members possess particular KSAs (knowledge, skills, attitudes; e.g., Cannon-Bowers & Bowers, 2011). Analyses of KSAs often distinguish between *taskwork* competencies, which allow individual team members to perform their specific tasks, and *teamwork* competencies, which allow members to interact effectively in seeking to achieve team-level goals (Lacerenza, Marlow, Tannenbaum, & Salas, 2018). Although extreme groups and work teams frequently require different taskwork competencies (e.g., bomb-making vs. surgical skills), this is not always the case (e.g., both kinds of groups frequently value computer skills). And, more often than not, extreme groups and work teams desire the same teamwork competencies (e.g., accurate knowledge of other members' abilities, capacity to resolve interpersonal conflicts). In light of these similarities, research on determinants of work team performance and strategies for increasing team members' competencies (e.g., Kozlowski & Bell, 2020; Lacerenza et al., 2018; Salas, Reyes, & Woods, 2020) can shed light on the functioning of extreme groups (e.g., Spitzmuller & Park, 2018).

In the socialization phase, the group and the individual continue to engage in reciprocal evaluation. If their commitment levels to rise to their acceptance criteria (AC), as depicted in Figure 4.1, the individual undergoes the role transition of acceptance and enters the maintenance phase as a full member. Like entry, acceptance is an inclusive transition that signals promotion to a higher status.

Maintenance Phase

During maintenance, the two parties negotiate the specialized role that the individual will play in the group, with the goal of maximizing both parties' outcomes. Depending on the group's structure, many such roles may be available. However, one role—that of leader—is important in all groups, and this is particularly true in extreme groups because of the high stakes of success/failure. Rather than discussing the process by which leadership roles are negotiated, we will make the simplifying assumption that group prototypicality is the key driver of negotiation (Hogg & Adelman, 2013; Hogg & van Knippenberg, 2003). Thus, we will assume that prototypical full members seek leadership roles and that groups want them to occupy these roles. Earlier we discussed how leaders can influence the creation of group narratives. In this section, we focus on how leaders can influence the implementation of these narratives.

Leadership is viewed as the sine qua non of group extremism. How to explain Catholic monasticism without Saint Benedict, the women's suffrage movement without Susan B. Anthony, the fledgling Nazi party without Adolf Hitler, the Unification Church without Reverend sun Myung Moon, al Qaeda without Osama bin Laden, and so on? Not surprisingly, numerous efforts have been made to explain why leadership matters and how it functions in extreme groups, particularly terrorist groups and religious cults (e.g., Filippidou, 2020; Freeman, 2014; Galanter, 1999; Galanter & Forest, 2006; Hogg & Adelman, 2013; Hoffman, 2015; Raphaeli, 2002; Seyranian, 2012; Zenn, 2019). Leadership is a complex phenomenon, and many theories have been developed to explain its origins and consequences (see Hogg, 2007b, for a review). A common assumption of these theories is that leadership in extreme groups serves two basic functions—motivation and management (cf. Freeman, 2014).[5]

Motivation involves inspiring members to engage in self-sacrificial acts on behalf of the group. Leaders can use several strategies to achieve this goal, including increasing members' collective identification, their liking and respect for colleagues, their commitment to group goals, and their respect for authority and willingness to obey superiors' orders. Many analyses of extreme groups highlight the role that 'charismatic' leaders play in motivating members (e.g., Hoffman, 2015; Galanter, 1999). Although the source of charisma varies from group to group, the leader's religious powers are often emphasized, as in the following description of a far-right Charismatic Christian terrorist group (The Covenant, the Sword, and the Arm of the Lord):

> In addition to the recognition of Ellison's extraordinary nature, there was a clear charismatization process in the CSA that focused on educating new members about his special powers and authority. Much of this process hinged on the semi-regular commission of miracles by Ellison and the recitation of stories about his supernatural powers among the group's membership. For example, Ellison claimed to have resurrected his son from the dead after he had been hit by a car in 1977.
>
> *(Hoffman, 2015, p. 724)*

It is worth noting that, notwithstanding the widespread belief in the importance of leader charisma in nonextreme as well extreme groups, serious questions have been raised about its theoretical and empirical status (van Knippenberg & Sitkin, 2013).

Management involves controlling factors that affect members' ability to accomplish their individual and joint tasks and directing members' actions. Management strategies that leaders can use include influencing the group's composition (e.g., staffing priorities, socialization practices); structure (e.g., authority hierarchy, reward system); internal dynamics (e.g., pressure for uniformity, cooperation); tactics (e.g., violence, voting); resources (e.g., firearms, money); and relations to other groups (e.g., competition, alliances). Leaders of extreme groups typically have wide latitude in employing these strategies and hence exert strong control over how the group operates. They thus often exhibit authoritarianism (Hogg & Adelman, 2013), which is one manifestation of 'group-centrism' based on members' desire for nonspecific cognitive closure (Kruglanski et al., 2006).

There is perhaps no clearer indicator of the assumed importance of terrorist leadership than the popularity of targeted assassinations designed to 'decapitate' terrorist networks. Although physically eliminating leaders is a common counterterrorism strategy, its efficacy is a matter of some debate (e.g., Abrams & Mireau, 2017; Johnston, 2012; Jordan, 2009; Price, 2012).

In the maintenance phase, the group and the individual (regardless of that person's role) continue to engage in reciprocal evaluation. In many cases, the group's and the individual's commitment levels decline during this phase. For example, the group might become disenchanted with the individual because the person fails to meet role expectations or dissents from consensus on opinion issues. Similarly, the individual might become disenchanted with the group because it fails to provide promised rewards or is disrespectful. If the commitment levels of the two parties fall to their divergence criteria (DC), as depicted in Figure 4.1, the individual undergoes the role transition of divergence and enters the resocialization phase as a marginal member. Divergence differs from entry and acceptance in two ways. First, it is an 'exclusive' transition that signals demotion to a lower status. Second, one party can sometimes force the transition even if the other party's commitment level has not fallen to its divergence criterion.

Resocialization Phase

During resocialization, the group attempts to increase the individual's contribution to group goals, and the individual attempts to increase the group's contribution to personal needs. As in the case of socialization, the latter type of influence is relatively rare (e.g., when a marginal member's skills are impossible to replace). For this reason, when successful, resocialization typically involves assimilation on the part of marginal members, rather than accommodation on the part of the group.

Because of their strong belief in the group narrative and ardent desire to achieve group goals, members of extreme groups are highly vigilant to norm transgressions. Their responses to such transgressions are influenced by several factors, including the seriousness of the violation, the violator's assumed motivation and competence, the violator's status in the group, the difficulty of replacing the violator, and the availability of particular sanctions. Extreme groups employ a range of tactics for 'reforming' marginal members, including re-education and therapy, loss of privileges, reduced status and power, ostracism, and threat and use of physical punishment (see Levine & Kerr, 2007; Levine & Moreland, 2002, for analyses of group reaction to deviance).

As Sageman (2016) notes, violations of group norms often elicit harsh responses:

> Other ingroup members subject deviant members to greater denigration and rejection than comparable out-group members and punish them disproportionately for acts committed against the in-group. This greater hostility to traitors in sometimes called the 'black sheep effect'.

An example is the treatment meted out to a member of a right-wing terrorist group for maintaining friendly relations with outgroup members:

> I would meet other people, and they [group members] didn't like them because they were black, or Mexican, or whatever. They threatened me all the time. . . . They were like, 'You can't hang out with them, and that's your warning.' Then they saw me again and I got a boot party [being knocked to the ground and then stomped and kicked].
>
> (Adam, New Order Skins, September 9, 2013; quoted in Simi et al., 2016, p. 76)

Resocialization can end in one of two ways. If the individual succumbs to group pressure and the commitment levels of the parties rise (back) to their DC, the individual undergoes a special role transition—convergence—and returns to full membership, though often under a cloud. However, if the individual resists group pressure and the commitment levels of the parties fall to their exit criteria (XC), as depicted in Figure 4.1, the individual undergoes the role transition of exit and enters the remembrance phase as an ex-member. Like divergence, exit is an exclusive transition signaling demotion to a lower status. Moreover, one party can sometimes 'force' the transition even if the other party's commitment level has not fallen to its exit criterion.

Substantial attention has been devoted to member-initiated exit from extreme groups, with the goal of identifying why individuals leave these groups (i.e., the reasons their commitment falls to their XC) (see Galanter, 1999; Kruglanski et al., 2019; Raets, 2017; Reinares, 2011; Rosenau, Espach, Ortiz, & Herrera, 2014; Windisch, Simi, Ligon, & McNeel, 2016). Many useful analyses of member-initiated exit, or disengagement, have been proposed. We highlight one that both integrates prior formulations and aligns with SQT (Kruglanski et al., 2019).

This analysis distinguishes (a) between 'Push' factors that drive people away from an extreme group and 'Pull' factors that attract them to the mainstream and (b) between specific mechanisms associated with the quest for significance (QfS), the narrative (Nar), and the network (Net). Push factors include difficulty coping with violence, burnout, and fear of being caught (QfS), loss of faith in ideology and disillusionment with strategy (Nar), and disillusionment with leaders or other members (Net). Pull factors include a desire to focus more on occupation or family (QfS), religious conversion (Nar), and social ties to, and pressure from, people outside the group (Net).

An example relates to dehumanization, which, as noted earlier, is a common feature of extreme group narratives. To the extent enemies are viewed as less than human, a particular Push-Nar mechanism, namely, 'rehumanization', can reduce faith in such narratives and thereby facilitate disengagement:

> We went to have Turkish food. It tasted good, and the Turk did not stink. I also chatted with an African, and they also didn't shout UGU or they moved their arms like normal people and not like chimpanzees. . . . And then we came to the next topic: What are German virtues? Order, discipline, punctuality. The African said according to that he is completely German. 'I keep things orderly. Do you want to see my apartment? I am disciplined. I am always on time. I am there 10 minutes before work starts. I am really German'. And he was right somehow. That does start to make sense after a while.
>
> *(Kruglanski et al., 2019, p. 169)*

Extreme groups often resist member-initiated exit because of the threats they pose. These include loss of ex-members' skills, concern that other members will be tempted to leave, and fear that ex-members will undermine the group by revealing secret information. For these reasons, extreme groups often seek to prevent exit by offering rewards for staying and/or threatening punishments for leaving. To the extent such incentives are not effective, groups seek to guarantee ex-members' loyalty by promises/threats of future rewards/punishments. However, such incentives are not always effective, and ex-members thought to pose significant risks to the group are often treated very harshly:

> eight German right-wing extremists 'held court' and sentenced a former leading member of their 'combat group' to severe punishment due to treason. The group broke into the victim's home at night, read the indictment to him, and severely beat him. The victim was water boarded and 'shot' with an unloaded weapon put to his head. After another round of beating up the person was brought to a forest to die but he luckily survived.
>
> *(Koehler, 2015, p. 44)*

In a relevant investigation, Koehler (2015) used interview and archival data to assess factors influencing the punishment meted out to ex-members of right-wing extremist groups in Germany and the United States. He found, consistent with Levine and Moreland's (2002) model of group responses to disloyalty, that punishment severity was affected by defectors' length of time in the group, former status, motivation for leaving, and betrayal of group information to outsiders.

Finally, in some cases, exit is initiated by the group rather than by the individual. For the reasons mentioned earlier, extreme groups are usually reluctant to expel marginal members. Sometimes, however, the threat such members pose to group effectiveness requires their expulsion. In such cases, groups often use harsh exit procedures (what Garfinkel, 1956, calls 'status degradation ceremonies'). In extreme cases, rather than being physically (and psychologically) banished from the group, recalcitrant marginal members are killed.

Group Development

In this section, we move from focusing on changes in the relationship between a group and its members (socialization) to changes in the group as a whole (development). Like socialization, development is a process that unfolds over time, beginning with formation and ending with demise. Moreover, just as all individuals do not have identical socialization experiences (e.g., some full members are demoted to marginal status whereas others are not), all groups do not have identical development experiences (e.g., some dissolve in months whereas others thrive for decades).

Much is known about the life cycles of terrorist groups (e.g., Gupta, 2008; Hoffman, 2006; Pearson, Akbulut, & Lounsbery, 2017; Perliger, 2014; Phillips, 2015, 2019; Tinnes, 2017b). Most of this work relies on some variant of social network analysis, which uses constructs such as nodes, links, hubs, size, density, and centralization, to explain the structure and dynamics of terrorist groups (e.g., Bright, Whelan, & Harris-Hogan, 2020; Kelly & McCarthy-Jones, 2019; Kilberg, 2012; Malthaner, 2018; Spitzmuller & Park, 2018; Tinnes, 2017a). For example, Sageman (2004) makes the case that the global Salafi jihad has a 'small-world' structure, in which nodes are linked to well-connected hubs. From a social evolutionary perspective, this structure ought to yield benefits to the jihad, and it does:

> Unlike a hierarchical network that can be eliminated through decapitation of its leadership, a small-world network resists fragmentation because of its dense interconnectivity. A significant fraction of nodes can be randomly removed without much impact on its integrity.
>
> *(p. 132)*

> Dense networks like cliques [which are the building blocks of the Salafi jihad] commonly produce social cohesion and a collective identity and foster solidarity, trust, community, political inclusion, identity-formation, and other valuable social outcomes.
>
> *(p. 145)*

Hubs receive the most communications from the more isolated nodes. Because of their larger numbers, innovations are more likely in nodes.

The nodes link to hubs, who, in turn, send the information along to their numerous other nodes.

(p. 154)

In addition to rapidly diffusing innovations, the topology of a small-world network is also able to adapt to changing circumstances and solve unforeseen obstacles in the execution of general plans. This flexibility is especially useful in terrorist operations. . . . These operations involve much uncertainty and many unanticipated obstacles . . . the mujahedin form a network of information processors, where the network handles large volumes of information efficiently without overloading any individual processor.

(p. 154)

Of course, no network configuration is perfect. Among the weaknesses of small-world networks are their vulnerability to targeted attacks on hubs, their inability to exert strong control over members, and the tendency for cliques to isolate themselves (Sageman, 2004). Thus, networks of all kinds must be vigilant to their weaknesses and prepared to make necessary adjustments. The injunction that 'form should follow function' is as relevant to terrorist groups as it is to surgical teams and jazz ensembles, and a group that discovers its configuration no longer serves its needs faces a stark choice—adapt or die.

Our earlier analyses of quest for significance and narratives addressed many issues relevant to the early stages of group development, formation, and growth. Therefore, in this section, we focus on the later stages, decline and demise. According to Cronin (2009), terrorist groups lose viability and eventually disappear for six reasons: (a) their leaders are assassinated or captured (decapitation); (b) they successfully negotiate settlements with governments; (c) they achieve their objectives; (d) they fail to achieve their objectives; (e) they are forcibly repressed; or (f) they transition to different means (reorientation). These reasons are not necessarily independent (e.g., decapitation can be used as a technique of repression), and the likelihood they will occur (either separately or in combination) is dependent on factors both internal and external to the group.

From our perspective, the most interesting determinant of group decline and demise is failure to achieve group objectives. According to Cronin (2009), this failure can arise for two reasons. The first is 'marginalization', which occurs when the group loses community support, because community members come to see the group ideology as irrelevant, lose contact with the group, or respond negatively to group 'targeting errors' (e.g., civilian casualties). The second reason is 'implosion', which occurs when the group falls apart for internal reasons, such as failure to pass the cause to the next generation, infighting and fractionalization, loss of operational control, and willingness to accept amnesty. Of the various determinants of implosion, infighting and fractionalization has received the most attention (e.g., Bencherif, 2020; Hafez, 2020; Ives, 2019; Mahoney, 2020; Zenn, 2019).

The infighting associated with fractionalization can have many sources, including disagreement about the group's operational goals and tactics and its leadership. Cronin (2009, p. 98) provides several examples:

> Sean O'Callaghan, who joined the PIRA [Provisional Irish Republican Army] at the age of 15, later recounted his disgust when fellow members wanted a murder victim to be pregnant so as to get 'two for the price of one.' Feelings of growing dissatisfaction catalyzed his decision to become a police informant.
>
> Ambitious individuals may struggle for predominance within a group hierarchy, as was the case with GIA (Groupe Islamique Arme) in Algeria in the mid-1990s, where the government actively encouraged factions within the group to proliferate.
>
> Bickering within a group can have deadly consequences, not only for individual members but also for the viability of the group, especially if revenge killings are used to maintain operational control. According to interviews with prisoners, members of left-wing groups in Italy began to be disgusted by the growing brutality of operations, particularly after the assassination of a very young member who was suspected of being a traitor.
>
> . . . the FLQ (Le Front de Liberation du Quebec) essentially self-destructed as a result of bitter disputes over operations.

Cronin's analysis does not stipulate that group demise means group death, in the sense that all vestiges of the group suddenly disappear. For example, groups that negotiate settlements with governments or transition to new means remain 'alive', albeit in a different form, for some period of time. A particularly interesting form of survival in the midst of demise is schisms, in which members of a subgroup withdraw from a parent group, either to form a breakaway group or to join a different group. The infighting that precipitates a schism involves more than disagreement about operational goals/tactics or leadership—rather, it concerns the parent group's fealty to a set of 'sacred' values. According to Sani (2005, 2009), schisms begin when a subset of group members perceives that those who speak for the group are endorsing values that subvert the core principles underlying the group's identity. This perception causes dissatisfied members to experience negative emotions, feel reduced identification with the group, and perceive the group as lower in entitativity. Weak identification and strong negative emotions, in turn, increase schismatic intentions, which are reduced to the extent dissatisfied members feel they have 'voice' in the group (see also Hart & Van Vugt, 2006; Wagoner & Hogg, 2016).

Conclusion

Our goal in this chapter was to analyze the psychological and social processes that underlie group extremism. Our analysis was informed by Kruglanski et al.'s (2021) general model of extremism, which views extremism as a special form of motivational imbalance in which one need dominates all others, justifying extreme actions in the service of the focal need. We based our analysis on SQT (Kruglanski et al., 2019; Kruglanski, Gelfand et al., 2014; Kruglanski et al., 2018), which is a special case of the general extremism model. Although designed to explain violent extremism (terrorism), SQT is also applicable to extreme groups of other kinds, such as cults, religious orders, and single-issue political groups.

SQT is based on three fundamental premises: (a) the dominant need underlying extremism is desire for personal significance; (b) this need is linked to action by an ideological narrative that justifies extreme means for achieving group goals; and (c) a social network is important in providing validation for the narrative and rewards for engaging in narrative-consistent behavior. Our analysis extended SQT by highlighting an unacknowledged, though crucial, role that networks play in extreme groups—providing the *means* by which groups engage in narrative-specified behavior designed to satisfy their quest for significance.

This chapter was organized in terms of the three basic components of SQT: quest for significance, narratives, and networks. In each section, we selectively reviewed the extensive literature on extreme groups, with an (unavoidable) emphasis on the most-studied category, namely, violent groups. The papers and books we consulted were written by scholars from a wide range of disciplines. Our literature search led us to conclude that, while social psychology is making important contributions to understanding individual-level (cognitive-emotional-motivational) processes associated with extremism, our discipline is 'not in the game' when it comes to studying relevant group-level processes. Although this state of affairs is perhaps not surprising, given the paucity of current social psychological work on intragroup (as opposed to intergroup) phenomena (Levine & Moreland, 2012), it is unfortunate. As we hope this chapter illustrates, group-level theory and research from social (as well as organizational) psychology, though not generated with extreme groups in mind, can be quite helpful in clarifying their dynamics.

In conclusion, let us return briefly to SQT, which served as the organizing framework for our review. In discussing the three components of the theory, we assumed that the quest for significance precedes an individual's involvement with a social network and acceptance of its narrative. But this temporal order is not written in stone (Bélanger et al., 2020). For example, under some conditions, discussion of a narrative might precede (and increase) network members' desire for significance and/or commitment to the network. These and other possible relationships between the three components deserve systematic attention.

Notes

1. This is not to say, of course, that extreme opinions invariably produce extreme behaviors (cf. McCauley & Moskalenko, 2017).
2. This does not mean that new members play a completely passive role during socialization. As suggested earlier, they can facilitate their integration into the group via information seeking and other proactive behaviors (e.g., Griffin, Colella, & Goparaju, 2000; Saks, Gruman, & Cooper-Thomas, 2011). We return to this issue when discussing networks.
3. Evidence indicates that frequently expressing one's opinions and hearing others' opinions can also contribute to polarization (Brauer & Judd, 1996).
4. Figure 4.1 presents an idealized representation of an individual's passage through a group and hence masks several complexities. For example, group and individual commitment levels may change abruptly rather than gradually; decision criteria may shift over time, thereby influencing how long individuals spend in particular membership phases; and the relative positions of some decision criteria may vary (e.g., the entry criterion may be higher or lower than the exit criterion). In addition, individuals may not pass through all five membership phases (e.g., a person might exit during the socialization phase or remain in the maintenance phase until death). Finally, individual and group commitment levels and decision criteria may not be identical, which can produce conflict about role transitions.
5. In recent years, substantial attention has been given to 'leaderless' groups espousing extreme agendas (e.g., Joose, 2007; Hoffman, 2006; Sageman, 2008). We would argue that leaders are in fact important in such groups. Although their management role is much reduced, it is often not zero, and they often play an important motivational role. As Gray (2013) argued, 'In leaderless resistance, the group provides ideological guidance—and possibly technical guidance—as well as targets. But it does not organize cells, provide logistical support, or indeed have any interaction with "street-level" agents' (p. 659) (see also Hoffman's, 2020, discussion of lone-actor terrorists).

References

Abrams, D., & Levine, J. M. (2017). Norm formation: Revisiting Sherif's autokinetic illusion study. In J. R. Smith & S. A. Haslam (Eds.), *Social psychology: Revisiting the classic studies* (2nd ed., pp. 58–76). London: Sage.

Abrams, M., & Mireau, J. (2017). Leadership matters: The effects of targeted killings on militant group tactics. *Terrorism and Political Violence, 29*, 830–851.

Atran, S. (2016). The devoted actor: Unconditional commitment and intractable conflict across cultures. *Current Anthropology, 57*(S13), S192–S203.

Atran, S. (2021). Psychology of transnational terrorism and extreme political conflict. *Annual Review of Psychology, 72*, 6.1–6.31.

Atran, S., & Sheikh, H. (2015). Dangerous terrorists as devoted actors. In V. Zeigler-Hill, L. Welling, & T. Shackelford (Eds.), *Evolutionary perspectives on social psychology* (pp. 401–416). New York: Springer.

Bandura, A. (1999). Moral disengagement in the perpetration of inhumanities. *Personality and Social Psychology Review, 3*, 193–209.

Baron, R. S. (2005). So right it's wrong: Groupthink and the ubiquitous nature of polarized group decision making. In M. P. Zanna (Ed.), *Advances in experimental social psychology* (Vol. 37, pp. 219–253). San Diego, CA: Elsevier.

Basra, R., & Neumann, P. R. (2016). Criminal pasts, terrorist futures: European jihadists and the new crime-terror nexus. *Perspectives on Terrorism, 10*, 25–40.

Becker, J. C., & Tausch, N. (2015). A dynamic model of engagement in normative and non-normative collective action: Psychological antecedents, consequences, and barriers. *European Review of Social Psychology, 26*, 43–92.

Bélanger, J. J., Caouette, J., Sharvit, K., & Dugas, M. (2014). The psychology of martyrdom: Making the ultimate sacrifice in the name of a cause. *Journal of Personality and Social Psychology, 107*, 494–515.

Bélanger, J. J., Robbins, B. G., Muhammad, H., Moyano, M., Nisa, C. F., Schumpe, B. M., & Blaya-Burgo, M. (2020). Supporting political violence: The role of ideological passion and social network. *Group Processes & Intergroup Relations, 23*, 1187–1203.

Bencherif, A. (2020). From resilience to fragmentation: Al Qaeda in the Islamic Maghreb and jihadist group modularity. *Terrorism and Political Violence, 32*, 100–118.

Betts, K. R., & Hinsz, V. B. (2013). Group marginalization: Extending research on interpersonal rejection to small groups. *Personality and Social Psychology Review, 17*, 355–370.

Block, G., & Drucker, M. (Eds.). (1992). *Rescuers: Portraits of moral courage in the holocaust*. New York: Holmes & Meier.

Bloom, M. (2017). Constructing expertise: Terrorist recruitment and "talent spotting" in the PIRA, Al Qaeda, and ISIS. *Studies in Conflict & Terrorism, 40*, 603–623.

Bongiorno, R., McGarty, C., Kurz, T., Haslam, A., & Sibley, C. G. (2016). Mobilizing cause supporters through group-based interaction. *Journal of Applied Social Psychology 2016, 46*, 203–215.

Borzekowski, D. L. G., Schenk, S., Wilson, J. L., & Peebles, R. (2010). e-Ana and e-Mia: A content analysis of pro-eating disorder web sites. *American Journal of Public Health, 100*, 1526–1534.

Brauer, M., & Judd, C. M. (1996). Group polarization and repeated attitude expressions: A new take on an old topic. In W. Stroebe & M. Hewstone (Eds.), *European review of social psychology* (Vol. 7, pp. 173–207). Chichester, UK: John Wiley & Sons.

Bright, D., Whelan, C., & Harris-Hogan, S. (2020). On the durability of terrorist networks: Revealing the hidden connections between *Jihadist* cells. *Studies in Conflict & Terrorism, 43*, 638–656.

Bromley, D. G., & Melton, J. G. (Eds.). (2004). *Cults, religion, and violence*. Cambridge: Cambridge University Press.

Buhrmester, M. D., Fraser, W. T., Lanman, J. A., Whitehouse, H., & Swann, W. B., Jr. (2015). When terror hits home: Identity fused Americans who saw Boston bombing victims as "family" provided aid. *Self and Identity, 14*, 253–270.

Cannon-Bowers, J. A., & Bowers, C. (2011). Team development and functioning. In S. Zedeck (Ed.), *APA handbook of industrial and organizational psychology: Vol. 1. Building and developing the organization* (pp. 597–650). Washington, DC: American Psychological Association.

Cialdini, R. B., & Trost, M. R. (1998). Social influence: Social norms, conformity, and compliance. In D. Gilbert, S. Fiske, & G. Lindzey (Eds.), *The handbook of social psychology* (4th ed., Vol. 2, pp. 151–192). Boston, MA: McGraw-Hill.

Cronin, A. K. (2009). *How terrorism ends: Understanding the decline and demise of terrorist campaigns*. Princeton, NJ: Princeton University Press.

della Porta, D. (1988). Recruitment processes in clandestine political organizations: Italian left-wing terrorism. *International Social Movement Research, 1*, 155–169.

Doosje, B., Loseman, A., & van den Bos, K. (2013). Determinants of radicalization of Islamic youth in the Netherlands: Personal uncertainty, perceived injustice, and perceived group threat. *Journal of Social Issues, 69*, 586–604.

Dugas, M., Bélanger, J. J., Moyano, M., Schumpe, B. M., Kruglanski, A. W., Gelfand, M. J., . . . Nociti, N. (2016). The quest for significance motivates self-sacrifice. *Motivation Science, 2*, 15–32.

Echterhoff, G., & Higgins, E. T. (2017). Creating shared reality in interpersonal and intergroup communication: The role of epistemic processes and their interplay. *European Review of Social Psychology, 28*, 175–226.

Echterhoff, G., Higgins, E. T., & Levine, J. M. (2009). Shared reality: Experiencing commonality with others' inner states about the world. *Perspectives on Psychological Science, 4*, 496–521.

Ellemers, N., Spears, R., & Doosje, B. (2002). Self and social identity. *Annual Review of Psychology, 53*, 161–186.

Filippidou, A. (2020). The oxymoron of a benevolent authoritarian leadership: The case of Lebanon's Hezbollah and Hassan Nasrallah. *Terrorism and Political Violence*. https://doi.org/10.1080/09546553.2020.1724967

Freeman, A. (2014). A theory of terrorist leadership (and its consequences for leadership targeting). *Terrorism and Political Violence, 26*, 666–687.

Futrell, R., Simi, P., & Tan, A. E. (2019). Political extremism and social movements. In D. A. Snow, S. A. Soule, H. Kriesi, & H. J. McCammon (Eds.), *The Wiley Blackwell companion to social movements* (2nd ed., pp. 618–634). Hoboken, NJ: Wiley Blackwell.

Galanter, M. (1999). *Cults: Faith, healing, and coercion* (2nd ed.). New York: Oxford University Press.

Galanter, M., & Forest, J. J. F. (2006). Cults, charismatic groups, and social systems: Understanding the transformation of terrorist recruits. In J. J. F. Forest (Ed.), *The making of a terrorist: Recruitment, training, and root causes* (pp. 51–70). Westport, CT: Praeger.

Garfinkel, H. (1956). Conditions of successful degradation ceremonies. *American Journal of Sociology, 61*, 420–424.

Gilbert, M. (1985). *The Holocaust: A history of the Jews of Europe during the second world war*. New York: Henry Holt and Company.

Giner-Sorolla, R., Leidner, B., & Castano, E. (2012). Dehumanization, demonization, and morality shifting: Paths to moral certainty in extremist violence. In M. A. Hogg & D. L. Blaylock (Eds.), *Extremism and the psychology of uncertainty* (pp. 165–182). Boston, MA: Wiley-Blackwell.

Golec de Zavala, A. G., Cichocka, A., Eidelson, R., & Jayawickreme, N. (2009). Collective narcissism and its social consequences. *Journal of Personality and Social Psychology, 97*, 1074–1096.

Gómez, A., Brooks, M. L., Buhrmester, M. D., Vázquez, A., Jetten, J., & Swann, W. B., Jr. (2011). On the nature of identity fusion: Insights into the construct and a new measure. *Journal of Personality and Social Psychology, 100*, 918–933.

Gomez, A., Chinchilla, J., Vazquez, A., Lopez-Rodriguez, L., Paredes, B., & Martinez, M. (2020). Recent advances, misconceptions, untested assumptions, and future research agenda for identity fusion theory. *Social and Personality Psychology Compass, 14*, e12531. https://doi.org/10.1111/spc3.12531

Gómez, A., Morales, J. F., Hart, S., Vázquez, A., & Swann, W. B., Jr. (2011). Rejected and excluded forevermore, but even more devoted: Irrevocable ostracism intensifies loyalty to the group among identity-fused persons. *Personality and Social Psychology Bulletin, 37*, 1574–1586.

Gray, P. W. (2013). Leaderless resistance, networked organization, and ideological hegemony. *Terrorism and Political Violence, 25*, 655–671.

Griffin, A. E. C., Colella, A., & Goparaju, S. (2000). Newcomer and organizational socialization tactics: An interactionist perspective. *Human Resource Management Review, 10*, 453–474.

Gupta, D. K. (2008). *Understanding terrorism and political violence: The life cycle of birth, growth, transformation, and demise.* New York: Routledge.

Hafez, M. M. (2020). Fratricidal rebels: Ideological extremity and warring factionalism in civil wars. *Terrorism and Political Violence, 32*, 604–629.

Hart, C. M., & Van Vugt, M. (2006). From fault line to group fission: Understanding membership changes in small groups. *Personality and Social Psychology Bulletin, 32*, 392–404.

Hasbrouck, J. (2020). *How needs, narratives, and networks promote a willingness to engage in extremism* (Unpublished doctoral dissertation). University of Maryland, College Park.

Haslam, N. (2006). Dehumanization: An integrative review. *Personality and Social Psychology Review, 10*, 252–264.

Haslam, N. (2015). Dehumanization and intergroup relations. In M. Mikulincer & P. R. Shaver (Eds.), *APA Handbook of personality and social psychology: Group processes* (J. F. Dovidio & J. A. Simpson, Assoc. Eds.) (Vol. 2, pp. 295–314). Washington, DC: American Psychological Association.

Higgins, E. T. (2019). *Shared reality: What makes us strong and tears us apart.* New York, NY: Oxford University Press.

Hoffman, B. (2006). *Inside terrorism: Revised and expanded edition.* New York: Columbia University Press.

Hoffman, D. C. (2015). Quantifying and qualifying charisma: A theoretical framework for measuring the presence of charismatic authority in terrorist groups. *Studies in Conflict & Terrorism, 38*, 710–733.

Hoffman, D. C. (2020). How "alone" are lone-actors? Exploring the ideological, signaling, and support networks of lone-actor terrorists. *Studies in Conflict & Terrorism, 43*, 657–678.

Hogg, M. A. (2007a). Uncertainty-identity theory. In M. P. Zanna (Ed.), *Advances in experimental social psychology* (Vol. 39, pp. 69–126). San Diego, CA: Academic Press.

Hogg, M. A. (2007b). Social psychology of leadership. In A. E. Kruglanski & E. T. Higgins (Eds.), *Social psychology: Handbook of basic principles* (2nd ed., pp. 716–733). New York: Guilford.

Hogg, M. A. (2012). Self-uncertainty, social identity, and the solace of extremism. In M. A. Hogg & D. L. Blaylock (Eds.), *Extremism and the psychology of uncertainty* (pp. 19–35). Boston, MA: Wiley-Blackwell.

Hogg, M. A. (2014). From uncertainty to extremism: Social categorization and identity processes. *Current Directions in Psychological Science, 23*, 338–342.

Hogg, M. A., & Adelman, J. (2013). Uncertainty-identity theory: Extreme groups, radical behavior, and authoritarian leadership. *Journal of Social Issues, 69*, 436–454.

Hogg, M. A., Kruglanski, A., & Van den Bos, K. (2013). Uncertainty and the roots of extremism. *Journal of Social Issues, 69*, 407–418.

Hogg, M. A., & van Knippenberg, D. (2003). Social identity and leadership processes in groups. In M. P. Zanna (Ed.), *Advances in experimental social psychology* (Vol. 35, pp. 1–52). San Diego, CA: Academic Press.

Holbrook, D., & Horgan, J. (2019). Terrorism and ideology: Cracking the nut. *Perspectives on Terrorism, 13*, 1–14.

Horgan, J. G., Taylor, M., Bloom, M., & Winter, C. (2017). From cubs to lions: A six stage model of child socialization into the Islamic State, *Studies in Conflict & Terrorism, 40*, 645–664.

Hull, C. L. (1951). *Essentials of behavior.* Westport, CT: Greenwood Press.

Hunter, S. T., Shortland, N. D., Crayne, M. P., & Ligon, G. S. (2017). Recruitment and selection in violent extremist organizations: Exploring what industrial and organizational psychology might contribute. *American Psychologist, 72*, 242–254.

Inoguchi, R., & Nakajima, T., & Pineau, R. (1958). *The divine wind: Japan's Kamikaze force in world war II*. Annapolis, MD: United States Naval Institute.

Isenberg, D. J. (1986). Group polarization: A critical review and meta-analysis. *Journal of Personality and Social Psychology, 50*, 1141–1151.

Ives, B. (2019). Ethnic external support and rebel group splintering. *Terrorism and Political Violence*. doi:10.1080/09546553.2019.1636035

Janis, I. L. (1982). *Groupthink: Psychological studies of policy decisions and fiascoes* (2nd ed.). Boston, MA; Houghton Mifflin.

Jasko, K., Szastok, M., Grzymala-Moszczynska, J., Maj, M., & Kruglanski, A. W. (2019). Rebel with a cause: Personal significance from political activism predicts willingness to self-sacrifice. *Journal of Social Issues, 75*, 314–349.

Jasko, K., Webber, D., Kruglanski, A. W., Gelfand, M., Taufiqurrohman, M., Hettiarachchi, M., & Gunaratna, R. (2020). Social context moderates the effects of quest for significance on violent extremism. *Journal of Personality and Social Psychology, 118*, 1165–1187.

Jimenez-Moya, G., Spears, R., Rodriguez-Bailon, R., & de Lemus, S. (2015). By any means necessary? When and why low group identification paradoxically predicts radical collective action. *Journal of Social Issues, 71*, 517–535.

Johnston, P. B. (2012). Does decapitation work? Assessing the effectiveness of leadership targeting in counterinsurgency campaigns. *International Security, 36*, 47–79.

Jong, J., Whitehouse, H., Kavanagh, C., & Lane, J. (2015). Shared negative experiences lead to identity fusion via personal reflection. *PLoS One, 10*(12), e0145611. doi:10.1371/journal.pone.0145611

Joose, P. (2007). Leaderless resistance and ideological inclusion: The case of the earth liberation front. *Terrorism and Political Violence, 19*, 351–368.

Jordan, J. (2009). When heads roll: Assessing the effectiveness of leadership decapitation. *Security Studies, 18*, 719–755.

Jost, J. T., & Napier, J. L. (2012). The uncertainty-threat model of political conservatism. In M. A. Hogg & D. L. Blaylock (Eds.), *Extremism and the psychology of uncertainty* (pp. 90–111). Boston, MA: Wiley-Blackwell.

Kanter, R. M. (1968). Commitment and social organization: A study of commitment mechanisms in utopian communities. *American Sociological Review, 33*, 499–517.

Kelly, M., & McCarthy-Jones, A. (2019). Mapping connections: A dark network analysis of neojihadism in Australia. *Terrorism and Political Violence*. doi:10.1080/09546553.2019.1586675

Kilberg, J. (2012). A basic model explaining terrorist group organizational structure. *Studies in Conflict & Terrorism, 35*, 810–830.

Koehler, D. (2015). Radical groups' social pressure towards defectors: The case of right-wing extremist groups. *Perspectives on Terrorism, 9*, 36–50.

Kozlowski, S. W. J., & Bell, B. S. (2020). Advancing team learning: Process mechanisms, knowledge outcomes, and implications. In L. Argote & J. M. Levine (Eds.), *The Oxford handbook of group and organizational learning* (pp. 195–230). New York: Oxford University Press.

Kruglanski, A. W. (2004). *The psychology of closed mindedness*. New York: Psychology Press.

Kruglanski, A. W., Belanger, J. J., & Gunaratna, R. (2019). *The three pillars of radicalization: Needs, narratives, and networks*. New York: Oxford University Press.

Kruglanski, A. W., Chernikova, M., Rosenzweig, E., & Kopetz, C. E. (2014). On motivational readiness. *Psychological Review, 121,* 367–388.

Kruglanski, A. W., Gelfand, M. J., Bélanger, J. J., Sheveland, A., Hetiarachchi, M., & Gunaratna, R. (2014). The psychology of radicalization and deradicalization: How significance quest impacts violent extremism. *Political Psychology, 35,* 69–93.

Kruglanski, A. W., Jasko, K., Chernikova, M., Dugas, M., & Webber, D. (2017). To the fringe and back: Violent extremism and the psychology of deviance. *American Psychologist, 72,* 217–230.

Kruglanski, A. W., Jasko, K., Webber, D., Chernikova, M., & Molinario, E. (2018). The making of violent extremists. *Review of General Psychology, 22,* 107–120.

Kruglanski, A. W., Molinario, E., Jasko, K., Webber, D., Leander, P., & Pierro, A. (in press). Significance quest theory. *Perspectives on Psychological Science.*

Kruglanski, A. W., & Orehek, E. (2012). The need for certainty as a psychological nexus for individuals and society. In M. A. Hogg & D. L. Blaylock (Eds.), *Extremism and the psychology of uncertainty* (pp. 3–18). Boston, MA: Wiley-Blackwell.

Kruglanski, A. W., Pierro, A., Mannetti, L., & De Grada, E. (2006). Groups as epistemic providers: Need for closure and the unfolding of group-centrism. *Psychological Review, 113,* 84–100.

Kruglanski, A. W., Raviv, A., Bar-Tal, D., Raviv, A., Sharvit, K., Ellis, S., . . . Mannetti, L. (2005). Says who? Epistemic authority effects in social judgment. *Advances in Experimental Social Psychology, 37,* 345–392.

Kruglanski, A. W., Szumowska, E., Kopetz, C. H., Vallerand, R. J., & Pierro, A. (2021). On the psychology of extremism: How motivational imbalance breeds intemperance. *Psychological Review, 128*(2), 264–289.

Kruglanski, A. W., Webber, D., & Koehler, D. (2019). *The radical's journey: How German neo-nazis voyaged to the edge and back.* New York: Oxford University Press.

Kteily, N., & Bruneau, E. (2017). Backlash: The politics and real-world consequences of minority group dehumanization. *Personality and Social Psychology Bulletin, 43,* 87–104.

Lacerenza, C. N., Marlow, S. L., Tannenbaum, S. I., & Salas, E. (2018). Team development interventions: Evidence-based approaches for improving teamwork. *American Psychologist, 73,* 517–531.

Leander, N. P., Agostini, M., Stroebe, W., Kreienkamp, J., Spears, R., Kuppens, T., . . . Kruglanski, A. W. (2020). Frustration-*affirmation*? Thwarted goals motivate compliance with social norms for violence and nonviolence. *Journal of Personality and Social Psychology, 119,* 249–271.

Levine, J. M. (2018). Socially-shared cognition and consensus in small groups. *Current Opinion in Psychology, 23,* 52–56.

Levine, J. M., Bogart, L. M., & Zdaniuk, B. (1996). Impact of anticipated group membership on cognition. In R. M. Sorrentino & E. T. Higgins (Eds.), *Handbook of motivation and cognition: The interpersonal context* (Vol. 3, pp. 531–569). New York: Guilford.

Levine, J. M., & Choi, H.-S. (2011). Minority influence in interacting groups: The impact of newcomers. In J. Jetten & M. Hornsey (Eds.), *Rebels in groups: Dissent, deviance, difference, and defiance* (pp. 73–92). Chichester, UK: Wiley-Blackwell.

Levine, J. M., Choi, H.-S., & Moreland, R. L. (2019). Newcomer influence and creativity in work groups. In P. B. Paulus & B. A. Nijstad (Eds.), *The Oxford handbook of group creativity and innovation* (pp. 51–72). New York: Oxford University Press.

Levine, J. M., & Higgins, E. T. (2001). Shared reality and social influence in groups and organizations. In F. Butera & G. Mugny (Eds.), *Social influence in social reality:*

Promoting individual and social change (pp. 33–52). Bern, Switzerland: Hogrefe & Huber Publishers.

Levine, J. M., & Kerr, N. L. (2007). Inclusion and exclusion: Implications for group processes. In A. E. Kruglanski & E. T. Higgins (Eds.), *Social psychology: Handbook of basic principles* (2nd ed., pp. 759–784). New York: Guilford.

Levine, J. M., & Moreland, R. L. (1994). Group socialization: Theory and research. In W. Stroebe & M. Hewstone (Eds.), *The European review of social psychology* (Vol. 5, pp. 305–336). Chichester: John Wiley & Sons.

Levine, J. M., & Moreland, R. L. (1998). Small groups. In D. Gilbert, S. Fiske, & G. Lindzey (Eds.), *The handbook of social psychology* (4th ed., Vol. 2, pp. 415–469). Boston, MA: McGraw-Hill.

Levine, J. M., & Moreland, R. L. (2002). Group reactions to loyalty and disloyalty. In S. R. Thye & E. J. Lawler (Eds.), *Group cohesion, trust and solidarity: Advances in group processes* (Vol. 19, pp. 203–228). Oxford, UK: Elsevier Science.

Levine, J. M., & Moreland, R. L. (2012). A history of small group research. In A. W. Kruglanski & W. Stroebe (Eds.), *Handbook of the history of social psychology* (pp. 383–406). New York: Psychology Press.

Levine, J. M., Moreland, R. L., & Hausmann, L. R. M. (2005). Managing group composition: Inclusive and exclusive role transitions. In D. Abrams, M. A. Hogg, & J. M. Marques (Eds.), *The social psychology of inclusion and exclusion* (pp. 137–160). New York: Psychology Press.

Levine, J. M., Resnick, L. B., & Higgins, E. T. (1993). Social foundations of cognition. *Annual Review of Psychology, 44*, 585–612.

Levine, J. M., & Tindale, R. S. (2015). Social influence in groups. In M. Mikulincer & P. R. Shaver (Eds.), *APA Handbook of personality and social psychology: Group processes* (J. F. Dovidio & J. A. Simpson, Assoc. Eds.) (Vol. 2). Washington, DC: American Psychological Association.

Lia, B. (2008). Doctrines for Jihadi terrorist training. *Terrorism and Political Violence, 20*, 518–542.

Lockwood, P., & Kunda, Z. (1997). Superstars and me: Predicting the impact of role models on the self. *Journal of Personality and Social Psychology, 73*, 91–103.

Lockwood, P., & Kunda, Z. (1999). Increasing the salience of one's best selves can undermine inspiration by outstanding role models. *Journal of Personality and Social Psychology, 76*, 214–228.

Lofland, J., & Stark, R. (1965). Becoming a world-saver: A theory of conversion to a deviant perspective. *American Sociological Review, 30*, 862–875.

Louis, W. R., Thomas, E., Chapman, C. M., Achia, T., Wibisono, S., Mirnajafi, Z., & Droogendyk, L. (2020). Emerging research on intergroup prosociality: Group members' charitable giving, positive contact, allyship, and solidarity with others. *Social and Personality Psychology Compass, 13*, e12436. https://doi.org/10.1111/spc3.12436

Lyons-Padilla, S., Gelfand, M. J., Mirahmadi, H., Farooq, M., & van Egmond, M. (2015). Belonging nowhere: Marginalization and radicalization risk among Muslim immigrants. *Behavioral Science & Policy, 1*, 1–12.

Mahoney, C. W. (2020). Splinters and schisms: Rebel group fragmentation and the durability of insurgencies. *Terrorism and Political Violence, 32*, 345–364.

Malthaner, S. (2018). Space, ties, and agency: The formation of radical networks. *Perspectives on. Terrorism, 12*, 32–43.

McCauley, C. (2002). Psychological issues in understanding terrorism and the response to terrorism. In C. Stout (Ed.), *The psychology of terrorism: Volume III, Theoretical understandings and perspectives* (pp. 3–29). Westport, CN: Praeger.

McCauley, C., & Moskalenko, S. (2011). *Friction: How radicalization happens to them and us.* New York: Oxford University Press.

McCauley, C., & Moskalenko, S. (2017). Understanding political radicalization: The two-pyramids model. *American Psychologist, 72,* 205–216.

McCauley, C., & Segal, M. (1987). Social psychology of terrorist groups. In C. Hendrick (Ed.), *Review of personality and social psychology: Group processes and intergroup relations* (Vol. 9, pp. 231–256). Newbury Park, CA: Sage.

McGregor, I., & Marigold, D. C. (2003). Defensive zeal and the uncertain self: What makes you sosSure? *Journal of Personality and Social Psychology, 85,* 838–852.

McGregor, I., Nash, K. A., & Prentice, M. (2012). Religious zeal after goal frustration. In M. A. Hogg & D. L. Blaylock (Eds.), *Extremism and the psychology of uncertainty* (pp. 147–164). Boston, MA: Wiley-Blackwell.

Meinhof, U. (1970, June 15). Natürlich kann geschossen werden. Ulrike Meinhof über die Baader-Aktion. *Der Spiegel, 25,* 74–75.

Moghaddam, F. M., & Love, K. (2012). Collective uncertainty and extremism: A further discussion on the collective roots of subjective experience. In M. A. Hogg & D. L. Blaylock (Eds.), *Extremism and the psychology of uncertainty* (pp. 246–262). Boston, MA: Wiley-Blackwell.

Molinario, E., Elster, A., Kruglanski, A. W., Webber, D., Jaśko, K., Leander, P., . . . Jaume, L. (2021). *Striving for significance: Development and validation of the quest for significance scale.* [Unpublished Manuscript]

Molinario, E., Jaśko, K., Kruglanski, A. W., Sensales, G., & Di Cicco, G. (2021). *Populism as a response to uncertainty and insignificance.* [Unpublished Manuscript]

Moreland, R. L., & Levine, J. M. (1982). Socialization in small groups: Temporal changes in individual-group relations. In L. Berkowitz (Ed.), *Advances in experimental social psychology* (Vol. 15, pp. 137–192). New York: Academic Press.

Moreland, R. L., & Levine, J. M. (1988). Group dynamics over time: Development and socialization in small groups. In J. E. McGrath (Ed.), *The social psychology of time: New perspectives* (pp. 151–181). Newbury Park, CA: Sage.

Obaidi, M., Bergh, R., Akrami, N., & Anjum, G. (2019). Group-based relative deprivation explains endorsement of extremism among Western-born Muslims. *Psychological Science, 30,* 596–605.

Olivola, C. Y., & Shafir, E. (2013). The martyrdom effect: When pain and effort increase prosocial contributions. *Journal of Behavioral Decision Making, 26,* 91–105.

Pearson, F. S., Akbulut, I., & Lounsbery, M. O. (2017). Group structure and intergroup relations in global terror networks: Further explorations. *Terrorism and Political Violence, 29,* 550–572.

Perliger, A. (2014). Terrorist networks' productivity and durability: A comparative multi-level analysis. *Perspectives on Terrorism, 8,* 36–52.

Phillips, B. J. (2015). Enemies with benefits? Violent rivalry and terrorist group longevity. *Journal of Peace Research, 52,* 62–75.

Phillips, B. J. (2019). Terrorist organizational dynamics. In E. Chenoweth, R. English, A. Gofas, & S. N. Kalyvas (Eds.), *The Oxford handbook of terrorism* (pp. 385–400). Oxford, UK: Oxford University Press.

Price, B. C. (2012). Targeting top terrorists: How leadership decapitation contributes to counterterrorism. *International Security, 36,* 9–46.

Raets, S. (2017). The we in me. Considering terrorist desistance from a social identity perspective. *Journal for Deradicaization, 13,* 1–28.
Raphaeli, N. (2002). Ayman Muhammad Rabi' Al-Zawahiri: The making of an archterrorist. *Terrorism and Political Violence, 14,* 1–22.
Reinares, F. (2011). Exit from terrorism: A qualitative empirical study on disengagement and deradicalization among members of ETA. *Terrorism and Political Violence, 23,* 780–803.
Rosenau, W., Espach, R., Ortiz, R. O., & Herrera, N. (2014). Why they join, why they fight, and why they leave: Learning from Colombia's database of demobilized militants. *Terrorism and Political Violence, 26,* 277–285.
Sageman, M. (2004). *Understanding terror networks.* Philadelphia, PA: University of Pennsylvania Press.
Sageman, M. (2008). *Leaderless Jihad: Terror networks in the twenty-first century.* Philadelphia, PA: University of Pennsylvania Press.
Sageman, M. (2016). *Misunderstanding terrorism.* Philadelphia, PA: University of Pennsylvania Press.
Sageman, M. (2017). *Turning to political violence: The emergence of terrorism.* Philadelphia, PA: University of Pennsylvania Press.
Saks, A. M., Gruman, J. A., & Cooper-Thomas, H. (2011). The neglected role of proactive behavior and outcomes in newcomer socialization. *Journal of Vocational Behavior, 79,* 36–46.
Salas, E., Reyes, D. L., & Woods, A. L. (2020). Team training in organizations: It works—when done right. In L. Argote & J. M. Levine (Eds.), *The Oxford handbook of group and organizational learning* (pp. 233–252). New York: Oxford University Press.
Sani, F. (2005). When subgroups secede: Extending and refining the social psychological model of schism in groups. *Personality and Social Psychological Bulletin, 31,* 1074–1086.
Sani, F. (2009). Why groups fall apart: A social psychological model of the schismatic process. In F. Butera & J. M. Levine (Eds.), *Coping with minority status: Responses to exclusion and inclusion* (pp. 243–266). New York, NY: Cambridge University Press.
Saucier, G., Akers, L. G., Shen-Miller, S., Knezevic, G., & Stankov, L. (2009). Patterns of thinking in militant extremism. *Perspectives on Psychological Science, 4,* 256–271.
Seyranian, V. (2012). Constructing extremism: Uncertainty provocation and reduction by terrorist leaders. In M. A. Hogg & D. L. Blaylock (Eds.), *Extremism and the psychology of uncertainty* (pp. 228–245). Boston, MA: Wiley-Blackwell.
Sheikh, H., Gómez, Á., & Atran, S. (2016). Empirical evidence for the devoted actor model. *Current Anthropology, 57*(S13), S204–S209.
Sherif, M. (1935). A study of some social factors in perception. *Archives of Psychology, 27*(187), 23–46.
Sherif, M. (1936). *The psychology of social norms.* New York: Harper and Row.
Silke, A. (Ed.). (2018). *Routledge handbook of terrorism and counterterrorism.* New York: Routledge.
Simi, P., Windisch, S., & Sporer, K. (2016). *Recruitment and radicalization among US far-right terrorists.* College Park, MD: START.
Skitka, L. J., Hanson, B. E., & Wisneski, D. C. (2017). Utopian hopes or dystopian fears? Exploring the motivational underpinnings of moralized political engagement. *Personality and Social Psychology Bulletin, 43,* 177–190.
Smith, H. J., Pettigrew, T. F., Pippin, G. M., & Bialosiewicz, S. (2012). Relative deprivation: A theoretical and meta-analytic review. *Personality and Social Psychology Review, 16,* 203–232.
Smith, L. G. E., Blackwood, L., & Thomas, E. F. (2020). The need to focus on the group as the site of radicalization. *Perspectives on Psychological Science, 15,* 327–352.

Smith, L. G. E., Thomas, E. F., & McGarty, C. (2015). "We must be the change we want to see in the world": Integrating norms and identities through social interaction. *Political Psychology, 36,* 543–557.

Snow, D. A., Zurcher, L. A. Jr., & Ekland-Olson, S. (1980). Social networks and social movements: A microstructural approach to differential recruitment. *American Sociological Review, 45,* 787–801.

Spence, K. W. (1956). *Behavior theory and conditioning.* New Haven, CT: Yale University Press.

Spitzmuller, M., & Park, G. (2018). Terrorist teams as loosely coupled systems. *American Psychologist, 73,* 491–503.

Swann, W. B., Jr., & Buhrmester, M. (2015). Identity fusion. *Current Directions in Psychological Science, 24,* 52–57.

Swann, W. B., Jr., Buhrmester, M., Gómez, Á., Jetten, J., Bastian, B., Vázquez, A., . . . Zhang, A. (2014). What makes a group worth dying for? Identity fusion fosters perception of familial ties, promoting self-sacrifice. *Journal of Personality and Social Psychology, 106,* 912–926.

Swann, W. B., Jr., Gómez, A., Buhrmester, M. D., López-Rodríguez, L., Jiménez, J., & Vázquez, A. (2014). Contemplating the ultimate sacrifice: Identity fusion channels pro-group affect, cognition, and moral decision making. *Journal of Personality and Social Psychology, 106,* 713–727.

Swann, W. B., Jr., Gómez, A., Dovidio, J. F., Hart, S., & Jetten, J. (2010). Dying and killing for one's group: Identity fusion moderates responses to intergroup versions of the trolley problem. *Psychological Science, 21,* 1176–1183.

Swann, W. B., Jr., Jetten, J., Gómez, A., Whitehouse, H., & Bastian, B. (2012). When group membership gets personal: A theory of identity fusion. *Psychological Review, 119,* 441–456.

Tajfel, H., & Turner, J. C. (1979). An integrative theory of intergroup conflict. In W. G. Austin & S. Worchel (Eds.), *The social psychology of intergroup relations* (pp. 33–47). Monterey, CA: Brooks/Cole.

Thomas, E. F., McGarty, C., & Louis, W. (2014). Social interaction and psychological pathways to political engagement and extremism. *European Journal of Social Psychology, 44,* 15–22.

Thomas, E. F., McGarty, C., & Mavor, K. I. (2009). Aligning identities, emotions, and beliefs to create commitment to sustainable social and political action. *Personality and Social Psychology Review, 13,* 194–218.

Tinnes, J. (2017a). Bibliography: Terrorist organizations: Cells, networks, affiliations, splits. *Perspectives on Terrorism, 11*(5), 67–107.

Tinnes, J. (2017b). Bibliography: Life cycles of terrorism. *Perspectives on Terrorism, 11*(5), 108–139.

Tinnes, J. (2018). Bibliography: Social aspects of terrorism. *Perspectives on Terrorism, 12*(1), 112–151.

Tucker, B. (1998). Deaf culture, cochlear implants, and elective disability. *Hastings Center Report, 28,* 6–14.

Turner, J. C., Hogg, M. A., Oakes, P. J., Reicher, S. D., & Wetherell, M. S. (1987). *Rediscovering the social group: A self-categorization theory.* Oxford: Wiley Blackwell.

van Knippenberg, D., & Sitkin, S. B. (2013). A critical assessment of charismatic-transformational leadership research: Back to the drawing board? *The Academy of Management Annals, 7,* 1–60.

van Stekelenburg, J., & Klandermans, B. (2017). Individuals in movements: A social psychology of contention. In C. Roggeband & B. Klandermans (Eds.), *Handbook of social movements across disciplines* (2nd ed., pp. 103–139). Cham, Switzerland: Springer.

van Stekelenburg, J., Klandermans, B., & Walgrave, S. (2019). Individual participation in street demonstrations. In D. A. Snow, S. A. Soule, H. Kriesi, & H. J. McCammon (Eds.), *The Wiley Blackwell companion to social movements* (2nd ed., pp. 371–391). Hoboken, NJ: Wiley Blackwell.

Van Zomeren, M. (2015). Psychological processes in social action. In M. Mikulincer & P. R. Shaver (Eds.), *APA Handbook of personality and social psychology: Group processes* (J. F. Dovidio & J. A. Simpson, Assoc. Eds.) (Vol. 2, pp. 507–533). Washington, DC: American Psychological Association.

van Zomeren, M., Kutlaca, M., & Turner-Zwinkels, F. (2018). Integrating who "we" are with what "we" (will not) stand for: A further extension of the *Social Identity Model of Collective Action*. *European Review of Social Psychology, 29*, 122–160.

Victoroff, J., Adelman, J. R., & Matthews, M. (2012). Psychological factors associated with support for suicide bombing in the Muslim diaspora. *Political Psychology, 33*, 791–809.

Wagoner, J. A., & Hogg, M. A. (2016). Normative dissensus, identity-uncertainty, and subgroup autonomy. *Group Dynamics: Theory, Research, and Practice, 20*, 310–322.

Waller, J. (2007). *Becoming evil: How ordinary people commit genocide and mass killing* (2nd ed.). New York: Oxford University Press.

Wasmund, K. (1983). The political socialization of terrorist groups in West Germany. *Journal of Political and Military Sociology, 11*, 223–239.

Webber, D., Klein, K., Kruglanski, A. W., Brizi, A., & Merari, A. (2017). Divergent paths to martyrdom and significance among suicide attackers. *Terrorism and Political Violence, 29*, 852–874.

Whitehouse, H. (2018). Dying for the group: Towards a general theory of extreme self-sacrifice. *Behavioral and Brain Sciences, 41*, e192, 1–62. doi:10.1017/S0140525X18000249

Willer, R. (2009). Groups reward individual sacrifice: The status solution to the collective action problem. *American Sociological Review, 74*, 23–43.

Williams, K. D. (2001). *Ostracism: The power of silence*. New York: Guilford Press.

Windisch, S., Logan, M. K., & Ligon, G. S. (2018). Headhunting among extremist organizations: An empirical assessment of talent spotting. *Perspectives on Terrorism, 12*, 44–62.

Windisch, S., Simi, P., Ligon, G. S., & McNeel, H. (2016). Disengagement from ideologically-based and violent organizations: A systematic review of the literature. *Journal for Deradicalization, 9*, 1–38.

Zenn, J. (2019). Boko Haram's factional feuds: Internal extremism and external interventions. *Terrorism and Political Violence*. doi:10.1080/09546553.2019.1566127

5

MASTERS OF BOTH

Balancing the Extremes of Innovation Through Tight–Loose Ambidexterity

Piotr Prokopowicz, Virginia K. Choi, and Michele J. Gelfand

What makes nations successful? In a time of globalization and rapidly advancing technologies, a widely championed answer to this critical question has become innovation or the development and implementation of new and useful ideas (Bledow, Frese, Anderson, Erez, & Farr, 2009; Camisón & Villar-López, 2014; Porter, 1990, 2011). Across the globe, rapid societal shifts and a fast-changing marketplace has galvanized both private and public sector enterprises to stay competitive primarily by cultivating an innovative edge.

Although innovation is commonly understood as the creation of new inventions, it actually involves two radically different, if not opposed, processes. Inherent to innovation is the need to execute both novel *idea generation* and *successful implementation* of these ideas (Rosing, Frese, & Bausch, 2011). The paradoxical nature of innovation manifests itself in the trade-off between exploration and exploitation (March, 1991), both of which, while in opposition, are needed for successful idea generation and implementation (Bledow et al., 2009). Exploration processes are characterized by risk-taking, experimentation, variation, and discovery. Exploitation processes focus on refinement, precision, efficiency, implementation, and execution. Either of these processes, when taken to their extremes, can produce ineffectual outcomes. When teams, nations, or organizations focus all their efforts on exploring new solutions without designating any resources toward implementation, their novel ideas fail to be translated into reality. On the flip side, overreliance on exploitation can put once thriving enterprises on a fast-track to extinction due to their unchecked rigidity and outdated systems. The aim of this chapter is to advance a theory-driven explanation of how these exploration–exploitation extremes are contingent on culture (Gelfand, 2018). Specifically, we focus on the strength of cultural norms as a key

DOI: 10.4324/9781003030898-7

determinant in understanding the dynamic capabilities that underlie innovation successes and failures.

With the growing demand to innovate, there has been an upsurge in research efforts to understand how to develop and manage innovation, as nations compete in areas as diverse as renewable energy, artificial intelligence, and biotechnology. Yet with few exceptions, there has been little attention to *cultural factors* in innovation (see Bukowski & Rudnicki, 2019; Hayton, George, & Zahra, 2002; Nakata & Sivakumar, 1996; Shane, 1993; Taylor & Wilson, 2012; Tellis, Prabhu, & Chandy, 2009 for notable exceptions). Much of the research has focused on economic variables such as *wealth* (Akçomak & Ter Weel, 2009; Dutta, Lanvin, & Wunsch-Vincent, 2019; Jacobs, 1984; Landes, 1969; Murphy, Shleifer, & Vishny, 1993; Porter, 1990), *R&D expenditures* (Chan, Martin, & Kensinger, 1990; Hall, 1993; Rosenbusch, Brinckmann, & Bausch, 2011; Schankerman & Pakes, 1986), and *human capital* (Diebolt & Hippe, 2019). Other, noneconomic predictors of innovation have also been investigated (Ghazinoory, Bitaab, & Lohrasbi, 2014; Gorodnichenko & Roland, 2017; Howaldt & Schwarz, 2010; OECD, 2018), with a particular focus on structural and institutional drivers such as intellectual property protection, openness to international trade, and the degree of technological specialization on the national level (Furman, Porter, & Stern, 2002; see also Baldridge & Burnham, 1975; Damanpour, 1991; Mohr, 1969; Pierce & Delbecq, 1977 for other structural drivers at the organizational level).

What is missing from both the theoretical and empirical investigations of determinants of innovative performance are the implicit norms that drive organizational behaviors. One promising construct to explore the link between culture and the extremes of underlying innovation is tightness–looseness (TL), or the extent to which a given social group can be characterized by a high number of clear social norms and low tolerance for deviant behaviors (Gelfand et al., 2011). It could be argued that an exclusive emphasis on tightness *or* on looseness would be detrimental to innovation. In highly loose cultures, with their weaker norms and higher tolerance for deviance, creative deviations and tolerance for risky ideas would help the exploration aspect of innovation, yet loose norms would make implementation and delivery very difficult. In contrast, extremely tight cultures would excel in exploitation processes through their focus on structure, synchronicity, and coordination, yet they would have difficulty with exploration processes that require creativity. Accordingly, we'd expect both high levels of cultural tightness and high levels of looseness to be detrimental to innovation. Only cultures that reach *a balance* of both tightness and looseness would be positioned to best manage the demands of both exploration and exploitation. In this chapter, we elaborate upon this theory and test it with data on over 30 nations and multiple years.

Anecdotally, one can see the interplay between TL and innovation exemplified in popular historical accounts. Thomas Edison, praised as 'America's greatest inventor', cultivated a working group of thinkers and scientists who assisted him in many projects, including the invention of the lightbulb and phonograph. Among his team of specialists employed at his laboratory in Menlo Park, New Jersey, he fostered a culture that illustrated this tight–loose balance (Morris, 2019). In his shop, Edison introduced high levels of worker autonomy and occupational mobility (Pretzer, 2002). Although he provided structure by informing the engineers on the general idea of the invention, he would sometime refuse to oversee their experiments, encouraging them to pursue their own ideas and solutions (Israel, 1998). At the same time, Edison would monitor the implementation process, guiding the assistants through testing, refinement, and patent application (Millard, 1990). Balancing exploration and exploitation, Edison's focus, and the reason for Menlo Park's success, can be attributed to his emphasis on the dual nature of the innovative process.

In what follows, we first describe the contradictory nature of exploration and exploitation, focusing on the role the two processes play in driving innovation. Next, we introduce cultural TL as a framework that offers unique insight into the relationship between social norms and innovation, arguing that through tight–loose ambidexterity organizations can balance exploration and exploitation processes to build a truly innovative edge. We then present data that examines the relationship between cultural TL and innovation indicators on a national level. We conclude by demonstrating ways in which balancing the extremes of cultural tightness and looseness allows for building innovative environments, focusing on the role leadership plays in developing cultures of tight–loose ambidexterity.

Exploration and Exploitation: Understanding the Extremes of Innovation

One critical challenge in studying complex phenomena such as innovation is finding a universally accepted definition of the term. Definitions from the literature that aim at grappling with the complexity of innovation are often oversimplified, focusing on one particular aspect of innovation (e.g., exclusively on intentional introduction of ideas, processes or products; West & Farr, 1990) or overly complex (e.g., encompassing both an outcome and activity, a product and a process, all on multiple levels of analysis and in multiple business domains; OECD, 2018). This definitional difficulty might be linked to the fact that, at its core, innovation is inherently multifaceted and contradictory, involving both the development and large-scale implementation of new and improved ways of doing things, and as such, is paradoxical (Anderson, Potočnik & Zhou, 2014). Understanding the tension between creativity and implementation is a key for studying innovation, and the most popular theoretical framework used to problematize and discuss it has been the exploration–exploitation trade-off (March, 1991).

The use of exploration and exploitation as terms denoting two opposite business processes originates in the oil industry (Hargadon, 2003), in which a constant balancing act between short-term and long-term costs and benefits of exploring new oil fields and extracting from existing wells needs to be maintained. These two contradictory pursuits of oil companies have historically led to the emergence of two separate divisions—exploration and exploitation—each with different goals, performance measures, and cultures (Hargadon, 2003). According to March, who applied the idea of these opposing processes to a wider set of contexts, exploration and exploitation form two mutually exclusive modes and goals of organizational behavior. In March's view, exploration is related to variation, risk-taking, experimentation, and discovery, whereas exploitation is related to refinement, efficiency, implementation, and execution. Moreover, exploration and exploitation compete for similar resources and therefore lead inevitably to conflicts (March, 1991). In order to successfully overcome these conflicts, any system trying to engage in both exploration and exploitation (individuals, teams, organizations, and nations) must find a way of balancing these dual pursuits (March, 1991, 1996). If they end up pursuing only one of these ends to the extreme, this will lead to suboptimal stable equilibria and failure to realize both of these goals well (March, 1996).

Since the publication of March's account, exploration and exploitation have been successfully linked to a variety of processes related to individual and organizational performance (Hoang & Rothaermel, 2010; Lavie, Kang, & Rosenkopf, 2011; Raisch, Birkinshaw, Probst, & Tushman, 2009). For example, Uotila, Maula, Keil, and Zahra (2009) measured the relative exploration versus exploitation orientations of 279 S&P companies by analyzing the vocabulary used in news articles describing their business activities. Uotila and colleagues found an inverted U-shaped relationship between the relative share of companies' explorative orientation and financial performance, and this relationship was especially pronounced in high R&D industries. The study also concluded that about 80% of the analyzed organizations *lacked the optimal balance* between exploration and exploitation, relying too much on the latter. For these companies, according to the authors, greater emphasis on exploration would provide immediate financial gain. By analyzing the dynamic interplay between the dominant modes and goals of organizational activities, Uotila et al. (2009) illustrate that extreme exploration and extreme exploitation lead to inferior performance and that achieving and maintaining performance requires organizations to stay *ambidextrous*.

The concept of ambidexterity has a long tradition in management scholarship. Ambidexterity usually refers to the ability to use both hands with equal ease. In management research, it has been used to denote the ability to engage in exploration and exploitation equally well (Benner & Tushman, 2003; Gibson & Birkinshaw, 2004; He & Wong, 2004; Raisch & Birkinshaw, 2008). Both exploration and exploitation are directly related to innovation, but different activities and behaviors engage these separate processes. Granting a group more autonomy, for instance, has been linked to their higher generation of new ideas (Shalley, Zhou, & Oldham, 2004), whereas initiating structure has been found to help groups implement their goals (Keller, 2006). For organizations to be successful and innovative, it seems necessary to enact norms that are both explorative and exploitative, that is, be ambidextrous. However, is it possible to build cultures that facilitate ambidexterity and foster innovation? A theoretical perspective that sheds light on this question comes from a body of research in cultural TL.

Cultural Tightness–Looseness

The TL framework provides insights about a group's culture based on whether its behavioral expectations are generally permissive or constraining. Tight cultures have strict norms and little tolerance for deviance, whereas loose cultures have more permissive norms and more tolerance (Gelfand et al., 2011; Gelfand, 2018). Research has found that the evolution of TL differences is related to collective threat. Nations with a history of ecological and human-made threats (e.g., natural disasters, disease prevalence, resource scarcity, and invasions) tend to be tight (i.e., had stricter norms and little tolerance for deviance), whereas groups with less threat tend to be loose (i.e., had weaker norms and more permissiveness) (Gelfand et al., 2011). From an evolutionary perspective, strict norms and punishments that deter free-riders are essential to helping groups coordinate their social action to survive and thus would evolve in times of threat (Roos, Gelfand, Nau, & Lun, 2015). These patterns have also been recently replicated in nonindustrial societies (Jackson, Gelfand, & Ember, 2020).

Notably, research has shown that as groups tighten to deal with coordination needs, they also experience a number of trade-offs associated with *order* versus *openness*. As we review in the following, loose cultures exhibit higher openness—to different people, ideas, and change, but they tend to have lower order. Conversely, tight cultures have more order and synchrony, yet they tend to have lower openness. This tight–loose trade-off is presented in Figure 5.1.

Exploring Looseness, Exploiting Tightness

Loose cultures, with their greater permissiveness, are more open when it comes to new ideas, people, and change. For instance, Chua, Roth, and Lemoine (2015) analyzed the field data accumulated from responses submitted to a series of creative contests organized by a global online crowdsourcing platform. Examples of the creative briefs for the contests included coming up with designs for a new shopping mall in Spain or rebranding instant coffee for Australians. The study found that individuals from loose cultures were more likely than their counterparts from tight cultures to engage in and succeed at coming up with novel and useful ideas for a foreign audience. In addition, judges from loose cultures were more receptive to foreign creative ideas. Longitudinal studies have also found that increasing looseness over the last 200 years is related to rising creativity (Jackson, Gelfand, De, & Fox, 2019).

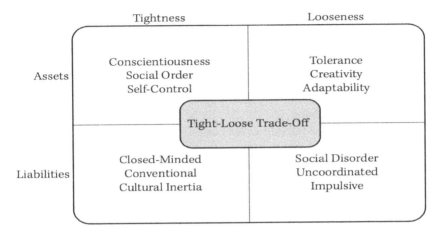

FIGURE 5.1 Tight–loose trade-off.

People from loose cultures are not only more creative but also demonstrate greater tolerance of people from commonly stigmatized backgrounds than people from tight cultures (Jackson, Van Egmond et al., 2019). In a set of four studies, Jackson and colleagues demonstrated how cultural tightness was associated with greater prejudices across nations, US states, and preindustrial societies. They found that stigma toward immigrants, minority religious groups, gay persons, and racial minorities were higher in tighter cultures. These findings demonstrated a global and historic association between TL and levels of societal tolerance for groups considered outside the norm. With greater tolerance found in loose cultures, these contexts provide more favorable conditions for eclectic ideas.

With loose cultures being more open and accepting to new groups and ideas, it is not surprising that social and cultural change in loose societies is more volatile (people in loose cultures are more likely to protest, see Gelfand, 2018) and more widespread. In an agent-based laboratory simulation, De, Nau, and Gelfand (2017) showed how introducing new, beneficial norms (such as better economic or social conditions) would be far more difficult in tight cultures, as they manifest high levels of cultural inertia, resisting the necessary changes for much longer than loose groups. Conversely, low need for coordination, characteristic for loose cultures, makes the members of loose groups more adaptive and more likely to seek out new strategies for societal change.

While loose groups corner the market on openness, tight groups more excel at processes that promote efficiency and coordination that are essential for implementation. Tight nations have higher levels of social order as manifested in lower crime, greater cleanliness, and more monitoring of behavior (Gelfand et al., 2011; Gelfand, 2018). They also tend to have more uniformity, even in the clocks on city streets (Gelfand, 2018). This high level of synchrony has also been shown in experimental studies. In contexts where threat is activated (which is more common in tight cultures), groups coordinate much faster both behaviorally and neurally as compared to groups that don't have threat (Mu, Han, & Gelfand, 2017). And at the individual level, research has shown that individuals in tight cultures have higher levels of self-monitoring, impulse control, and prevention focus as compared to loose cultures.

The TL trade-off of order versus openness has also been found at the state level (Harrington & Gelfand, 2014). Tighter states have higher trait conscientiousness (John, Naumann, & Soto, 2008), greater social organization (e.g., lower mobility, less divorce), and greater self-control (e.g., less drug and alcohol abuse and debt) (Gelfand, 2018). Looser states, in contrast, have higher trait openness (John et al., 2008), less discrimination (e.g., lower rates of Equal Employment Opportunity Commission claims and more women and minority-owned businesses), and higher creativity (e.g., more utility patents and artists per capita). While looser states are more disorganized and have more self-control failures, they are also more innovative and tolerant as compared to tighter states. Similar patterns have been found in China (Chua et al., 2019).

Finally, neuroscience research also illustrates the order-openness trade-off of TL. In a study on Chinese and US participants, Mu et al. (2015) studied cross-cultural differences in the neural markers of detecting norm violations. In their experiment, subjects read vignettes about norm-abiding or norm-defiant behaviors while connected to an electroencephalogram monitor. An example of a norm-abiding behavior was a description of a hypothetical person dancing during a tango lesson, whereas in the norm-defiant description, a person was dancing in an art museum. While both the American and Chinese subjects registered the norm violations in the central-parietal brain region that is associated with processing surprising events, Chinese subjects registered a significantly stronger response to norm violations in the frontal region, which is associated with theory of mind and punishment decisions, than Americans. They also found that differences in frontal responses to norm violations mediated cultural differences in self-control and creativity. People who didn't notice norm violations at the neural level were much more creative but reported having lower self-control, whereas the reverse was the case for people who had a high degree of neural activity when viewing norm violations.

The tight–loose trade-off is especially easy to observe in relation to the dual characteristics of innovation. Highly loose cultures, with their weaker norms and higher tolerance for deviance, can be expected to promote exploration, for example, greater variation of ideas, creative deviations, experimentation, and have more tolerance for risky endeavors. Highly tight cultures, with strong norms and emphasis on rule obedience should facilitate exploitation, for example, refinement, efficiency, execution, and implementation (Table 5.1).

TABLE 5.1 Paradoxical Behaviors and Processes Involved in Innovation.

Loose Cultures	Tight Cultures
Exploration	**Exploitation**
Variation	Refinement
Risk-taking	Efficiency
Experimentation	Implementation
Discovery	Execution

As innovation depends on the contradictory processes of exploration and exploitation, we expect high levels of cultural tightness to be positively related to greater exploitation competencies, but negatively related to exploration processes, and vice versa. Taking both relationships into consideration, we would expect a *curvilinear* relationship between cultural TL and innovation, with extreme tightness and extreme looseness leading to lower innovation and tight–loose balance leading to the highest levels of innovation. Cultures capable of balancing TL and therefore maximizing innovation would manifest what could be called tight–loose ambidexterity. This is what we refer to as the Goldilocks principle of TL, discussed in the next section.

The Goldilocks Principle of Innovation

'The Goldilocks principle' refers to the principle of moderation and has been used by previous researchers to describe phenomena ranging from teacher cognition (Kagan, 1990) to calcium release restitution in heart cells (Liu, Lederer, & Sobie, 2012). This principle is based on the title of the 19th-century children's fairy tale about a girl named Goldilocks, who encounters the home of three bears. She enters the house and discovers that the family of bears has left three plates of oatmeal on the table. The largest plate of oatmeal is too hot, and Goldilocks burns her tongue. The smallest plate of oatmeal is too cold and she spits it out. Finally, the middle plate is found to be 'just right', the ideal temperature.

In astrobiology, the 'Goldilocks zone' describes a zone around a star that is habitable, where conditions are 'just right' for life to develop. In medicine, the principle refers to the perfect dose–exposure balance (Leeder, Brown, & Soden, 2014). In psychology, the Goldilocks Principle has been successfully applied to describe work stress (Hargrove, Becker, & Hargrove, 2015) and decision-making (Schwarz, 2004), among other phenomena. The idea of moderation as an optimal strategy for satisfying multiple, often contradictory goals and needs is also echoed in the psychological body of work on that links extreme behaviors to motivational imbalance (Kruglanski, Szumowska, Kopetz, Vallerand, & Pierro, 2021). According to this perspective, extremism emerges when a certain need dominates other basic concerns, often leading to suboptimal outcomes. Motivational balance, conversely, leads to moderation, which results in different needs of an individual being equitably satisfied, producing a system of checks and balances that inhibits extreme actions (Kruglanski et al., 2021).

More recently the Goldilock's Principle has been applied to the study of social norms. Evidence suggests that a curvilinear relationship exists between TL and societal outcomes such as happiness, depression, and suicide (Harrington, Boski, & Gelfand, 2015). While cultures may need to veer tight or loose depending on their ecologies, those that get too extreme become dysfunctional. Cultures that are exceedingly loose experience what Durkheim called *anomie* or normlessness, whereas cultures that are exceedingly tight have high levels of repression, making both less functional for different reasons. Here, we examine the Goldilocks principle of innovation, wherein looseness is positively related to exploration outputs and tightness is beneficial for exploitation outputs, but the most optimal innovation levels are achieved by a cultural balance of TL. Accordingly, we sought to explore whether the relationship between TL and innovation is curvilinear.

Methods

To test the positive effect of tightness on indicators of *exploitation* and the negative effect of tightness on indicators of *exploration*, measures for our outcome variables were collected from several databases with global coverage. Exploitation measures represented features of efficiency and precision, and exploration indicators were selected based on their exemplification of processes involving creativity and discovery. Values for nation-level TL were from Gelfand and colleagues (2011)'s cross-national study. Tight–loose differences in exploitation and exploration were examined from two separate vantage points.

First, we investigated how the culture–innovation link impacts *the nature of work* in tight versus loose countries. We predicted individuals in tighter cultures would be less likely to be involved in work tasks involving entrepreneurial ventures and more likely to describe their everyday work tasks as routine. Furthermore, we tested TL's association with macro-level indicators of exploration and exploitation using data on *nation-level outputs*. We expected national tightness to be associated with more exploitation-related production strengths (value added in manufacturing and agriculture production) and looseness to be related to more exploration-based outputs (numbers of startups and scientific publications). Data included in this study were samples from nations that overlapped with countries found in the TL index (Gelfand et al., 2011) that could be matched with a tightness value. Detailed information on the data sources of these indicators is provided in the Appendix.

We examined our curvilinear hypotheses regarding TL and innovation with data from the Global Innovation Index (GII). This index is based on measures that reflect both exploration and exploitation (Dutta et al., 2019). Prior research using the GII found the index to be a robust source of data for comparing regional and socio-cultural variance in innovation (Bukowski & Rudnicki, 2019; Janger, Schubert, Andries, Rammer, & Hoskens, 2017; Rinne, Steel, & Fairweather, 2012). We expected an inverted U-relationship with TL, such that nations on the extreme end of TL would have lower rates of innovation, while nations that have more balance would be more innovative.

Results

Multilevel modeling (MLM) was performed for the individual-level data we collected on work task characteristics. By using MLM procedures with intercepts modeled as randomly varying across countries, we could simultaneously account for both the between-group and within-group variance in our estimation of the effect of tightness on the nature of work tasks (i.e., entrepreneurial or routine).

The measure asking employees whether they participated in entrepreneurial activities at their jobs was coded as 0 for 'no' and 1 for 'yes'. A multilevel logistic regression was fitted with the level-one binary outcome variable regressed on the level-2 predictor TL. Individuals were nested within countries. We used the same MLM analysis strategy for the measure on employee impressions of work tasks as routine, with higher values indicating work tasks are mostly routine. In both models, we controlled for nation-level gross domestic product (GDP) per cap, as well as a set of variables known to influence employees' workplace autonomy and, by extension, their general impressions of their work. These level-one controls included the subject's age and their relative income group, both of which testified to their job experience and workplace status. Tightness was significantly associated with our nature of work measures, correlating positively with descriptions of tasks as more routine and correlating negatively with entrepreneurial tasks. Table 5.2 summarizes the coefficients from our two mixed models.

In our next set of analyses, we investigated the degree to which cultural tightness was associated with *national outputs* indicative of either exploitation or exploration advantages. Both our TL index and the following sources of data were

TABLE 5.2 Multilevel Regression Models for Work Task Characteristics.

Models	Tightness		GDP Per Cap		Income		Age	
	γ_{01} (P)	95% CI	γ_{02} (P)	95% CI	β_{01} (P)	95% CI	β_{02} (P)	95% CI
Routine	0.09 (0.01)	0.03 to 0.15	−0.31 (0.00)	−0.49 to 0.12	−0.29 (0.00)	−0.30 to −0.27	0.00 (0.30)	−0.03 to 0.04
Entrepreneurial	−0.18 (0.00)	−0.25, to −0.11	0.45 (0.00)	0.27 to 0.62	0.63 (0.00)	0.60 to 0.67	−0.10 (0.00)	−0.13 to −0.08

values aggregated at the national level, thereby establishing ordinary least squares (OLS) models as a more appropriate means of fitting the data than MLM. Our control variable was GDP per capita because we expected that products of both exploitation and exploration processes could vary based on differing levels of wealth and economic development.

We first performed OLS analyses on data from 2019, which was the most recently available annual value for our measures. Results from our regressions showed a significant negative relationship with cultural tightness and number of scientific publications ($b = -2.98$, 95% confidence interval [CI] [−5.41, −0.56], $P = 0.02$, $R^2 = 0.29$), as well as number of startups per cap ($b = -0.006$, 95% CI [−0.01, −0.002], $P < 0.01$, $R^2 = 0.25$). By contrast, cultural tightness was significantly positively related to value added as percentage of GDP for the following industries: manufacturing ($b = 0.82$, 95% CI [0.073, 1.563], $P = 0.03$, $R^2 = 0.11$) and agriculture ($b = 0.77$, 95% CI [0.332, 1.199], $P < 0.01$, $R^2 = 0.62$). In Figure 5.2, we illustrate the partial regression plots for these national output indicators.

As a robustness check, we performed another set of analyses including data from the last 5 years with years modeled as fixed effects. The results of our analyses showed tightness remained significantly negatively associated with scientific publications ($b = -3.08$, 95% CI [−5.01, −1.16], $P < 0.01$, $R^2 = 0.35$) and startups per cap ($b = -0.002$, 95% CI [−0.005, −0.0002], $P = 0.03$, $R^2 = 0.16$). In addition, cultural tightness was positively related to the value-added measures in manufacturing ($b = 0.81$, 95% CI [0.059, 1.553], $P = 0.03$, $R^2 = 0.16$) and agriculture ($b = 0.72$, 95% CI [0.125, 1.313], $P = 0.02$, $R^2 = 0.62$). Due to possible heteroscedasticity and within-country correlations in the error terms of these models, these analyses were run using robust standard errors clustered by nation.

Finally, we tested the curvilinear relationship between our indices of cultural tightness and innovation. For these analyses, we used stepwise multiple regression to first examine a linear effect of TL and cumulative innovation scores from the GII. We then included a quadratic term to our regression equation to test for a curvilinear effect, using Akaike information criterion (AIC) estimates to compare models. Our results showed a significant curvilinear relationship when we analyzed the 2019 GII scores ($b = -0.50$, 95% CI [−0.835, −0.102], $P = 0.01$, $R^2 = 0.16$), with the quadratic model providing a better fit of the data than the linear model ($AIC_{linear} = 230.04$, $AIC_{quadratic} = 225.25$). We also analyzed all 5 years of recent GII data employing robust standard errors clustered by countries in our regressions, wherein adding the quadratic TL term also improved the initial linear model ($AIC_{linear} = 1171.18$, $AIC_{quadratic} = 1137.93$). Findings from this quadratic model showed that generally speaking, extremely tight and loose cultures have the lowest innovation values ($b = -0.51$, 95% CI [−0.853, −0.171], $P < 0.01$, $R^2 = 0.18$). See Figure 5.3.

All analyses were conducted using R version 3.6.2 (R Core Team, 2019). Data and code are available at https://osf.io/pqh5e/.

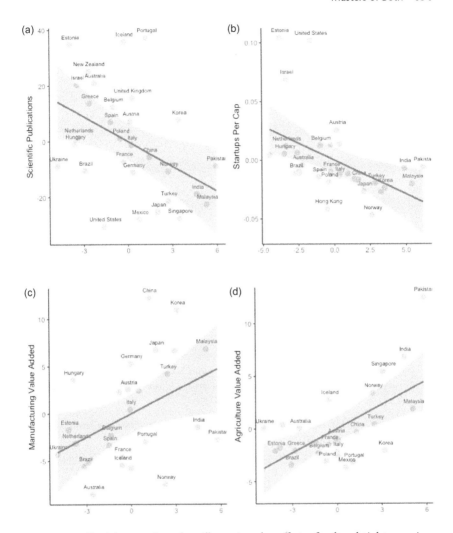

FIGURE 5.2 Partial regression plots illustrating the effect of cultural tightness given the effect of GDP per cap on (a) relative number of scientific publications, (b) startups per cap, (c) manufacturing value added, and (d) agriculture value added. Data for these response variables were collected in 2019.

Discussion

Innovation is critical for nations worldwide, yet to date, we have little understanding of which nations are better able to explore, exploit, or do both. We showed that loose cultures are better at exploration and tight cultures at exploitation on a number of measures. We also provided evidence for the Goldilocks principle of

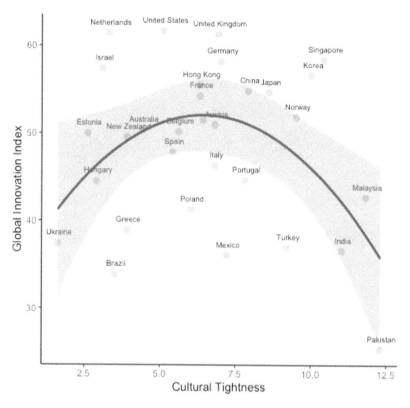

FIGURE 5.3 Curvilinear relationship found between nation-level tightness and GII scores from 2019.

innovation such that cultures that had more balance in TL had higher innovation. From a theoretical point of view, this study advances the study of innovation by specifically illustrating the effect of cultural norms on a society's capacity to innovate. Ambidextrous cultures, enabling the right balance between freedom and structure, are able to boost creativity while simultaneously driving implementation. This finding builds upon previous work in the paradox (Miron-Spektor, Erez, & Naveh, 2011; Miron-Spektor & Erez, 2017; Smith, Lewis & Tushman, 2016;) and ambidexterity literatures (Gibson & Birkinshaw, 2004; O'Reilly & Tushman, 2008; Zacher & Wilden, 2014; Zacher, Robinson, & Rosing, 2016) and shows that cultural norms are an important aspect of societies that foster exploration and exploitation simultaneously.

A key question is how organizations can achieve TL balance in order to maximize innovation. One key mechanism for doing so is through leadership. For example, tight cultures tend to value leaders who are independent and who are better able to provide order and closure, which is helpful for implementation, while loose cultures value leaders who are charismatic, team oriented, and are more open minded—facilitating creativity (Aktas, Gelfand, & Hanges, 2016). Yet leadership needs to be *ambidextrious* (Rosing et al., 2011), wherein leaders are able to simultaneously promote *order* and *openness* in groups. Leaders who emphasize openness encourage experimentation, give room for independent thinking and acting, and support challenges to established approaches. Leaders who help provide order set specific guidelines, monitor goal attainment, and take corrective actions when necessary. Since innovation requires maximizing two opposing requirements (exploration and exploitation), leaders need to deploy behaviors that help to promote both openness and order to foster team creativity and implementation, respectively (Lewis, Welsh, Dehler, & Green, 2002).

Some companies are already doing this. For instance, Baidu, known as 'China's Google', has tried to balance both openness and order through a number of practices. Recognizing the extent to which China's cultural tightness might stifle creativity in the innovation process, representatives of Baidu intentionally created a set of rules and anecdotes, which they titled the 'Baidu Analects', in reference to the Analects of Confucius. According to the company, the main goal of this book is to encourage organizational dissent and teach the employees how to break from conformity, while at the same time delivering tasks to the next team or person only when they've been perfected (Gelfand, 2018). This combination of loose norms with high accountability—which encourages both exploration and exploitation—has helped Baidu to be highly innovative.

Indeed, when facilitating workshops and conducting classes in leadership and innovation management, one of the authors of this chapter (P.P.) uses an exercise in which he invites participants to reflect on people who have empowered them to innovate and the environments that made it possible. When they reflect on their stories and examples, the participants usually recognize that innovative cultures and leaders share an interesting pattern—they exhibit often seemingly *opposite behaviors*. Leaders who enabled them to be creative and to implement their ideas not only provided freedom but also imposed restrictions. They encouraged experimentation, but only within certain deadlines. The results from the present study confirm that cultures that drive innovation are able to succeed in navigating between the Scylla and Charybdis of tight–loose extremes through ambidexterity; they facilitate creativity and experimentation and build resources and structures for implementation.

There are other ways to produce TL balance in organizations. For example, leaders can temporarily switch between exploration and exploitation across time (Gupta, Smith, & Shalley, 2006; Tushman & O'Reilly, 1996; Turner, Swart, & Maylor, 2013). One instance of such *temporal ambidexterity* is the approach used by the Navy Seals, who follow a strict hierarchy during their missions, but during mission debriefings, they get rid of rank and democratically discuss their successes and failures (Greer, Van Bunderen, & Yu, 2017). Another way to create TL balance is through *structural ambidexterity*, wherein leaders separate units who perform exploratory and exploitative activities to a large extent independently (Bower & Christensen, 1995; Gupta et al., 2006; O'Reilly & Tushman, 2004). In this approach, leaders can design organizations to capitalize on both creativity and implementation, by either recruiting or training some organizational members to emphasize the former and others emphasize the latter, or by designing 'teams of teams', each designated to perform the creativity or implementation function. Leaders should highlight superordinate goals to all teams, noting that both tightness and looseness are necessary for innovation, and that tight and loose divisions of the company help offset each other's weaknesses. The example of *USA Today*, made widely popular by O'Reilly and Tushman (2004), shows that a path to balancing exploration and exploitation might lead through separating tight and loose units with different objectives and modes of operations, while cultivating integration and mutual respect between them.

Our study only provides a 'proof of concept' regarding TL balance and innovation. Future research should extend this work to other nations, examine within-country variation in TL, and examine whether these investigations produce similar effects. New measures of exploration and exploitation—or measures that capture both concepts—will need to be developed, given that existing measures have imperfect construct validity (Potočnik & Anderson, 2016). Future research should account for the interplay between national and organizational TL in order to examine closely the relationship between tight–loose ambidexterity and innovation, as well as study these dynamics in teams. In addition, future research can also investigate other social and psychological mechanisms through which national culture impacts both exploration and exploitation.

We note that as our research illustrates a robust relationship between TL ambidexterity and innovation, it also implies that balancing exploration and exploitation brings about optimal innovation-related outcomes for nations and organizations alike. However, this integration of exploitation and exploration does not have to mean that each exists in equal measure. Certain occupations and industries, for instance, may require a dominant focus on exploration or exploitation to meet their organizational goals. The military's tight culture enables their ability to safely complete dangerous missions and complex operations. Although these are exploitation-relevant skills; however, the military still benefits from integrating some exploration capabilities. As there is never the guarantee of a foolproof military operation, unexpected complications will necessitate soldiers to be inventive and adaptive. In all, the balance of tightness and looseness, and in turn, exploration and exploitation, must align with the unique strategic and economic goals of different nations and organizations.

In conclusion, we present the integral role of cultural TL in shaping how different ventures exhibit varying focus on exploration or exploitation processes. However, both processes are essential to the advancement of innovations. Importantly, the management of cultural TL facilitates the circumvention of excess prioritization on just exploration or exploitation. By leveraging the strength of cultural norms, organizations can strike a balance between these often-conflicting objectives. At present, innovation is one of the most important forces driving growth and market success. For leaders, this cultural framework offers guidance in establishing suitable norms and organizational processes that can help them successfully drive innovative results.

Funding

This research was funded in part by Office of Naval Research grant N000141912407 (MJG). The information in this Article does not imply or constitute an endorsement of the views therein by the Office of Naval Research, US Navy, or Department of Defense.

Appendix

Tightness–Looseness. Cultural TL scores were derived from a large-scale, cross-cultural study by Gelfand and colleagues (2011), which compared 32 nations on the strength of their social norms. In each country, a representative sample of participants completed a well-validated measure of TL that featured questions on the normative expectations found within each respective nation. Countries were assigned an overall tightness value, with high scores indicating greater cultural tightness and lower scores representing more looseness. Nations highest on tightness were found to be Malaysia, India, and Norway. The loosest nations from the study included Hungary, Israel, and Brazil. We used this index as the basis of our country-level tightness scores.

Nature of Work. Two of the most recently available waves of the Adult Population Survey (APS)—years 2014 and 2015—were collected at the time of our data collection for this study (Reynolds et al., 2005). APS is part of the Global Entrepreneurship Monitor (GEM)'s annual assessment of entrepreneurial activities around the world. The years selected were based on GEM's report on recent improvements to their sampling methodology. The APS measure selected asked employed individuals a yes-or-no question on whether they had participated in developing new and entrepreneurial business activities for their employers in the last 3 years. Our final sample from GEM included 63,301 people from 25 different countries.

The World Values Survey (WVS) provided an additional sample of individual-level data from numerous countries (Inglehart et al., 2014). We collected the data for the fifth (2005–2009) and sixth (2010–2014) waves of the WVS. In particular, we prioritized these waves because they included a measure well suited for our research focus on work task characteristics. The WVS item of interest asked respondents on a ten-point scale to rate whether they would describe their overall work tasks as either routine or creative. We combined responses to this question from both waves, only including currently employed respondents, which resulted in a final sample of 29,963 subjects from 25 countries.

National Outputs. By analyzing overall national output estimates, these measures allowed us to observe macro-level signs of exploitation and exploration. We gathered several years of data (2015–2019) from the World Bank on a country's value added, reported as a percentage of GDP, for manufacturing and agriculture sectors. The past research has found that national cultures that stress high adherence to strong norms and rules are better situated for advancing their manufacturing industries (Campante & Chor, 2017). In addition, we chose the agriculture sector based on research that has found societies that manage crop fields develop strong norms in order to execute the routines and group cooperation needed to secure a good harvest (Barry, Child, & Bacon, 1959). In turn, strong norms are expected to be beneficial toward future production capabilities and success in large-scale agricultural activities.

Exploration indicators for the study included data counting the annual (a) scientific publications and (b) startups per capita for each country in our sample from the years 2015 to 2019. Both of these indicators are reflective of creativity through novel contributions and inventive business endeavors. Data on scientific publications were tabulated from the Science Citation Index and the Social Sciences Citation Index platforms, archival databases of scientific journals worldwide published by Clarivate Analytics and extracted by the GII (Dutta et al., 2019). Each countries' article numbers were based on the location of a publication's author(s) per given year. The publications encompassed a number of diverse scientific fields (engineering, chemistry, astronomy, physics, computer science, mathematics, medicine, etc.). Final counts were scaled with GDP based on purchasing power parity per billion international dollars. Then, min–max normalization was applied to create final scores in the 0–100 range per country. Data on startups were based on the annual number of registered start-up companies on Startup Ranking, a website that maintains a record of new businesses from over 190 countries (Startup Ranking, 2019). This series was scaled by annual country population data provided by the World Bank's World Development Indicators (2019c).

References

Akçomak, I. S., & Ter Weel, B. (2009). Social capital, innovation and growth: Evidence from Europe. *European Economic Review, 53*(5), 544–567.

Aktas, M., Gelfand, M. J., & Hanges, P. J. (2016). Cultural tightness—looseness and perceptions of effective leadership. *Journal of Cross-Cultural Psychology, 47*(2), 294–309.

Anderson, N., Potočnik, K., & Zhou, J. (2014). Innovation and creativity in organizations: A state-of-the-science review, prospective commentary, and guiding framework. *Journal of Management, 40*(5), 1297–1333.

Baldridge, J. V., & Burnham, R. A. (1975). Organizational innovation: Individual, organizational, and environmental impacts. *Administrative Science Quarterly*, 165–176.

Barry, H., Child, I. L., & Bacon, M. K. (1959). Relation of child training to subsistence economy. *American Anthropologist, 61*(1), 51–63.

Benner, M. J., & Tushman, M. L. (2003). Exploitation, exploration, and process management: The productivity dilemma revisited. *Academy of Management Review, 28*(2), 238–256.

Bledow, R., Frese, M., Anderson, N., Erez, M., & Farr, J. (2009). A dialectic perspective on innovation: Conflicting demands, multiple pathways, and ambidexterity. *Industrial and Organizational Psychology, 2*(3), 305–337.

Bower, J. L., & Christensen, C. M. (1995). Disruptive technologies: Catching the wave. *Harvard Business Review, 73*(1), 43–53.

Bukowski, A., & Rudnicki, S. (2019). Not only individualism: The effects of long-term orientation and other cultural variables on national innovation success. *Cross-Cultural Research, 53*(2), 119–162.

Camisón, C., & Villar-López, A. (2014). Organizational innovation as an enabler of technological innovation capabilities and firm performance. *Journal of Business Research, 67*(1), 2891–2902.

Campante, F. R., & Chor, D. (2017). *"Just do your job": Obedience, routine tasks, and the pattern of specialization* (ERIA Discussion Paper No. DP-2016–35). Jakarta: Economic Research Institute for ASEAN and East Asia.

Chan, S. H., Martin, J. D., & Kensinger, J. W. (1990). Corporate research and development expenditures and share value. *Journal of Financial Economics, 26*(2), 255–276.

Chua, R. Y., Huang, K. G., & Jin, M. (2019). Mapping cultural tightness and its links to innovation, urbanization, and happiness across 31 provinces in China. *Proceedings of the National Academy of Sciences, 116*(14), 6720–6725.

Chua, R. Y., Roth, Y., & Lemoine, J. F. (2015). The impact of culture on creativity: How cultural tightness and cultural distance affect global innovation crowdsourcing work. *Administrative Science Quarterly, 60*(2), 189–227.

Damanpour, F. (1991). Organizational innovation: A meta-analysis of effects of determinants and moderators. *Academy of Management Journal, 34*(3), 555–590.

De, S., Nau, D. S., & Gelfand, M. J. (2017, May). Understanding norm change: An evolutionary game-theoretic approach. In *Proceedings of the 16th conference on autonomous agents and MultiAgent systems* (pp. 1433–1441). Richland, SC: International Foundation for Autonomous Agents and Multiagent Systems.

Diebolt, C., & Hippe, R. (2019). The long-run impact of human capital on innovation and economic development in the regions of Europe. *Applied Economics, 51*(5), 542–563.

Dutta, S., Lanvin, B., & Wunsch-Vincent, S. (2019). *The global innovation index 2019: Creating healthy lives—the future of medical innovation*. Geneva: World Intellectual Property Organization.

Furman, J. L., Porter, M. E., & Stern, S. (2002). The determinants of national innovative capacity. *Research Policy, 31*(6), 899–933.

Gelfand, M. J. (2018). *Rule makers, rule breakers: How tight and loose cultures wire our world*. New York: Scribner.

Gelfand, M. J., Raver, J. L., Nishii, L., Leslie, L. M., Lun, J., Lim, B. C., . . . Aycan, Z. (2011). Differences between tight and loose cultures: A 33-nation study. *Science, 332*(6033), 1100–1104.

Ghazinoory, S., Bitaab, A., & Lohrasbi, A. (2014). Social capital and national innovation system: A cross-country analysis. *Cross Cultural Management, 21*(4), 453–475.

Gibson, C. B., & Birkinshaw, J. (2004). The antecedents, consequences, and mediating role of organizational ambidexterity. *Academy of Management Journal, 47*(2), 209–226.

Gorodnichenko, Y., & Roland, G. (2017). Culture, institutions, and the wealth of nations. *Review of Economics and Statistics, 99*(3), 402–416.

Greer, L. L., Van Bunderen, L., & Yu, S. (2017). The dysfunctions of power in teams: A review and emergent conflict perspective. *Research in Organizational Behavior, 37*, 103–124.

Gupta, A. K., Smith, K. G., & Shalley, C. E. (2006). The interplay between exploration and exploitation. *Academy of Management Journal, 49*(4), 693–706.

Hall, B. H. (1993). The stock market's valuation of R&D investment during the 1980's. *The American Economic Review, 83*(2), 259–264.

Hargadon, A. (2003). *How breakthroughs happen: The surprising truth about how companies innovate*. Boston, MA: Harvard Business Press.

Hargrove, M. B., Becker, W. S., & Hargrove, D. F. (2015). The HRD eustress model: Generating positive stress with challenging work. *Human Resource Development Review, 14*(3), 279–298.

Harrington, J. R., Boski, P., & Gelfand, M. J. (2015). Culture and national well-being: Should societies emphasize freedom or constraint? *PloS One, 10*(6), e0127173.

Harrington, J. R., & Gelfand, M. J. (2014). Tightness—looseness across the 50 United States. *Proceedings of the National Academy of Sciences, 111*(22), 7990–7995.

Hayton, J. C., George, G., & Zahra, S. A. (2002). National culture and entrepreneurship: A review of behavioral research. *Entrepreneurship Theory and Practice, 26*(4), 33–52.

He, Z. L., & Wong, P. K. (2004). Exploration vs. exploitation: An empirical test of the ambidexterity hypothesis. *Organization Science, 15*(4), 481–494.

Hoang, H. A., & Rothaermel, F. T. (2010). Leveraging internal and external experience: Exploration, exploitation, and R&D project performance. *Strategic Management Journal, 31*(7), 734–758.

Howaldt, J., & Schwarz, M. (2010). *Social innovation: Concepts, research fields and international trends*. Aachen: IMA/ZLW.

Inglehart, R., Haerpfer, C., Moreno, A., Welzel, C., Kizilova, K., Diez-Medrano, J., . . . Puranen, B. (2014). *World values survey: All rounds-country-pooled datafile version*. Madrid: JD Systems Institute.

Israel, P. (1998). *Edison: A life of invention*. New York: John Wiley.

Jackson, J. C., Gelfand, M., De, S., & Fox, A. (2019). The loosening of American culture over 200 years is associated with a creativity—order trade-off. *Nature Human Behaviour, 3*(3), 244–250.

Jackson, J. C., Gelfand, M., & Ember, C. R. (2020). A global analysis of cultural tightness in non-industrial societies. *Proceedings of the Royal Society B, 287*(1930), 20201036.

Jackson, J. C., Van Egmond, M., Choi, V. K., Ember, C. R., Halberstadt, J., Balanovic, J., . . . Fulop, M. (2019). Ecological and cultural factors underlying the global distribution of prejudice. *PLoS One, 14*(9), e0221953.

Jacobs, J. (1984). *Cities and the wealth of nations: Principles of economic life*. New York: Vintage.

Janger, J., Schubert, T., Andries, P., Rammer, C., & Hoskens, M. (2017). The EU 2020 innovation indicator: A step forward in measuring innovation outputs and outcomes? *Research Policy, 46*(1), 30–42.

John, O. P., Naumann, L. P., & Soto, C. J. (2008). Paradigm shift to the integrative big five trait taxonomy. *Handbook of Personality: Theory and Research, 3*(2), 114–158.

Kagan, D. M. (1990). Ways of evaluating teacher cognition: Inferences concerning the Goldilocks principle. *Review of Educational Research*, *60*(3), 419–469.

Keller, R. T. (2006). Transformational leadership, initiating structure, and substitutes for leadership: A longitudinal study of research and development project team performance. *Journal of Applied Psychology*, *91*(1), 202.

Kruglanski, A. W., Szumowska, E., Kopetz, C. H., Vallerand, R. J., & Pierro, A. (2021). On the psychology of extremism: How motivational imbalance breeds intemperance. *Psychological Review*, *128*(2), 264–289.

Landes, D. S. (1969). *The unbound Prometheus: Technological change and industrial development in Western Europe from 1750 to the present*. Cambridge and New York: Press Syndicate of the University of Cambridge.

Lavie, D., Kang, J., & Rosenkopf, L. (2011). Balance within and across domains: The performance implications of exploration and exploitation in alliances. *Organization Science*, *22*(6), 1517–1538.

Leeder, J. S., Brown, J. T., & Soden, S. E. (2014). Individualizing the use of medications in children: Making Goldilocks happy. *Clinical Pharmacology & Therapeutics*, *96*(3), 304–306.

Lewis, M. W., Welsh, M. A., Dehler, G. E., & Green, S. G. (2002). Product development tensions: Exploring contrasting styles of project management. *Academy of Management Journal*, *45*(3), 546–564.

Liu, O. Z., Lederer, W. J., & Sobie, E. A. (2012). Does the goldilocks principle apply to calcium release restitution in heart cells? *Journal of Molecular and Cellular Cardiology*, *52*(1), 3–6.

March, J. G. (1991). Exploration and exploitation in organizational learning. *Organization Science*, *2*(1), 71–87.

March, J. G. (1996). Continuity and change in theories of organizational action. *Administrative Science Quarterly*, 278–287.

Millard, A. J (1990). *Edison and the business of innovation*. Baltimore: Johns Hopkins University Press.

Miron-Spektor, E., & Erez, M. (2017). Looking at creativity through a paradox lens. *The Oxford Handbook of Organizational Paradox*, 434.

Miron-Spektor, E., Erez, M., & Naveh, E. (2011). The effect of conformists and attentive-to-detail members on team innovation: Reconciling the innovation paradox. *Academy of Management Journal*, *54*(4), 740–760.

Mohr, L. B. (1969). Determinants of innovation in organizations. *American Political Science Review*, *63*(1), 111–126.

Morris, E. (2019). *Edison* (1st ed.). New York: Random House.

Mu, Y., Han, S., & Gelfand, M. J. (2017). The role of gamma interbrain synchrony in social coordination when humans face territorial threats. *Social Cognitive and Affective Neuroscience*, *12*(10), 1614–1623.

Mu, Y., Kitayama, S., Han, S., & Gelfand, M. J. (2015). How culture gets embrained: Cultural differences in event-related potentials of social norm violations. *Proceedings of the National Academy of Sciences*, *112*(50), 15348–15353.

Murphy, K. M., Shleifer, A., & Vishny, R. W. (1993). Why is rent-seeking so costly to growth? *The American Economic Review*, *83*(2), 409–414.

Nakata, C., & Sivakumar, K. (1996). National culture and new product development: An integrative review. *Journal of Marketing*, *60*(1), 61–72.

O'Reilly III, C. A., & Tushman, M. L. (2004). The ambidextrous organization. *Harvard Business Review*, *82*(4), 74–83.

O'Reilly III, C. A., & Tushman, M. L. (2008). Ambidexterity as a dynamic capability: Resolving the innovator's dilemma. *Research in Organizational Behavior, 28*, 185–206.

Organisation for Economic Co-operation and Development, & Statistical Office of the European Communities. (2018). *Oslo Manual 2018: Guidelines for collecting, reporting and using data on innovation.* Paris: OECD Publishing.

Pierce, J. L., & Delbecq, A. L. (1977). Organization structure, individual attitudes and innovation. *Academy of Management Review, 2*(1), 27–37.

Porter, M. E. (1990). The competitive advantage of nations. *Harvard Business Review, 68*(2), 73–93.

Porter, M. E. (2011). *Competitive advantage of nations: Creating and sustaining superior performance.* New York: Simon and Schuster.

Potočnik, K., & Anderson, N. (2016). A constructively critical review of change and innovation-related concepts: Towards conceptual and operational clarity. *European Journal of Work and Organizational Psychology, 25*(4), 481–494.

Pretzer, W. S. (Ed.). (2002). *Working at inventing: Thomas A. Edison and the Menlo Park experience.* London: Johns Hopkins University Press.

Raisch, S., & Birkinshaw, J. (2008). Organizational ambidexterity: Antecedents, outcomes, and moderators. *Journal of Management, 34*(3), 375–409.

Raisch, S., Birkinshaw, J., Probst, G., & Tushman, M. L. (2009). Organizational ambidexterity: Balancing exploitation and exploration for sustained performance. *Organization Science, 20*(4), 685–695.

R Core Team. (2019). *R: A language and environment for statistical computing.* R Foundation for Statistical Computing, Vienna, Austria. Retrieved from www.R-project.org/

Reynolds, P., Bosma, N., Autio, E., Hunt, S., De Bono, N., Servais, I., . . . Chin, N. (2005). Global entrepreneurship monitor: Data collection design and implementation 1998–2003. *Small Business Economics, 24*(3), 205–231

Rinne, T., Steel, G. D., & Fairweather, J. (2012). Hofstede and Shane revisited: The role of power distance and individualism in national-level innovation success. *Cross-Cultural Research, 46*(2), 91–108.

Roos, P., Gelfand, M., Nau, D., & Lun, J. (2015). Societal threat and cultural variation in the strength of social norms: An evolutionary basis. *Organizational Behavior and Human Decision Processes, 129*, 14–23.

Rosenbusch, N., Brinckmann, J., & Bausch, A. (2011). Is innovation always beneficial? A meta-analysis of the relationship between innovation and performance in SMEs. *Journal of Business Venturing, 26*(4), 441–457.

Rosing, K., Frese, M., & Bausch, A. (2011). Explaining the heterogeneity of the leadership-innovation relationship: Ambidextrous leadership. *The Leadership Quarterly, 22*(5), 956–974.

Schankerman, M., & Pakes, A. (1986). Estimates of the value of patent rights in European countries during the post-1950 period. *Economic Journal, 96*, 1052–1076.

Schwarz, B. (2004). *The paradox of choice. Why more is less.* New York: Harper Collins.

Shalley, C. E., Zhou, J., & Oldham, G. R. (2004). The effects of personal and contextual characteristics on creativity: Where should we go from here? *Journal of Management, 30*(6), 933–958.

Shane, S. (1993). Cultural influences on national rates of innovation. *Journal of Business Venturing, 8*(1), 59–73.

Smith, W. K., Lewis, M. W., & Tushman, M. L. (2016). "Both/and" leadership. *Harvard Business Review, 94*(5), 62–70.

Startup Ranking. (2019). Retrieved from www.startupranking.com
Taylor, M. Z., & Wilson, S. (2012). Does culture still matter? The effects of individualism on national innovation rates. *Journal of Business Venturing, 27*(2), 234–247.
Tellis, G. J., Prabhu, J. C., & Chandy, R. K. (2009). Radical innovation across nations: The preeminence of corporate culture. *Journal of Marketing, 73*(1), 3–23.
Turner, N., Swart, J., & Maylor, H. (2013). Mechanisms for managing ambidexterity: A review and research agenda. *International Journal of Management Reviews, 15*(3), 317–332.
Tushman, M. L., & O'Reilly III, C. A. (1996). Ambidextrous organizations: Managing evolutionary and revolutionary change. *California Management Review, 38*(4), 8–29.
Uotila, J., Maula, M., Keil, T., & Zahra, S. A. (2009). Exploration, exploitation, and financial performance: Analysis of S&P 500 corporations. *Strategic Management Journal, 30*(2), 221–231.
West, M. A., & Farr, J. L. (1990). Innovation at work. In M. A. West & J. L. Farr (Eds.), *Innovation and creativity at work: Psychological and organizational strategies*. Chichester: Wiley Blackwell.
World Bank, World Development Indicators. (2019a). *Agriculture, forestry, and fishing, value added (% of GDP)* [API_NV.AGR.TOTL.ZS_DS2_en_excel_v2_1218009]. Retrieved from https://data.worldbank.org/indicator/NV.AGR.TOTL.ZS?view=chart
World Bank, World Development Indicators. (2019b). *Manufacturing, value added (% of GDP)* [API_NV.IND.MANF.ZS_DS2_en_excel_v2_1218297]. Retrieved from https://data.worldbank.org/indicator/NV.IND.MANF.ZS
World Bank, World Development Indicators. (2019c). *Population, total* [API_SP.POP.TOTL_DS2_en_excel_v2_1217749]. Retrieved from https://data.worldbank.org/indicator/SP.POP.TOTL
Zacher, H., Robinson, A. J., & Rosing, K. (2016). Ambidextrous leadership and employees' self-reported innovative performance: The role of exploration and exploitation behaviors. *The Journal of Creative Behavior, 50*(1), 24–46.
Zacher, H., & Wilden, R. G. (2014). A daily diary study on ambidextrous leadership and self-reported employee innovation. *Journal of Occupational and Organizational Psychology, 87*(4), 813–820.

6
THE EVOLUTION OF EXTREMISM

William von Hippel and Nadia Fox

The focus of this chapter is on the evolution of extremism, by which we mean two inter-related concepts. First and foremost, we follow Kruglanski, Fernandez, Factor, and Szumowska's (2019) *psychological* definition of extremism as the single-minded pursuit of a particular goal, with all or nearly all other goals subordinated to it. Secondarily, we also consider Kruglanski et al.'s (2019) *statistical* definition of extremism as attitudes and behaviors that reside on the tail end of the normal distribution on any particular domain. Our conflation of these two approaches to extremism is based on the observation that attitude extremity correlates with attitude strength (Petty & Krosnick, 1995). Statistically extreme attitudes are held with greater certainty (Krosnick, Boninger, Chuang, Berent, & Carnot, 1993), are regarded as more important (Boninger, Krosnick, Berent, & Fabrigar, 1995), and are more resistant to change ('their bulldog nature'; Abelson, 1995). These features of statistically extreme attitudes ensure that they are more self-defining than moderate ones and are held with greater zeal (Klein & Kruglanski, 2013). As a consequence of these converging factors, statistically extreme attitudes and behaviors are often associated with the single-minded pursuit of relevant goals.

With these definitions in mind, this chapter is organized around four questions regarding extremism:

1. How might extremism have evolved?
2. What function might extremism serve?
3. Is extremism unique to humans?
4. What role does morality play in extremism?

These questions are clearly inter-related in the sense that the function of extremism tells us something about how and why it evolved, and a possible role for

DOI: 10.4324/9781003030898-8

morality tells us something about the uniqueness of extremism to humans (as other animals do not have the capacity for self-reflective moral reasoning; Boehm, 2012; De Waal, Macedo, & Ober, 2006; Suddendorf, 2013). Nonetheless, organizing the material with regard to these four questions provides a clear and easy structure, while also allowing readers direct access to the material of greatest interest to them.

How Might Extremism Have Evolved?

When social scientists began studying the role of genes in complex traits, it was widely expected that genes would map onto traits in a relatively simple fashion—there would be a few genes for this and a few genes for that. Based on this belief, the field of behavioral genetics took a deep dive into candidate gene studies, examining whether various genes (e.g., those known to be important from knockout studies in mice) might influence human behavior and traits in a predictable manner. These studies appeared promising at first, but their findings proved hard to replicate (Hewitt, 2012). This replication crisis preceded that of the larger field of psychology (Nosek, Aarts, Anderson, Kappes, & Collaboration, 2015) and led to two important realizations. First, just because a gene can be shown to play a critical role in a trait or a behavior by artificially removing it from a mouse or other animals does not mean that natural variations in that gene (i.e., different alleles) play the same role. Second, as the field moved from candidate gene studies to genome-wide association studies (GWAS), in which the effects of a vast number of genes on a particular trait were examined simultaneously, it became evident that most complex traits are influenced by a very large number of genes, and individual genes often have (very small) effects on a large number of complex traits (Plomin, DeFries, Knopik, & Neiderhiser, 2016).

This knowledge of genetics has changed our understanding of what it means to have a psychological disorder, as it clarifies that the genetic component of abnormality is really just an extreme position on a bell curve rather than a difference of type. As Plomin (2018) explains in *Blueprint*:

> What we call disorders are merely the extremes of the same genes that work throughout the normal distribution. That is, there are no genes 'for' any psychological disorder. Instead, we have many of the DNA differences that are related to disorders . . . the more we have, the more likely we are to have problems. In other words, the genetic causes of what we call disorders are quantitatively, not qualitatively, different from the rest of the population.
>
> (p. 58)

Before exploring the implications of this statement for extremism, it is worth noting that the overall influence of genes on complex traits varies substantially as a function of the trait in question (e.g., see Tesser, 1993), but averages around 50%

(Polderman et al., 2015). In other words, although any individual gene accounts for only a fraction of the variance in most complex traits, our entire genome accounts for about half the variance. In contrast, only 4% of the variance in these traits is from the shared environment of our parents' home (on average), as the preponderance of the environmental variance is accounted for by our experiences elsewhere (Polderman et al., 2015).

These conclusions about genetic influences on complex traits are based on studies of fraternal and identical twins, which allow researchers to statistically parse the effects of genes and the environment on any particular trait without actually examining the genes in question. As of this writing, the molecular genetics data from GWAS have not come close to accounting for the large genetic effects that emerge from twin studies, with GWAS typically only accounting for a small portion of the expected genetic variance (Young, 2019).

To return to Plomin's (2018) point that disorders represent quantitative rather than qualitative differences, it stands to reason that the genetics of extremism are similarly quantitative. Indeed, Plomin's definition of disorder invokes extremity on the genetic distribution as the underlying cause. Because extreme positions on the genetic distribution are an inevitable but rare consequence of a continuous distribution, a small percentage of people will always have a genetic proclivity for extremism for no other reason than the random shuffling of DNA that takes place during sexual reproduction (the process of Mendelian segregation during meiosis). An important implication of this genetic perspective is that there need not be any evolutionary advantage of extremism (and there could even be notable costs) for it to emerge at very low frequency in every generation. This perspective does not preclude an evolved role for extremism, in which extreme behaviors enhance fitness under certain circumstances, but it suggests that extremism will continue to exist even if it only leads to costs. Such a genetic proclivity toward extremism would likely manifest in a variety of individual differences, such as addiction, engagement in extreme sports or violence, holding extreme political or religious attitudes, etc.

Perhaps surprisingly, the second implication of this genetic perspective is that environmental factors will always play a major role in the emergence of extremism. As noted earlier, the environment impacts the emergence of all complex traits, as genes and environments jointly influence attitudes and behaviors (Plomin et al., 2016). And that, in turn, leads to the question of what environmental factors might facilitate extremism.

When we think of extremism as the single-minded pursuit of a particular goal, with all or nearly all other goals subordinated to it (Kruglanski et al., 2019), it suggests that extremism is more likely to emerge when environmental forces favor such single-minded pursuit. Furthermore, whenever such environmental forces are a regular occurrence across evolutionary time, and whenever they predictably emerge in certain situations, extremism is likely to evolve as a genetic response to those environmental regularities. One situation that fits both of these requirements is the presence of intergroup conflict, which has such severe consequences for group members that it can cause a single goal or approach to become dominant across an entire society (Kruglanski, Jasko, & Friston, 2020).

Several lines of evidence suggest that intergroup conflict has been a prevalent feature of our existence since well before the onset of agriculture 10,000 years ago, and probably for the entire 200,000 to 300,000 years that we have been *Homo sapiens* (Pinker, 2011; Rusch, 2014). Indeed, once our ancestors developed the capacities for division of labor and planning for unfelt needs, both of which date back to *Homo erectus* over a million years ago (Domínguez-Rodrigo, 2002; Shipton, 2013; Shipton & Nielsen, 2015), the greatest predatory threat to our hominin ancestors was no longer the large animals that roamed the savannah, but rather other groups of hominins.[1] Other hominins were the only species that shared our capacities for planning and coordination and hence the incredible lethality of the groups in which we lived (von Hippel, 2018).

Nonetheless, encounters with other groups were not inevitably conflictual. Interactions with outgroups also led to numerous benefits, for example, in the form of trade, knowledge transmission, and reduced inbreeding (e.g., Smith, 2008). As a consequence, we have not evolved an automatic negativity toward outgroups (Perdue, Dovidio, Gurtman, & Tyler, 1990), but we have evolved a tendency for negativity to emerge rapidly at any sign of conflict or intergroup vulnerability (McDonald, Navarrete, & van Vugt, 2012; Schaller & Neuberg, 2012). The result of these evolutionary processes is that we feel an automatic affiliation with members of our own group but a sense of volatile ambivalence (which often appears as neutrality) toward members of other groups (Brewer, 1999). Cooperation with outgroups can create affinities, but competition quickly leads to intergroup hostility (Sherif, Harvey, White, Hood, & Sherif, 1961).

Because intergroup conflict was an existential threat to our ancestors (Keeley, 1997), it selected for extreme behavior by rewarding those ancestors who prioritized winning the conflict above all other possible goals. Those ancestors who engaged in extreme behaviors that supported the ingroup and were hostile to the outgroup in situations of intergroup conflict would have often been the most effective in such conflicts (Rusch, 2014) and hence would have passed on the heritable aspects of this proclivity to their offspring. This line of reasoning suggests that *humans have evolved a tendency toward extremism whenever they are in conflict with other groups.*

In contrast to this reasoning, it is easy to imagine ancestral humans who could reflect dispassionately on their conflicts with other groups, recognize that both sides share some blame, realize that conflict is best resolved by compromise, and understand that violence is unlikely to serve either group's interests over the long run. No doubt our ancestors engaged in such reasoning processes on occasion, and when that happened, conflict was lessened or avoided altogether. Nonetheless, we also evolved a tendency toward self-deceptive hypocrisy and a resultant belief in the unique righteousness of our cause (Trivers, 2011; von Hippel, 2017;

von Hippel & Trivers, 2011). This tendency would have supported extremism, as it would justify an extreme response to our enemies, and would have made compromise less likely. This tendency would also be exacerbated by the fact that our logical abilities are often more focused on supporting positions that benefit us than on finding the truth (Kruglanski et al., 2020; Mercier & Sperber, 2011).

What Function Might Extremism Serve?

Beyond its basic function in making one goal ascendant above all others, and hence benefitting the individual for whom achieving that particular goal is paramount, extremism also has the potential to provide other benefits. First and foremost, *extremism can be a form of virtue signaling to one's group* that communicates the importance of group membership. From outside the group, we tend to view extremism with a fair bit of humor or horror, depending on the attitude or behavior in question, but extreme behaviors are interpreted very differently by fellow group members. For example, a report prepared for the Nuremberg Trials for Nazi war crimes (Alexander, 1948) found that the violent and sadistic behaviors of SS officers were often exhibitions of loyalty intended to display to other SS officers that the extreme individual was highly committed to the cause and a valuable group member. The same principle no doubt holds for countless other examples of genocide and intergroup violence, from the rape and slaughter of Chinese civilians in Nanjing by Japanese soldiers to the Rwandan genocide of Tutsis by Hutu civilians. Eyewitness accounts of these atrocities record not only numerous examples of reluctant participation in the slaughter but also numerous examples of wanton violence and barbarity (Buckley, 1997). These latter cases were not met with censure, but often provoked similarly extreme acts of barbarism by other group members, consistent with the idea that group members were in some sort of competition of extremism to demonstrate their value and righteousness to each other.

The intended audience of such virtue signaling is other members of one's own group, but extremism can also serve as a warning to those outside the group or as a signal to rival group members. In such cases, *extremism is more aptly considered a commitment device* that communicates to an adversary or outgroup that if they come into conflict with the extreme individual, the individual will subordinate all other goals to that of winning the conflict. In this sense, extremism serves as an honest signal of an individual's willingness to engage in further extreme behaviors. Anger has been theorized to play exactly this sort of role (Frank, 1988), as on the surface it seems like a misguided emotion that only gets people into unnecessary trouble. Indeed, anger does lead people to engage in all sorts of self-destructive and costly behaviors (Denson, DeWall, & Finkel, 2012; Zillmann, 1994). But because people know that individuals who are angry are likely to seek retribution, even when such behaviors are irrational, people go out of their way to avoid angering others and to mollify people who are angered. In this way, anger can lead people to act in a manner that is inconsistent with their self-interests in the moment, but nonetheless promotes their long-term self-interests (Burnham, 2007; Nowak, Page, & Sigmund, 2000).

By analogy, extremism works much like anger, as people will avoid provoking someone with a reputation for extremism, as such provocations are likely to be costly. Consider the assassination by the US military of Qassem Soleimani, the Major General of the Quds Force of Iran's Revolutionary Guard. Despite Soleimani's long-term support for terrorism and violence throughout the Middle East, previous US administrations chose not to assassinate him. An important reason for their prior reticence was to avoid retribution and the escalation of ongoing conflict (e.g., Panetta, 2020). Given that the United States' defense budget of $716 billion compares favorably to Iran's $6.3 billion (Global Fire Power, 2020), it is unlikely that previous US administrations feared losing a war with Iran. Rather, the possibility of extreme retaliation from Iranian forces, directed at military personnel or civilians, evidently stood as a substantial deterrent for such an assassination for many years.

Finally, because extremism involves subordinating all other goals to one dominant goal, *extremism can be a highly effective leadership strategy*. Extremists are, by definition, the people most committed to a particular goal. When that goal coincides with an important group goal or is adopted by the extremist because of his/her group membership, an extreme leader can be of enormous value to a group. Such a leader will sacrifice all else in service of that goal and hence can be relied on by group members to provide goal-oriented and group-serving leadership. Mahatma Gandhi is an example of such a leader, whose single-minded pursuit of equality and self-rule, often at great personal sacrifice, hastened India's independence from Great Britain.

Is Extremism Unique to Humans?

If extreme attitudes and behavior among humans were simply a product of genetics, we would expect other animals to be equally likely to display the single-minded pursuit of a particular goal. After all, our extension of Plomin's (2018) point regarding extremism applies equally well to nonhuman animals, in that extremism exists on a continuous genetic distribution for all species, suggesting that other animals are likely to have the same genetic proclivities toward extremism as humans. Nonetheless, there are many aspects of extremism that will be unique to human beings. First and foremost, humans are unique in the animal kingdom in the degree to which survival relies on learning of social rules, norms, and culture (Baumeister, 2005; Henrich, 2015; Tomasello, Kruger, & Ratner, 1993). Despite being born knowing almost nothing, humans have an enormous advantage over other animals in that our culture is cumulative, as our incredible communicative abilities allow us to pass on and then build upon the achievements of prior generations (Baumeister, 2005; Tennie, Call, & Tomasello, 2009;

Tomasello et al., 1993). This ability, in turn, ensures that our attitudes and proclivities are intertwined with learning and experience. A chimpanzee could be extremely violent or kind, but it could not be extremely religious or ideological. Nor could it hold extreme attitudes toward abstract concepts such as country, liberty, or respect. Humans can and often do.

Processes of social and ideological extremism in humans are enabled by the fact that our dependence on learning, norms, and culture also ensure that we are highly cognitively and attitudinally flexible. We can learn to love things that we are not biologically predisposed to love and to hate things that we are not biologically predisposed to hate. As a consequence, learning processes can combine with extreme experiences to create a particularly potent and unique form of human extremism. Rapid and cumulative cultural change then plays a critical role in the manifestation of extremism, and indeed whether it is perceived as extremism at all.

If you are fiercely devoted to an ideology that everyone in your culture shares, your extremism would largely pass unnoticed, as the status quo provides ongoing support for your social or ideological goal. For example, Berserkers were apparently regarded as exemplary Vikings despite the fact that they were prone to high levels of violence and rage (Dunbar, Clark, & Hurst, 1995; Fatur, 2019). Because violence and rage were their occupational requirements, a psychotically extreme Berserker would be largely undifferentiable from his comrades. In contrast, an individual inciting violence within a community of Quakers, known for their commitment to peace and harmony (Robson, 2010), would quickly be identified as an extreme outlier. Conversely, in the slaveholding South of pre-Civil War America, Quakers were viewed as extremists, by virtue of their unpopular views that condemned slavery and advocated for equality (Soderlund, 1985). Such views were in stark contrast to the majority of the South, where slavery was deemed a morally and socially acceptable enterprise (Fox-Genovese & Genovese, 2005).

The fact that extreme individuals can easily disappear into the extremeness of their culture (when their culture is motivationally imbalanced; Kruglanski et al., 2020) suggests that extremism is often more readily identified in heterogenous societies. In such societies, where there is substantial attitudinal and ideological pluralism, anyone who holds extreme positions is likely to stand out in comparison to their comrades, as there will always be others who do not hold such extreme views. In contrast, when one is a member of a homogenous society, one's extreme views will only be noticeable if they differ from those of everyone else, such as would be the case if a Berserker found himself in a community of Quakers.

As noted earlier, evolutionary pressures toward extremism—such as virtue signaling and leadership aspirations—are likely to lead people to hold more extreme versions of the views held by members of their own society. Thus, it is unlikely in homogenous societies that extremism will emerge in a direction that is counter to the attitudes of other group members. When counter-normative extremism does emerge in homogenous societies, it will be even more noticeable than it would be in a heterogenous society, but such occurrences will be exceedingly rare. In contrast, because heterogenous societies include a multiplicity of perspectives, people with a tendency toward extremism in such societies will often form extreme views that counter one another, and those positions will be readily noticed by the large majority of people who do not share their extremism. By way of example, consider the United States, a country that is high in heterogeneity and individualism (Hamamura, 2012). In such a culture, diversity of perspective and deviation from social norms and attitudes are met with relative tolerance (Gelfand, Lafree, Fahey, & Feinberg, 2013; Triandis, 2001), making extreme attitudes more likely to emerge. In contrast, in homogenous and collectivist cultures like Japan, swift and strict sanctions act to enforce the established social norms among those who are detected to deviate (Gelfand et al., 2013), thereby reducing instances of extremism that challenge these norms.

Cultural values also influence the directionality of extreme views. Collectivist cultures prioritize ingroup goals, interdependence, and harmony, so an individual holding extreme attitudes that clash with those of their ingroup would essentially be challenging the inherent values of their society, sparking disapproval (Hornsey, Jetten, McAuliffe, & Hogg, 2006; Triandis, 2001). That same individual in a society that promotes autonomy, independence, and freedom of expression, as in individualistic cultures (Triandis, 1995), would be perceived more leniently for their divergence of thought.

Similarly, at times of great societal change, when new ideas are being offered but many people do not (yet) follow them, extremism is likely to be more apparent. For example, in 1821, the British parliament debated whether to outlaw the abuse of working animals. Parliamentarians who opposed the new law retorted (in jest) that it would only be a matter of time before laws were offered to protect dogs and cats (Pinker, 2011). They were right; by the end of the 19th century, Britain had established laws to prevent animal cruelty, including the protection of cats and dogs. Such ideas are perceived as extreme during the early stages of societal change, but when they are successful, they are absorbed by the culture, thereby becoming mainstream and no longer demanding motivational imbalance from their adherents. This process is illustrated by the establishment of the world's first society for the protection of animals in a London coffee shop in 1824, which assuredly seemed extreme (and downright odd) at the time, but soon was entirely mainstream. Indeed, were someone today to propose live vivisection in their lecture hall or arrange bear-baiting for a friend's dinner party, these everyday activities of the past would be regarded as extreme. The forces that create extremism are largely the same across these widely varying situations, but the outcome is more identifiably extreme in one set of circumstances than in the other.

How Does Morality Attenuate and Exacerbate Extremism?

Morality, by which we mean a set of principles that guide thought and action by distinguishing between right and wrong, evolved as a system for internalizing and thereby coordinating cooperation norms (Boehm, 2012). Chimpanzees have some of the basic building blocks of morality, but they lack the capacity to self-reflect on the morality of their behavior (Boehm, 2012; Suddendorf, 2013). Rather chimpanzee behavior is guided by the benefits afforded by their situation or relationship and by hierarchy. This strategy is computationally efficient, as it enables them to dominate or relinquish a resource depending on their relative holding power and the costs and benefits of cooperation. But this strategy has important costs, as it disrupts group coordination, only rarely promotes cooperation, and indeed tends to enhance competition.

In contrast to our chimpanzee cousins, once our ancestors moved to the savannah and became highly interdependent, the evolution of morality enabled people to agree on a set of rules that facilitated harmonious and cooperative relationships and hence benefitted everyone (Boehm, 2012; Tomasello, 2016). In this sense, the evolution of morality enabled our ancestors to domesticate themselves, as they could now police their own behavior and that of their fellow group members to keep costly forms of extremism in check. The evolution of morality also enabled people to represent their own plans and actions in moral terms, which had two notable advantages. First, they could predict how others would likely respond to their behavior and hence could rely on internal standards to guide their actions rather than suffering the retribution of others. Second, they could attempt to couch their actions in moral terms, making them seem more appropriate than they might otherwise.

The evolution of morality might appear to be an unmitigatedly positive development for humanity, by making us less violent and more considerate of others. Indeed, the basic idea of the Golden Rule—due unto others as you would have them do unto you—is engraved into the moral system of nearly every society on earth (Gensler, 2013). But there is an important caveat that limits the scope of our moral reasoning: *morality evolved for coordination within groups, not between groups* (Boehm, 2012; Böhm, Thielmann, & Hilbig, 2018). Because members of other groups were often a mortal threat to our ancestors, we did not evolve moral concern for outgroup members and they do not have inherent moral standing (Choi & Bowles, 2007). As a consequence, our actions toward outgroup members need not be considerate or even peaceful. Indeed, moral reasoning can cause extreme behavior toward outgroups and can lead people who would not otherwise behave extremely to engage in such behaviors (Giner-Sorolla, Leidner, & Castano, 2012; Staub, 1990).

To begin with religions, which at least in their modern instantiations typically codify societies' moral systems (Alexander, 1987; Norenzayan et al., 2016;

Whitehouse et al., 2019), history is replete with examples of religious justification and even provocation for extreme behavior. In such cases, religion creates a reason to harm others that otherwise did not exist. Indeed, someone's worship of a god other than one's own can often be sufficient justification for extreme behavior, as religious moral codes frequently demand adherence to their particular doctrine. Consider the Qur'an, the religious text of Islam, which encourages the killing of infidels (individuals who do not follow Islamic faith): 'Slay them wherever ye find them, and drive them out of the places whence they drove you out, for persecution is worse than slaughter . . . slay them; such is the recompense of the unbelievers' (Qur'an 2:191).

Lest it seem that Islam is a uniquely violent religion, it is worth noting that religious instruction to persecute outgroup members reflects a common theme across the monotheistic religions. For example, the Torah describes a story in which Moses led a battle against the Midianite people, in which all the men were to be slain. As for the women and children, Moses proclaimed 'Kill every male among the little ones, and kill every woman that hath known man by lying with him. But all the women children, that have not known a man by lying with him, keep alive for yourselves' (31: 17–18).

It is possible that these ancient texts are not to be taken literally, but are just efforts to signal their group's strength and willingness to levy total destruction, but history suggests otherwise. Consider the Massacre of Beziers in 1209, part of the Crusade against Catharism (a sect of Christianity not recognized by the Catholic Church) initiated by Pope Innocent III. On the discovery that the city of Béziers contained Catholics mixed with Cathars, the Abbot of Cîteaux is said to have declared, 'Kill them all, let God sort them out' (2 Tim. ii. 19). The Crusaders did their best to follow this advice, slaughtering an estimated 20,000 men, women, and children when they breached the city walls.

Historical and textual examples such as these might give the impression that religions are uniquely violent and extreme moral systems, but unfortunately they are not. The Rwandan genocide, the Nazi Holocaust, the treatment of prisoners of war by the Japanese during World War II, the stolen generation of Australian Aboriginal and Torres Strait Islander people, and a long list of other atrocities invoked ideologies other than religious ideals, but nevertheless were based on strong moral convictions that justified their actions (Anderson, 2017; Bialas, 2013).

Such extreme moral convictions might also appear to be the result of an aberrant combination of individual personalities and historical forces, but they are probably more accurately perceived as the tail end of the distribution of everyday moral systems. Consider the fact that self-identified conservative and liberal Americans rarely show a desire to kill one another, but nonetheless display an equal and ready willingness to deny members of the other party their constitutional rights (Crawford, 2014; Crawford & Pilanski, 2014). In other words, everyday Americans often perceive their political conflicts in terms of moral imperatives that justify extreme responses.

Implications and Conclusions

In this chapter, we have argued that a variety of evolved predilections can lead to extremism. The underlying source of much of the pressure toward extremism is based in our evolved tribalism, which leaves outgroup members outside our 'moral circles' (Singer, 1981) and can even cause us to enlist our moral faculties in service of their destruction or exploitation (Böhm, Thielmann, & Hilbig, 2018). We have also argued that humans have a genetic tendency toward extremism that is completely independent of our tribalism. Because all complex traits are polygenic, or influenced by a large number of genes, a tendency toward extremism is more likely to emerge whenever people have a genetic makeup that is on the far end of the distribution with regard to the importance of their attitudes or ideology. In other words, an extreme position on a genetic distribution will underlie a tendency toward extreme attitudes and behaviors (cf. Plomin, 2018). As a consequence, every generation will have a small number of individuals who have inherited a strong proclivity for extremism, which may be expressed in wide variety of ways that may or may not have an intergroup component.

In combination with our evolved tribalism, the inevitable emergence of people with a proclivity for extremism in every generation might lead to pessimism regarding human violence and conflict. Pessimism might also be warranted simply by reading the news; countless examples from everyday life (and the lab; e.g., Giner-Sorolla et al., 2012) show that when conflicts emerge between groups, or between kin and strangers, clear differences in moral mattering emerge. These daily reminders of our moral limitations can make societal progress seem like a thin veneer over an evolved favoritism toward family and ingroup, with that favoritism rising to the surface to displace our post-Enlightenment ideals whenever the two come into any real conflict. Indeed, such concerns regarding the implications of evolutionary principles in psychology appear to underlie a great deal of the hesitation that many scholars show in accepting an evolutionary approach (Buss & von Hippel, 2018; von Hippel & Buss, 2017).

Nevertheless, one need only reflect on the ghastly history of human cruelty toward each other and toward other animals to realize that enormous progress has been made over the last few thousand years, few hundred years, and even few decades (Pinker, 2011, 2018). The mistreatment and lack of concern toward outgroups that emerges all too easily inside and outside the laboratory today is both qualitatively and quantitatively different from that of a few hundred years ago, when the killing and torture of humans and other animals was a common source of entertainment (Pinker, 2011). Thus, pessimism about the human capacity to live in harmony with each other and with nature represents a misunderstanding of the lessons of our evolutionary history.

If humans evolved tendencies toward extremism to deal with environmental regularities, such as hostile outgroup members, then removing these environmental triggers removes the tendency toward extremism as well. In the absence of hostile outgroups, there is no corresponding need for extremism. Similarly, if extremism is adaptive when it acts as a virtue signal, a commitment device, or an effective leadership strategy, then removing the need for virtue signaling, commitment devices, and extreme leaders removes the impetus for extremism. A brief perusal of the scope of virtue signaling on social media might suggest that such an argument is utopian, but the basic principle is sound. If extremism delivers adaptive benefits, then changes to the environment will dramatically change levels of extremism. Furthermore, even those tendencies toward extremism that are a product of extreme polygenic scores are highly sensitive to environmental contingencies, as attitudes and behaviors are influenced jointly by genes and environments (Plomin et al., 2016). The lesson of our evolutionary history is that the frequency and forms of extremism will change as a function of what predominant social norms punish and reward.

Because we live our lives over such brief periods of time, and because those brief windows in time do not seem so brief as we live them, it is easy to lose sight of the massive changes in society that have occurred over larger time scales. From the vantage point of our everyday lives, the effects of the environment often seem dwarfed by the effects of individual personalities or groups. Consider terrorism as an example. Fifty years ago, terrorism was a common political problem in Europe, and to the degree that it interfaced with religion, it was largely Christian in origin (Catholic vs. Protestant in Northern Ireland). For the past 20 years, terrorism in Europe has been primarily tied to radical forms of Islamic fundamentalism, making it easy to forget that this was not always the case. Twenty years seems like a long time—so long that our collective amnesia regarding Christian terrorism emerges despite the fact that terrorism in Europe in the 1970s took many more lives than it did in the 2010s (see www.start.umd.edu/gtd/). And our collective amnesia, in turn, ensures that we fail to notice that societal incentives for extreme behavior, including dominant ideologies, play the larger role in determining when and where extreme behavior will emerge.

If we zoom out a little further in time from modern Europe, the effect of societal incentive structures on extreme behavior is readily apparent. For example, the Vikings of Denmark, Sweden, and Norway were some of the most violent people on earth at the turn of the first millennium, but by the turn of the second millennium, they have become some of the most peaceful people. Their genes may have changed only a little (Krzewińska et al., 2015), but their way of making a living has changed a lot, as raiding has become a lot less lucrative than selling build-it-yourself furniture or really safe automobiles.

Along with changes in environmental contingencies, evolutionary forces also play an important role in moderating extremism. Perhaps the most important counter-pressure to violent extremism is our evolved cooperative nature. Altruism and cooperation tend to be positive-sum across the animal kingdom, as the benefits to the recipients typically outweigh the costs borne by the providers. The positive-sum nature of altruism and cooperation emerges in part not only due to strategic decisions made by those who offer help but also due to the decreasing marginal utility of each additional resource item. Help becomes mutually beneficial via reciprocal altruism (Trivers, 1971), which is essentially cooperation. Our human success story goes beyond most forms of reciprocal altruism elsewhere in the animal kingdom, as human helping is firmly based in widespread group-level cooperation (von Hippel, von Hippel, & Suddendorf, 2021).

Human helping and cooperation are no exception to the positive-sum rule, as both helpers and recipients agree that the costs to the helper are far outweighed by the benefits to the recipient (Ent, Sjåstad, von Hippel, & Baumeister, 2020). Indeed, human helping has the potential to be far more positive sum than the helping of other animals. Once our ancestors began to exploit the cognitive niche (Tooby & DeVore, 1987), which relies on information sharing as our primary strategy for survival, human cooperation achieved a positive-sum quality that is unattainable for other animals. The key to our highly effective form of cooperation is that information is immensely valuable to those who do not have it and easily transmitted by those who do. By way of example, if you stop to give someone directions, 1 minute of your time can save hours of theirs, as the value of the information received is orders of magnitude greater than the cost of providing it. This was the principle that made our ancestors so effective in colonizing the entire planet as they migrated out of Africa nearly 100,000 years ago. Henrich (2015) has aptly referred to the spread of cultural knowledge as 'the secret of our success', but the key to our cultural heritage lies in the immensely positive-sum quality that marks its transfer from one person to another.

Due to the strongly positive-sum quality of human helping and cooperation, groups of humans who are helpful and cooperative with one another easily outcompete groups of humans who are violent and extreme. Hominin cooperation originally evolved to make us more effective killers (von Hippel, 2018), so cooperation and extremism are not mutually exclusive, but extremism tends to have enormous costs that are not created by cooperation. In contrast to helping and cooperation, acts of violence and evil are strongly negative-sum, with the costs borne by the recipient far outweighing the gains to the perpetrator (Baumeister, 1997).[2] From this perspective, the decreasing violence that is evident across most of the world (Pinker, 2011) reflects in part the greater efficiency and productivity of groups and societies that are based on kindness and cooperation rather than extremism and violence. The mathematics of positive-sum interaction all but assure that societies that are cooperative and kind will eventually dominate those that are not. The meek may or may not inherit the Earth, but the cooperative and interdependent most certainly will.

Notes

1. *Hominin* is a term for modern humans and our extinct ancestors back to our common ancestor with chimpanzees. *Hominid* is a term for all existing and extinct great apes, including modern gorillas, orangutans, chimpanzees, and humans.
2. One apparent consequence of this asymmetry is that retribution is much more focused than positive forms of reciprocity. By the age of four, children directly punish those who harm them, but it is not until the age of seven that they start to return favors directly (Chernyak, Leimgruber, Dunham, Hu, & Blake, 2019). Until then, children are just as likely to pay it forward as they are to reciprocate to the helper.

References

Abelson, R. P. (1995). Attitude extremity. In R. E. Petty & J. A. Krosnick (Eds.), *Attitude strength: Antecedents and consequences* (pp. 25–41). Mahwah, NJ: Lawrence Erlbaum Associates, Inc.

Alexander, L. (1948). Sociopsychologic structure of the ss: Psychiatric report of the nurnberg trials for war crimes. *Archives of Neurology And Psychiatry, 59*(5), 622–634. https://doi.org/10.1001/archneurpsyc.1948.02300400058003

Alexander, R. (1987). *The biology of moral systems.* Piscataway, NJ: Transaction Publishers.

Anderson, K. (2017). 'Who was i to stop the killing?' Moral neutralization among rwandan genocide perpetrators. *Journal of Perpetrator Research, 1*(1).

Baumeister, R. F. (1996). *Evil: Inside human cruelty and violence.* New York: Henry Holt & Co.

Baumeister, R. F. (2005). *The cultural animal: Human nature, meaning, and social life.* Oxford: Oxford University Press.

Bialas, W. (2013). Nazi ethics: Perpetrators with a clear conscience. *Dapim: Studies on the Holocaust, 27*(1), 3–25. https://doi.org/10.1080/23256249.2013.812821

Boehm, C. (2012). *Moral origins: The evolution of virtue, altruism, and shame.* New York: Basic Books.

Böhm, R., Thielmann, I., & Hilbig, B. E. (2018). The brighter the light, the deeper the shadow: Morality also fuels aggression, conflict, and violence. *Behavioral and Brain Sciences, 41*, e98. doi:10.1017/S0140525X18000031

Boninger, D. S., Krosnick, J. A., Berent, M. K., & Fabrigar, L. R. (1995). The causes and consequences of attitude importance. *Attitude Strength: Antecedents and Consequences, 4*(7), 159–189.

Brewer, M. B. (1999). The psychology of prejudice: Ingroup love or outgroup hate? *Journal of Social Issues, 55*(3), 429–444. https://doi.org/10.1111/0022-4537.00126

Buckley, S. (1997, December). Rekindling the horror in rwanda. *The Washington Post National Weekly Edition*, 15.

Burnham, T. C. (2007). High-testosterone men reject low ultimatum game offers. *Proceedings of the Royal Society B: Biological Sciences, 274*(1623), 2327–2330.

Buss, D. M., & von Hippel, W. (2018). Psychological barriers to evolutionary psychology: Ideological bias and coalitional adaptations. *Archives of Scientific Psychology, 6*, 148–158.

Chernyak, N., Leimgruber, K. L., Dunham, Y. C., Hu, J., & Blake, P. R. (2019). Paying back people who harmed us but not people who helped us: Direct negative reciprocity precedes direct positive reciprocity in early development. *Psychological Science, 30*(9), 1273–1286.

Choi, J. K., & Bowles, S. (2007). The coevolution of parochial altruism and war. *Science, 318*(5850), 636–640.

Crawford, J. T. (2014). Ideological symmetries and asymmetries in political intolerance and prejudice toward political activist groups. *Journal of Experimental Social Psychology, 55*, 284–298. https://doi.org/10.1016/j.jesp.2014.08.002

Crawford, J. T., & Pilanski, J. M. (2014). Political intolerance, right and left. *Political Psychology, 35*(6), 841–851. https://doi.org/10.1111/j.1467-9221.2012.00926.x

De Waal, F., Macedo, S. E., & Ober, J. E. (2006). *Primates and philosophers: How morality evolved* (S. Macedo & J. Ober, Eds.). Princeton, NJ: Princeton University Press.

Denson, T. F., DeWall, C. N., & Finkel, E. J. (2012). Self-control and aggression. *Current Directions in Psychological Science, 21*(1), 20–25.

Domínguez-Rodrigo, M. (2002). Hunting and scavenging by early humans: The state of the debate. *Journal of World Prehistory, 16*, 1–54.

Dunbar, R. I. M., Clark, A., & Hurst, N. L. (1995). Conflict and cooperation among the vikings: Contingent behavioral decisions. *Ethology and Sociobiology, 16*(3), 233–246. https://doi.org/10.1016/0162-3095(95)00022-D

Ent, M. R., Sjåstad, H., von Hippel, W., & Baumeister, R. F. (2020). Helping behavior is non-zero-sum: Helper and recipient autobiographical accounts of help. *Evolution and Human Behavior.* https://doi.org/10.1016/j.evolhumbehav.2020.02.004

Fatur, K. (2019, August). Sagas of the solanaceae: Speculative ethnobotanical perspectives on the norse berserkers. *Journal of Ethnopharmacology, 244*, 112151. https://doi.org/10.1016/j.jep.2019.112151

Fox-Genovese, E., & Genovese, E. D. (2005). *The mind of the master class: History and faith in the southern slaveholders worldview.* New York: Cambridge University Press.

Frank, R. H. (1988). *Passions within reason: The strategic role of the emotions.* New York: W. W. Norton & Co.

Gelfand, M. J., Lafree, G., Fahey, S., & Feinberg, E. (2013). Culture and extremism. *Journal of Social Issues, 69*(3), 495–517. https://doi.org/10.1111/josi.12026

Gensler, H. J. (2013). *Ethics and the golden rule.* New York: Routledge.

Giner-Sorolla, R., Leidner, B., & Castano, E. (2012). Dehumanization, demonization, and morality shifting: Paths to moral certainty in extremist violence. In M. A. Hogg & D. L. Blaylock (Eds.), *Extremism and the psychology of uncertainty* (pp. 165–182). Hoboken, NJ: Wiley-Blackwell.

Global Fire Power. (2020). *2020 Military strength ranking.* Retrieved from www.globalfirepower.com/countries-listing.asp

Hamamura, T. (2012). Are cultures becoming individualistic? A cross-temporal comparison of individualism-collectivism in the united states and Japan. *Personality and Social Psychology Review, 16*(1), 3–24. https://doi.org/10.1177/1088868311411587

Henrich, J. (2015). *The secret of our success: How culture is driving human evolution, domesticating our species, and making us smarter.* Princeton, NJ: Princeton University Press.

Hewitt, J. K. (2012). Editorial policy on candidate gene association and candidate gene-by-environment interaction studies of complex traits. *Behavior Genetics, 42*(1), 1.

Hornsey, M. J., Jetten, J., McAuliffe, B. J., & Hogg, M. A. (2006). The impact of individualist and collectivist group norms on evaluations of dissenting group members.

Journal of Experimental Social Psychology, 42(1), 57–68. https://doi.org/10.1016/j.jesp.2005.01.006

Keeley, L. H. (1997). *War before civilization.* Oxford: Oxford University Press.

Klein, K. M., & Kruglanski, A. W. (2013). Commitment and extremism: A goal systemic analysis. *Journal of Social Issues, 69*(3), 419–435.

Krosnick, J. A., Boninger, D. S., Chuang, Y. C., Berent, M. K., & Carnot, C. G. (1993). Attitude strength: One construct or many related constructs? *Journal of Personality and Social Psychology, 65*(6), 1132.

Kruglanski, A. W., Fernandez, J. R., Factor, A. R., & Szumowska, E. (2019). Cognitive mechanisms in violent extremism. *Cognition, 188,* 116–123.

Kruglanski, A. W., Jasko, K., & Friston, K. (2020). All thinking is 'wishful' thinking. *Trends in Cognitive Sciences, 24*(6), 413–424. https://doi.org/10.1016/j.tics.2020.03.004

Krzewińska, M., Bjørnstad, G., Skoglund, P., Olason, P. I., Bill, J., Götherström, A., & Hagelberg, E. (2015). Mitochondrial DNA variation in the Viking age population of Norway. *Philosophical Transactions of the Royal Society B: Biological Sciences, 370*(1660), 20130384.

McDonald, M. M., Navarrete, C. D., & van Vugt, M. (2012). Evolution and the psychology of intergroup conflict: The male warrior hypothesis. *Philosophical Transactions of the Royal Society B: Biological Sciences, 367*(1589), 670–679. https://doi.org/10.1098/rstb.2011.0301

Mercier, H., & Sperber, D. (2011). Why do humans reason? Arguments for an argumentative theory. *Behavioral and Brain Sciences, 34*(2), 57–111. doi:10.1017/S0140525X10000968

Norenzayan, A., Shariff, A. F., Gervais, W. M., Willard, A. K., McNamara, R. A., Slingerland, E., & Henrich, J. (2016). The cultural evolution of prosocial religions. *Behavioral and Brain Sciences, 39.*

Nosek, B. A., Aarts, A. A., Anderson, J. E., Kappes, H. B., & Collaboration, O. S. (2015). Estimating the reproducibility of psychological science. *Science, 349*(6251).

Nowak, M. A., Page, K. M., & Sigmund, K. (2000). Fairness versus reason in the ultimatum game. *Science, 289*(5485), 1773–1775.

Panetta, G. (2020, January 4). Here's why neither George W. Bush nor Barack Obama killed Iranian commander Qassem Soleimani, who the US just took out in an airstrike. *Business Insider Australia.* Retrieved from www.businessinsider.com.au/why-neither-bush-or-obama-killed-iranian-general-qassem-soleimani-2020–1?r=US&IR=T

Perdue, C. W., Dovidio, J. F., Gurtman, M. B., & Tyler, R. B. (1990). Us and them: Social categorization and the process of intergroup bias. *Journal of Personality and Social Psychology, 59*(3), 475.

Petty, R. E., & Krosnick, J. A. (1995). *Attitude strength: Antecedents and consequences.* Mahwah, NJ: Lawrence Erlbaum Associates, Inc.

Pinker, S. (2011). *The better angels of our nature: Why violence has declined.* New York, NY: Penguin Group.

Pinker, S. (2018). *Enlightenment now: The case for reason, science, humanism, and progress.* New York, NY: Penguin.

Plomin, R. (2018). *Blueprint: How DNA makes us who we are.* Cambridge, MA: The MIT Press.

Plomin, R., DeFries, J. C., Knopik, V. S., & Neiderhiser, J. M. (2016). Top 10 replicated findings from behavioral genetics. *Perspectives on Psychological Science, 11*(1), 3–23.

Polderman, T. J., Benyamin, B., De Leeuw, C. A., Sullivan, P. F., Van Bochoven, A., Visscher, P. M., & Posthuma, D. (2015). Meta-analysis of the heritability of human traits based on fifty years of twin studies. *Nature Genetics, 47*(7), 702–709.

Robson, S. (2010). Grasping the nettle: Conflict and the quaker condition. *Quaker Studies*, *15*(1), 67–83. https://doi.org/10.3828/quaker.15.1.67

Rusch, H. (2014). The evolutionary interplay of intergroup conflict and altruism in humans: A review of parochial altruism theory and prospects for its extension. *Proceedings: Biological Sciences*, *281*(1794), 1–9. https://doi.org/www.jstor.org/stable/43601730

Schaller, M., & Neuberg, S. L. (2012). Danger, disease, and the nature of prejudice(s). In *Advances in experimental social psychology* (pp. 1–54). https://doi.org/10.1016/B978-0-12-394281-4.00001-5

Sherif, M., Harvey, O. J., White, B. J., Hood, W. R., & Sherif, C. W. (1961). *Intergroup conflict and cooperation: The robbers cave experiment*. Norman, OK: University Book Exchange.

Shipton, C. (2013). *A million years of hominin sociality and cognition: Acheulean bifaces in the Hunsgi-Baichbal Valley, India*. Oxford: Archaeopress.

Shipton, C., & Nielsen, M. (2015). Before cumulative culture. *Human Nature*, *26*, 331–345.

Singer, P. (1981). *The expanding circle: Ethics, evolution, and moral progress*. Princeton, Oxford: Princeton University Press.

Smith, R. L. (2008). *Premodern trade in world history*. Oxford: Routledge.

Soderlund, J. (1985). *Quakers and slavery: A divided spirit*. Princeton University Press. https://doi.org/10.2307/j.ctt7ztwkr

Staub, E. (1990). Moral exclusion, personal goal theory, and extreme destructiveness. *Journal of Social Issues*, *46*(1), 47–64. https://doi.org/10.1111/j.1540-4560.1990.tb00271.x

Suddendorf, T. (2013). *The gap: The science of what separates us from other animals*. New York: Basic Books.

Tennie, C., Call, J., & Tomasello, M. (2009). Ratcheting up the ratchet: On the evolution of cumulative culture. *Philosophical Transactions of the Royal Society B: Biological Sciences*, *364*, 2405–2415.

Tesser, A. (1993). The importance of heritability in psychological research: The case of attitudes. *Psychological Review*, *100*(1), 129.

Tomasello, M. (2016). *A natural history of human morality*. Cambridge, MA: Harvard University Press.

Tomasello, M., Kruger, A. C., & Ratner, H. H. (1993). Cultural learning. *Behavioral and Brain Sciences*, *16*(3), 495–511.

Tooby, J., & DeVore, I. (1987). The reconstruction of hominid evolution through strategic modeling. In W. G. Kinzey (Ed.), *The evolution of human behavior: Primate models*. Albany, NY: SUNY Press.

Triandis, H. C. (1995). *Individualism & collectivism*. New York: Routledge.

Triandis, H. C. (2001). Individualism-collectivism and personality. *Journal of Personality*, *69*(6), 907–924. https://doi.org/10.1111/1467-6494.696169

Trivers, R. L. (1971). The evolution of reciprocal altruism. *The Quarterly Review of Biology*, *46*(1), 35–57. https://doi.org/10.1086/406755

Trivers, R. L. (2011). *The folly of fools: The logic of deceit and self-deception in human life*. New York: Basic Books (AZ).

von Hippel, W. (2017). Evolutionary psychology and global security. *Science & Global Security*, *25*, 28–41.

von Hippel, W. (2018). *The social leap*. New York: HarperCollins.

von Hippel, W., & Buss, D. M. (2017). Do ideologically driven scientific agendas impede the understanding and acceptance of evolutionary principles in social psychology? In L. Jussim & J. Crawford (Eds.), *The politics of social psychology* (Frontiers in psychology). New York: Psychology Press.

von Hippel, W., & Trivers, R. (2011). The evolution and psychology of self-deception. *Behavioral and Brain Sciences*, *34*, 1–16.

von Hippel, W., von Hippel, F. A., & Suddendorf, T. (2021). Evolutionary foundations of social psychology. In P. Van Lange, E. T. Higgins, & A. Kruglanski (Eds.), *Social psychology: The handbook of basic principles*. New York: Guilford.

Whitehouse, H., François, P., Savage, P. E., Currie, T. E., Feeney, K. C., Cioni, E., ... Turchin, P. (2019). Complex societies precede moralizing gods throughout world history. *Nature*, *568*(7751), 226–229. https://doi.org/10.1038/s41586-019-1043-4

Young, A. I. (2019). Solving the missing heritability problem. *PLoS Genetics*, *15*(6), e1008222.

Zillmann, D. (1994). Cognition-excitation interdependencies in the escalation of anger and angry aggression. In M. Potegal & J. F. Knutson (Eds.), *The dynamics of aggression: Biological and social processes in dyads and groups*. Mahwah, NJ: Lawrence Erlbaum Associates, Inc.

PART 2
Motivational Imbalance Across Domains of Human Endeavor

7
THE PSYCHOLOGY OF EXTREME SPORTS

Eric Brymer and Pierre Bouchat

Introduction

Research into extreme sports has evolved rapidly over the last few decades perhaps in response to a significant uptake in participation rates, which seem to have overtaken growth rates in more traditional sports (American Sports Data, 2002; Pain & Pain, 2005). These changes suggest a sustained evolution (Puchan, 2004) arguably stemming from a need for more meaningful activities not constrained by artificial rules, regulations, or physical boundaries that serve only to sterilize and 'disconnect people from their human potential' (Brymer & Houge Mackenzie, 2017, p. 3).

Despite this development, the lack of common terminology and operational definitions has negatively impacted our understanding of the psychology of extreme sports (Lebeau & Sides, 2015). The term extreme sport has been used as a synonym for solo or team adventure experiences such as white-water rafting, mountaineering, skydiving and surfing, and activities requiring little knowledge or skill, such as bungee jumping. Furthermore, the psychological processes and outcomes linked to successful extreme sport participation are often assumed to be the same as nontraditional competitive youth sports, such as skateboarding and BMX (Brymer & Schweitzer, 2017).

These conceptual cross-pollinations have facilitated the development of imprecise definitions, models, and theories that are too general and/or do not reflective of the lived experience of participants (Cohen, Baluch, & Duffy, 2018). For example, researchers might assume that BASE jumping and commercial raft trips require similar psychological processes or that findings from a study of teenage skateboarders that identifies thrill-seeking as a primary motivation can be extrapolated to assume that BASE jumping is just about thrills.

DOI: 10.4324/9781003030898-10

In this chapter, we consider the most extreme of extreme sports defined as *independent adventure activities where a mismanaged mistake or accident is most likely to result in death* (Brymer, 2005). Typical activities include BASE and proximity flying, waterfall kayaking, big-wave surfing, extreme skiing, high-level mountaineering, and 'free solo' climbing. BASE jumpers use parachutes to safely land after jumping from solid structures (e.g., bridges, buildings, cliffs) that are only a few hundred feet from the ground (Bouchat & Brymer, 2019; Celsi, Rose, & Leigh, 1993; Soreide, Ellingsen, & Knutson, 2007). Proximity BASE jumpers wear winged suits to aid forward movement when jumping from similar solid structures (Holmbom, Brymer, & Schweitzer, 2017). Waterfall kayakers tackle waterfalls often higher than 30 meters and rated as either 'portages' or the most difficult to navigate on the international white-water grading system (Stookesberry, 2009). In extreme skiing, participants navigate sheer cliffs where a slip or fall would likely result in an uncontrollable tumble. Big-wave surfers ride waves over 20 feet tall (Warshaw, 2000), and extreme mountaineers often venture above the death zone (8,000 meters) where bodies are tested to their limits (Schneider, 2002). Free solo climbers ascend cliffs without a rope or other aids as protection (Perkovik & Rata, 2008).

Traditional Perspectives on the Psychology of Extreme Sports

Traditional assumptions on the psychology of extreme sports stemmed from a pathological model that presupposed participants were abnormal individuals. Psychologically, participation was assumed to stem from a 'death wish', desire for thrills and risk-taking, or some other inherent pathological trait not found in the general population. Until recently participants were most often portrayed as self-centered young males, 'fascinated with the individuality, risk, and danger of [extreme] sports' (Bennett, Henson, & Zhang, 2003, p. 98). Media and advertising representations have mirrored these presuppositions (Davidson, 2008; Pollay, 2001; Puchan, 2004; Rinehart, 2005). These traditional perspectives stemmed from several fundamental, but until recently untested, assumptions that participants in extreme sports must be pathological, unskilled risk-seekers, who search out opportunities to be seen by others as glamourous and exciting, why else would someone willingly undertake a leisure activity where death is a potential outcome? Researchers have even extrapolated these to examine assumptions that extreme sport participation is akin to pharmacological addiction or other socially deviant behavior (Heirene, Shearer, Roderique-Davies, & Mellalieu, 2016). Motivations for participation have most often been attributed to innate characteristics where nonparticipants label participants as 'crazy people' with 'deviant' traits that predisposed them to aberrant risk-taking behaviors, due to deep, unfulfilled psychological needs and/or adrenaline addictions (Brymer, 2002; Delle Fave, Bassi, & Massimini, 2003; Lambton, 2000; Olivier, 2006; Pizam,

Reichel, & Uriely, 2002; Rinehart, 2000; Self, Henry, Findley, & Reilly, 2007; Simon, 2002).

In recent years, extreme sports have developed into a worldwide phenomenon and, perhaps as a consequence, there has been an evolution of approaches to understanding extreme sports psychology (Brymer, Feletti, Monasterio, & Schweitzer, 2020). However, traditional frameworks and models seeking to explain these seemingly 'paradoxical' pursuits are still adhered to by many researchers (Brymer et al., 2020). Some of the dominant theories that are still used in popular circles to address extreme sport motivations, performance, and outcomes include sensation seeking (Rossi & Cereatti, 1993; Zuckerman, 2000), psychoanalysis (Hunt, 1995a), Type 'T' personality (Self et al., 2007), reversal theory (Apter, 1992; Kerr & Houge Mackenzie, 2012), and edgework (Lyng, 2008). The following sections briefly outline and critique these theoretical assumptions.

Sensation Seeking

Sensation-seeking theory posits that extreme sports participants choose extreme sports and perform effectively in extreme environments because of a personality trait that drives some individuals to seek out thrills and high novel sensations. Individuals are 'born with a general sensation-seeking motive', which means that they are more likely to seek opportunities for risk-taking, such as those found in extreme sports (Zuckerman, 1979). Sensation-seeking theory has been presented as a theory that explains why some individuals innately need to continually search for risky, complex, or novel experiences (Rossi & Cereatti, 1993; Schroth, 1995). Zuckerman defined sensation seeking as 'the seeking of varied, novel, complex and intense sensations and experiences, and the willingness to take physical, social, legal, and financial risks for the sake of such experiences' (Zuckerman, 1979, p. 27). This theory postulates that individuals with the sensation-seeking trait need higher arousal levels than nonsensation seekers to maintain an optimal level of stimulation (Zuckerman, 1984; Zuckerman, Bone, Neary, Mangelsdorff, & Brustman, 1972).

Sensation-seeking theory proposes four related subcategories: (a) experience seeking (ES), (b) thrill and adventure seeking (TAS), (c) boredom susceptibility (BS), and (d) disinhibition (DIS). The sensation-seeking scale (Zuckerman, Eysenck, & Eysenck, 1978) was developed to measure these traits and also to determine risk preference. An overall sensation-seeking score is also determined by summing the scores for each of the individual subscales (Schroth, 1995). The ES scale measures the preference for experiences that involve the mind or senses, travel, and a nonconforming lifestyle. The DIS subscale attempts to determine the desire for sexual and social disinhibition. The BS scale refers to an individual's aversion to monotony and is often associated with an individual's general experience of restlessness (Zuckerman, 1984). The TAS scale was designed to measure preferences for sports and other physical activities involving speed or danger. The TAS scale is the one most often associated with extreme sport participation.

In the early days, sensation-seeking theory was probably most associated with the explication of activities claimed as extreme sports (Schrader & Wann, 1999; Zarevski, Marusic, Zolotic, Bunjevac, & Vukosav, 1998). While the sensation-seeking notion still seems to be attractive to researchers as a framework to guide studies (e.g., Frenkel et al., 2019), two major factors often impact how researchers interpret results. The first is that diverse definitions of what constitutes an extreme sport has meant that studies can often not be compared and, in some cases, studies claiming to examine extreme sports miss the point entirely. Many of the sports in these studies do not meet the definition highlighted earlier and those that broadly fit this definition often have inconclusive findings. The second issue is that traditional notions of what an extreme sport is supposed to be have created a priori assumptions that participation must be linked to risk-taking. Both these issues highlight flaws in defining extreme sports and also ignore many of the broader variables associated with extreme sport participation (Buckley, 2018; Cohen et al., 2018). A close examination of many of these studies indicates that the personality assumptions do not stand up to scrutiny as participants do not fit neatly into a homogenous sensation-seeking group (Monasterio & Brymer, in press). As Monasterio and colleagues concluded from their study on BASE jumpers:

> the large standard deviations across the measures of temperament and character indicate that there isn't a tightly defined personality profile among BASE jumpers and that factors other than personality, such as opportunity and social and cultural influences, contribute to engagement in the sport.
> *(Monasterio, Mulder, Frampton, & Mei-Dan, 2012, p. 398)*

Further indications that personality more broadly is not sufficient to explain participant choices include Goma's (1991) study that utilized the sensation-seeking scale, among other scales, to compare alpinists ($n = 27$), mountaineers ($n = 72$), general sportspeople ($n = 221$), and those who were categorized as not participating in any 'risky activity' ($n = 54$). Alpinists were labeled as extreme sport participants because death during this activity was a possibility. A close examination of the sensation-seeking scores suggests no significant difference between Alpinists and either the mountaineer group or the general sport group. The mountaineer group recorded high sensation-seeking scores for both TAS and ES when compared to the general sport group. Overall, the results suggest that extreme sports, as defined in this chapter, do not necessarily attract participants high in sensation-seeking traits. Furthermore, as the mountaineering group returned sensation-seeking scores significantly higher in TAS and ES than the general sports group, it may also be that if sensation seeking has any value when examining extreme sport participants, there may be a ceiling effect. This is supported by a study carried out by Slanger and Rudestam (1997) who found no significant differences among extreme-, high-, or low-risk sporting groups. They concluded that even if sensation-seeking theory was broadly useful at the very least, it was not suitable for examining extreme sports participants as sensation seeking was not capable of differentiating between extreme sports and nonextreme sports.

In summary, sensation-seeking theory might be useful to explain some patterns of behavior, but, as Buckley (2018) points out, when used to examine motivations for activities such as extreme sports participation the arguments tend to be circular and therefore not able to account for the motivations of extreme athletes. The mixed results of sensation-seeking research suggest that participation in extreme sports may be motivated by a range of factors in addition to, or separate from, sensation seeking.

Edgework

Edgework theory was developed to take into account the social psychology of voluntary risk-taking across various contexts including crime, self-harm, violent behavior, gaming, and sport (Lyng, 2005) and attempts to provide a framework to understand an individual's desire to explore the limits (or edge) of personal control (Lois, 2001). Edgework theory proposes that voluntary risk-taking might be in response to limitations in modern societies and voluntary risk-taking activities provide opportunities for physical mastery, personal control, and self-efficacy. Individuals who might be considered edgeworkers 'experience intense highs' (p. 3) and greater control over their lives, immediately following challenging psychological and physical events. Edgework activities are those that involve a 'clearly observable threat to one's physical or mental well-being or one's sense of an ordered existence' (Lyng, 1990, p. 857). The edge in edgework theory denotes a conceptual line between opposing states such as those that might be found in an extreme sport context; life and death, chaos and order, consciousness and unconsciousness (Lyng, 2005).

Research using edgework theory in adventure and outdoor contexts indicates that a primary motivation for participation was the capacity to negotiate various 'edges' (e.g., edge of competency/control). In these contexts, edgeworkers push personal physical and mental capacities to determine 'performance limits' (Lyng, 1990). As individuals develop skills and capacities, the edge is pushed further to maintain the desired sensations and feelings (Lyng, 1990; Stranger, 1999). Reports from participants deemed to be edgeworkers suggest that the experience is characterized by self-actualization, altered perceptions, sense of being at one with key objects, distortion of time, and the inability to fully articulate the experience (Fletcher, 2018; Lyng, 1990). Conceptually, voluntary risk-takers in an extreme sport context use a highly tuned skillset developed over many years to flirt with death and navigate the challenges of extreme sport participation (Tofler, Hyatt, & Tofler, 2018). While research utilizing edgework in an extreme sports context is scarce, this has not stopped theorists extrapolating edgework theory to extreme sport contexts, and assuming motivations for extreme sport participation are the same as those for stock-trading, unprotected sex, and sadomasochism (Lyng, 2005). A major issue with the edgework perspective in an extreme sport context is that even those using an edgework framework recognize that participants would rather walk away than walk the edge (Brymer & Schweitzer, 2017; Tofler et al., 2018). Almost two decades ago, Celsi et al. (1993) recognized that extreme sport participants did not push the edge of their control. Rather, they preferred to stop or postpone an activity if they felt that they were overextending their physical or psychological limits. Edgework theory has added to our understanding of extreme sports by adding sociological factors and emphasizing positive outcomes; however, it also turns out to be an oversimplification of extreme sport psychology.

Reversal Theory

Reversal theory (Apter, 1992, 2001) has been used to examine extreme sport psychology and the psychology of extreme sport athletes after injury (Pain & Kerr, 2004). Like the sensation seeking theory, reversal theory was not designed exclusively to examine extreme sports or even sports more broadly. Instead, reversal theory is a general theoretical model of motivation. Reversal theories posit that metamotivational states or frames of mind are involved in how a person interprets their motives at any given time. According to Apter (1992), a metamotivational state can be likened to a proverbial pair of rose-colored glasses where an experience is perceived in relation to a dominant metamotivational state. Broadly, reversal theory suggests that a participant can only enjoy dangerous extreme sports if the arousal that comes with participation is experienced as enjoyable. The main state that has been linked to extreme sports is the paratelic state that facilitates a feeling of safety even when the context is dangerous. Changes in an individual's mood, motivations, and emotional experiences are instigated via regular alternations, or *reversals*, between opposing metamotivational states. The majority of reversal theory research into extreme sport has focused on the telic/paratelic states. In the extreme sport context, these states are considered relevant because they help understand why people interpret intense emotional arousal in different ways such as when the same event can trigger excitement in one person but anxiety in another.

The *telic state* is characterized by seriousness arousal avoidance, whereas the *paratelic state* is characterized by spontaneity and arousal-seeking (Frey, 1999). In the telic state, activity is associated with future focus and an important end beyond the present moment (Apter, 1982). While low arousal is relaxing and pleasant, above a certain point, arousal triggers anxiety or fear. The paratelic state is direct, and activities are more in the moment and pursued as ends within themselves. High arousal is exciting and low arousal is boring. Individuals who spend more time in serious, telic states are 'telic dominant', whereas individuals who tend to operate in playful, paratelic states are 'paratelic dominant'. Reversals between the telic and paratelic states are dependent on so-called protective frames (Apter, 1993, p. 28), which provide feelings of protection from the presence of danger (e.g., due to confidence in skill level or equipment). When the protective frame is active (in the paratelic state), heightened arousal and challenge is experienced as exciting; when the protective frame is lacking (in the telic state), heightened arousal is experienced as anxiety. Although reversal theory is more state than trait-based, individuals are thought to have dominant states that they tend to reverse into more often than others. For example, investigations of paratelic dominance among athletes have supported the validity of this model. Studies by Kerr and colleagues (Kerr, 1991) generally support the hypothesis that participants who regularly participate in sports deemed risky are paratelic dominant. For example, surfers, sailboarders, motor-cycle racers, and skydivers demonstrated significantly lower levels of arousal avoidance compared to marathon runners, weight trainers, or the general public (Kerr, 1991). Low arousal avoidance (i.e., paratelic dominance) among skydivers, rock and mountain climbers, and deep-sea divers ($n = 72$) was also found in the study by Shoham, Rose, and Kahle (2000). However, as with sensation seeking, definitional issues exist in these studies, and there was no attempt to examine sports that could be deemed extreme. As such, extrapolating findings to extreme sport populations as defined in this chapter is likely inappropriate. These studies also raise another important issue. Due to the nature of the processes undertaken, it is impossible to determine whether participants chose to participate in extreme sports because of the characteristics highlighted by reversal theory or whether they learned to enjoy sensations that accompany extreme sports as a result of participation. Reasons for participation may not be as simple as suggested by a single theory, especially one that was not developed to understand extreme sports and might mask other more appropriate motivations such as goal achievement, connection to the natural environment, social motives, and pleasurable kinesthetic sensations such as those felt when moving through water or air (Brymer & Gray, 2010a, 2010b; Brymer & Schweitzer, 2017; Kerr & Houge Mackenzie, 2012).

Other Psychological Theories Employing the Risk Narrative

Other theoretical frameworks that have, in previous years, been employed to examine extreme sports also follow the risk narrative include Type 'T' personality theory and psychoanalysis. Type T personality theory assumes that extreme sports are a type of deviant behavior where participation leads to subcultures. The theory suggests that a person with a personality labeled type 'T' will search out thrills, risks, arousal, and novel sensations. Extreme sports are one of many such outlets. On the other hand, those with 't' personalities avoid thrill and risk (Farley, 1991). The type T framework recognizes that risk-taking behaviors can be both destructive (e.g., destruction, crime) and constructive (e.g., creativity, invention). Extreme sport participation is assumed to be a positive outlet (Self et al., 2007). However, while theorists and media outlets have made these connections, very little research on extreme sport populations as defined here has been undertaken.

One other approach that was previously used to examine extreme sports stems from case studies on particular extreme sport participants who have demonstrated dangerous behaviors within their sport. Psychoanalytic theory assumes pathological desires drive extreme sport participation, particularly a subelement known as the 'death wish' (Pain & Pain, 2005). Examples of earlier employment of psychoanalysis include Hunt's work on deep-sea diving. Hunt (1995a) extrapolated findings from her work with a particular male deep-sea diver to develop a psychoanalytic understanding of participation in what she termed 'risk sports'. Hunt reasoned that participation in dangerous sports stem from pathological concerns in everyday life, particularly concerns about masculinity and bisexuality, aggressive fantasies, and lack of power that might stem from childhood experiences (Hunt, 1993, 1995a, 1995b, 1996, 1998): 'the more risky and violent the sport, the more likely do issues of bisexuality, masculinity, aggression, and sadomasochism appear to influence an individual's sport participation' (Hunt, 1996, p. 620). However, Hunt (1996) also noted that the link between pathological concerns and extreme sport participation might not be straight forward as individuals react differently to childhood patterns. Hunt's findings might only apply to some male individuals, and how those individuals behave while participating. Using psychoanalysis in this way to explain extreme sport participation, in general, is tenuous at best (Brymer & Schweitzer, 2017).

In summary, the previous sections outlined some of the main psychological theories used to explain and understand extreme sports, whether in terms of motivation, process, or outcomes. These theories all stem from a traditional understanding of extreme sport psychology that assumes an inherent link to risk or thrill seeking. However, as Farley (1991) recognized almost three decades ago, these frameworks might be limited:

> too much energy is spent trying to pathologize? those that are not like us. I sometimes think psychologists see too much pathology out there . . . To the contrary, these are people who are pushing the envelope and that's their life. They would not want the life of someone who never pushes the envelope. To them, that is an unlived life.
>
> *(cited in Terwilliger, 1998, p. 4)*

These traditional perspectives might be limited not only because these assumptions have potentially masked more accurate appreciations but also because there has been an underlying assumption that participation must be abnormal and pathological.

Beyond the Risk-Taking Narrative

Risk-focused perspectives on extreme sports that attempt to explain motivation in terms of desire for risk or performance as the capacity to enjoy risk are overly simplistic and most likely based on naïve nonparticipant viewpoints (Brymer & Schweitzer, 2017; Milovanovic, 2005). Accounts of participation driven by the risk narrative mostly follow deficit models of behavior. Research indicates that these models do not reflect participant reports of their lived experience (Brymer, 2005, 2009, 2010; Brymer, Downey, & Gray, 2009; Brymer & Oades, 2009; Brymer & Schweitzer, 2013b; Brymer & Schweitzer, 2012; Celsi et al., 1993; Willig, 2008). This was recognized even in the early days of extreme sport research with, for example, reports that extreme sport participants are very well prepared and do not look for opportunities to participate beyond their capabilities, preferring instead to defer participation to a later date if they felt the limits of their control were being overextended (Celsi et al., 1993). Extreme athletes conduct extensive planning, spend considerable effort developing profound knowledge of the environmental and task variables of their particular activity, and work hard to hone skills (Pain & Pain, 2005). Epidemiological studies also suggest that extreme sports might be less dangerous that many socially accepted activities, such as motorcycle riding (Brymer & Feletti, in press; Soreide et al., 2007; Storry, 2003). The risk-taking narrative is at the very least overemphasized:

> Despite the public's perception, extreme sports demand perpetual care, high degrees of training and preparation, and, above all, discipline and control. Most of those involved are well aware of their strengths and limitations in the face of clear dangers. Findings of extensive research in climbers suggest that individuals do not want to put their lives in danger by going beyond personal capabilities.
>
> *(Pain & Pain, 2005, p. 34)*

The myopic focus on the risk narrative cannot explain activity choice, for example, why a person chooses skiing or climbing above surfing or BASE jumping, which are most often purposeful choices requiring years of skill development (Brymer et al., 2020; Brymer & Schweitzer, 2017; Celsi et al., 1993; Ogilvie, 1974). Extreme sports require considerable training, expertise, and planning. If risk-taking were the sole aim of these activities, it is questionable whether participants would spend years preparing to ensure relative 'safety' before undertaking their chosen pursuit (Ogilvie, 1974). Research suggests it can take years to plan a BASE jump (Swann & Singleman, 2007) and 14 years to plan one expedition (Muir, 2003). The idea that participants are somehow deceiving themselves while searching for thrills and uncertainty does not match more contemporary research findings (Brymer et al., 2020). The traditional risk focus narrative is more likely a response to the modern 'pathological' aversion to risk and the obsessive desire to be liberated from uncertainty. The risk narrative that has swamped extreme sport research has also meant a focus on negative outcomes. However there are many problems with this approach, including (a) contemporary research is finding characteristics and statistics that do not reflect traditional assumptions of risk motivations (e.g., Celsi et al., 1993; Soreide et al., 2007; Storry, 2003); (b) the focus on risk has largely obscured other aspects of the extreme sport experience (e.g., Brymer et al., 2009; Brymer & Oades, 2009; Kerr & Houge Mackenzie, 2012; Willig, 2008); and (c) traditional theory-driven perspectives often do not match the lived experiences of participants (Brymer & Oades, 2009; Holmbom et al., 2017; Willig, 2008). In the following sections, we highlight contemporary approaches to understanding motivations, processes, and outcomes of extreme sport participation.

Positive Psychology Perspectives

Several relevant positive psychology perspectives stem from research into the lived experience of those who participate in extreme environments. These perspectives do not support the risk-focused narrative.

Flow and Peak Experience

Flow and peak experience are positive psychology notions that describe related lived experiences (Privette, 1983). Both flow and peak experience describe positive subjective experiences that have striking similarities to research findings on the lived experience of extreme sport participants. The peak experience framework describes sensations and moments of intense joy, and the flow framework focuses on the intrinsically rewarding immersion experiences. According to Csikszentmihalyi (1988, 2000), flow denotes experiences that are so intensely enjoyable that people search for ways to repeat them, for their own sake. Flow states are found across a range of activities, such as music and surgery, not just extreme sports. Early studies investigating rock climbing found a range of positive experiences (see Table 7.1) (Csikszentmihalyi, 1975). Over the years, the flow model has evolved to include nine dimensions: challenge-skill balance, merging of action and awareness, clear goals, unambiguous feedback, concentration on the task at hand, paradox of control, loss of self-consciousness, transformation of time, and autotelic experience (Jackson, 1996). Flow theory has also been developed to include 'Deep Flow' and 'Deep Play' found in activities such as climbing (Csikszentmihalyi, 1975; p. 74). Other terms associated with flow include 'being in the zone' (Young & Pain, 1999) and 'fun' (Jackson, 1995, 1996; Kimiecik & Harris, 1996).

Peak experiences, while also enjoyable, include highly impactful, sometimes mystical, experiences (Maslow, 1977). Peak experience events are often assumed to be more commonly encountered by people considered to be self-actualized

TABLE 7.1 Flow in Rock Climbing as Compared to Normative Life Experiences (Csikszentmihalyi, 2000, pp. 96–97).

Normative Life	Rock Climbing Life
Informational noise: distraction and confusion of attention	One-pointedness of mind
Nebulosity of limits, demands, motivation, decisions, feedbacks	Clarity and manageability of limits, demands, decisions, feedbacks
Severing of action and awareness	Merging of action and awareness
Hidden, unpredictable dangers: unmanageable fears	Obvious danger subject to evaluation and control
Anxiety, worry, confusion	Happiness, health, vision
Slavery to the clock; life lived in spurts	Time out of time: timelessness
Carrot-and-stick preoccupation with exotelic, extrinsic material and social reward; orientation toward ends	Process orientation; concern for autotelic, intrinsic rewards; conquest of the useless
Dualism of mind and body	Integration of mind and body
Lack of self-understanding; false self-consciousness; wars between the selves	Understanding of the true self, self-integration
Miscommunication with others; masks, statuses, and roles in an inegalitarian order; false independence or misplaced dependency	Direct and immediate communication with others in an egalitarian order; true and welcomed dependency on others
Confusion about man's place in nature or the universe; isolation from the natural order; destruction of the earth	Sense of man's place in the universe; oneness with nature; congruence of psychological and environmental ecology
Superficiality of concerns; thinness of meaning in the flatland	Dimension of depth 'up there'; encounter with ultimate concerns

(Maslow, 1977). For example, Lipscombe (1999) found that veteran skydivers reported many peak experience characteristics (see Table 7.2). For Lipscombe, skydiving provided an opportunity to consistently invite peak experiences characterized by 'acute well-being, peace, calm and stillness, detachment, uniqueness, freedom, floating, flying and weightlessness, ecstasy, being in the present, immersed in the moment, immortality, unity, altered perceptions of time and space, self-validation, and awareness of other' (Lipscombe, 1999, p. 269).

Research into the lived experience of participating in extreme sports has found that participants report very similar experiences. For example, studies by Brymer and Schweitzer (2012, 2013a, 2017) found that participants reported extraordinary experiences characterized by transcendence, ineffability, freedom, and time slowing down. Extreme sport participation has changed the perceived relationship between humanity and the natural environment by enhancing a sense of integration and connection (Brymer et al., 2009; Brymer & Gray, 2010a, 2010b). Recent studies indicate that the link between flow and participation in extreme environments is more complex. Rather than achieving a single flow state, participants may experience a range of flow states. These states are dependent on attentional focus and goals and linked to varying arousal levels, perceived challenge,

TABLE 7.2 Maslow's 19 Peak Experience Characteristics (Lipscombe, 1999, p. 270).

Characterizations	Meaning
Experience/object unification	Total harmony
Total attention★	Complete absorption in the experience
Nature of the object in itself	Feeling of insignificance
Rich perception★	Lost in the experience
Awe, reverence of the experience★	The most blissful moment, ecstasy
Unity of the world	Feeling the world is unified
Abstract perception	Transcend the present situation
Fusion of dichotomies★	The person and the experience merge
Feeling Godlike	Fullest potential/total control
Nonclassifying perception	A new kind of viewing
Ego transcendence★	They are the activity
Self-justifying moment	The experience as an end in itself
No consciousness of time and space	Lack of spatio-temporal consciousness
Experience is intrinsically perfect★	Everything is perfect, beautiful, lasting
Awareness of the absolute	The ultimate truth is experienced
Effortlessness	No conscious deliberation in executing skills
Loss of fear	Momentary loss of psychological defenses
Unique being of the individual	Experiences the totality of one's unique self
Fusion of the individual★	Feeling integrated or together

★Observed in veteran skydivers, by Lipscombe (1999).

skill levels, and activity phases (Houge Mackenzie, Hodge, & Boyes, 2011, 2013). Using reversal theory, Houge Mackenzie and colleagues (2011, 2013) indicated that participants experience *telic flow* (a serious, outcome-oriented state) and *paratelic flow* (a playful, process-oriented state). Conceptually, these flow states are qualitatively distinct experiences. For example, participants most often described telic flow for situations of high challenge, where lowering arousal levels was the aim, and enjoyment was linked to goal attainment, rather than excitement or thrills. Paratelic flow represents the experience of enjoyment and thrill from a heightened challenge and high arousal. Houge Mackenzie and colleagues found that the telic flow was reported more frequently than the paratelic flow, suggesting that the flow was a more appropriate motivator than sensation seeking or thrill.

Combined, peak experience and flow seem to encompass a considerable chunk of the extreme sport experience. However, research has also found that extreme sport athletes report characteristics that seem unrelated. For example, Brymer (2009) reported that participants experience lasting transformations that are arguably in contrast to the more fleeting peak experience notion (Lipscombe, 1999; Ravizza, 1977). Peak experience and flow theory cannot account for the reports of high expectations of repeated sensations, which contrasts with the idea that peak experiences are very rare, or once-in-a-lifetime, occurrences (Maslow, 1971). However, experiences such as these have been reported in transpersonal psychology as extraordinary and transcendent experiences.

Extraordinary and Transcendent Experiences

An 'extraordinary experience' is an unusual event involving 'high levels of emotional intensity and experience' (Arnould & Price, 1993, p. 25). Arnould and Price (1993) found that adventure activities such as river rafting triggered positive extraordinary experiences characterized by 'absorption and integration, personal control, joy and valuing, and spontaneous letting-be of the process' (Arnould & Price, 1993, p. 39). In this study, feelings of fear and reduced safety developed *communitas* (a sense of intimacy that results from a rite of passage type experience), personal growth, and a sense of communion with nature. Arnould and colleagues (1999) also found that consumers and guides alike experienced 'transcendence and transformation' (p. 40) during white-water rafting trips. The authors described these events as 'spiritual' states. White (1993) argued that exceptional experiences such as these foster a sense of community and can reconnect people with their environments and inner self.

Research over the past 25 years substantiates these claims (Brymer et al., 2009; Brymer & Gray, 2010a, 2010b; Brymer & Schweitzer, 2013a, 2013b). Murphy and White (1995) categorized extraordinary experiences into three varieties: (a) mystical sensations (acute well-being, peace, calm, freedom, floating or flying, being in the present, instinctive action, surrender, and unity); (b) altered perceptions (time perception and awareness of alterity); and (c) extraordinary feats (exceptional energy). These transcendent or exceptional experiences might be related to life-threatening or near-death events (White, 1993). Several additional studies support these findings that extreme sports facilitate transcendence through altered perceptions of time and a sense of being at one with the surroundings. Brymer (2005) reported that many participants perceived altered perceptions such as time 'slowing down', which allowed for a greater perception of the environment:

> on every BASE jump, you will experience something interesting in that your awareness of one second expands enormously. So, what we would normally perceive in one second is very little compared to what you perceive in one second on a BASE jump. When you're doing it, it feels like it's in slow motion, whereas when you watch it back on footage you look and go 'whoa,' that's over in a blip clicks fingers. But when you're doing it you know you can see the tiny little creases in the rock and different colors in the sky, and you're totally aware of where your body is in space and how its moving and. . . . It's very surreal.
>
> *(Geoff)*

When people participate in extreme sports, their perceptions of reality seem to change. The extreme sport experience seems to be a medium for altered perceptions and an enhanced state of sensual awareness. Extreme sports appear to open up perceptions and possibilities, as opposed to narrowing them as in tunnel vision, which is often assumed when these activities are associated with adrenalin (Brymer, 2005).

In summary, far from the traditional risk-focused assumptions, extreme sport experiences appear to facilitate more positive psychological experiences and even lead to altered states of consciousness such as changes in time perception and increased sensory awareness. These extraordinary experiences are both transcendent and transforming. For extreme sport participants, the opportunity to transcend the everyday experience into the realm of the sacred often provides far more motivation and inspiration than experiencing short-term thrills through risk-taking.

The psychology of extreme sports and sports undertaken in extreme environments is more complex than traditional theories propose. Motivations for continuing to participate are more profound and transformational. Motivations for starting are varied and might involve focused decisions to develop skills or an affiliation with traditional perspectives (Brymer & Schweitzer, 2017). While motivation change for those who decide to continue participating (Brymer et al., 2009), it may still be the case that some initial decisions to try extreme sports are based on the assumed extreme nature of the sport and the desire of novice athletes to experience the extreme. This brings us to the last point of this discussion, which explores the links between extreme sports and the psychology of extremism. Recently, Kruglanski and colleagues have developed an integrative model of extremism based on the concept of motivational imbalance (Kruglanski, Szumowska, Kopetz, Vallerand, & Pierro, 2020). Drawing on strong theoretical bases and on the results of numerous empirical studies, the authors propose that extreme behaviors are the result of a motivational imbalance, wherein a given need overrides other basic concerns. This motivational imbalance has a series of motivational, cognitive, affective, behavioral, and social consequences, all presumably proportionate to impact on the imbalance (Kruglanski et al., 2020). This psychological model of extremism deepens the understanding of a set of human practices referred to as extremes: extreme diets, addictions, violent extremism, as well as extreme sports. Regarding extreme sports, the model proposed by Kruglanski and colleagues provides a much clearer understanding of the risk-taking motivation that might be behind the behaviors of a large number of extreme sportsmen and women and novice athletes who desire to try extreme sports. However, in our opinion, one of the main merits of the work of Kruglanski and colleagues lies in two other aspects.

First, their model helps us to distinguish between 'extreme sports' and 'extreme practices' of these sports. Indeed, until recently, the use of the term 'extreme' has blurred our understanding of the processes at work when practicing these sports. Or to put it another way, the question of terminology has perhaps overshadowed the appreciation of processes at work when practicing an extreme sport. Indeed, for many researchers working on extreme sports, these processes were reduced to extreme practices instead of distinguishing the type of (extreme) sport from the types of practices (extreme/extremist and 'nonextreme').

This psychological approach to extremism and the new approaches to extreme sports might help understand why some extreme sportsmen and women behave in an extremist way (e.g., sacrifice everything to their passion, take huge risks for the thrill of it), and this approach also supports the idea that that a majority of extreme sport athletes do behave in a balanced way, far different from extremism (e.g., very long and graduate progression, participants concerned by the balance between their family life and their passion). Extreme sports are therefore not necessarily the domain of extremists but extremists might use extreme sports as an outlet.

Second, the psychological approach to extremism leads us to question the label we attribute to 'extreme' sports. Indeed, this label seems to blur the understanding of the processes involved in the practice of these sports. Moreover, new perspectives in sports research show that not all sports labeled as extreme are extreme as defined in this chapter. Conceptual work may therefore be needed to improve the quality and relevance of research in this area.

Finally, just as research on the psychology of extremism sheds light on some aspects of extreme sports, we believe that recent research on extreme sports allows us to bring a series of nuances to other aspects of our lives (e.g., practices related to religion, food, work) that may seem extreme but are not caused by a motivational imbalance. For this reason, extreme sports and their less extreme cousins could be vital public health interventions to rebalance motivations for those intent on extreme behavior.

References

American Sports Data. (2002, August 1). "Generation Y" drives increasingly popular "extreme" sports. *Sector Analysis Report*. Retrieved from www.americansportsdata.com/pr-extremeactionsports.asp

Apter, M. J. (1982). *The experience of motivation: The theory of psychological reversals*. London: Academic Press.

Apter, M. J. (1992). *The dangerous edge: The psychology of excitement*. New York: Free Press.

Apter, M. J. (1993). Phenomenological frames and the paradoxes of experience. In J. H. Kerr, S. J. Murgatroyd, & M. J. Apter (Eds.), *Advances in reversal theory* (pp. 27–39). Amsterdam: Swets & Zeitlinger.

Apter, M. J. (2001). *Motivational styles in everyday life: A guide to reversal theory*. Washington, DC: American Psychological Association.

Arnould, E. J., & Price, L. L. (1993, June). River magic: Extraordinary experience and the extended service encounter. *Journal of Consumer Research, 20*(1), 24–45. doi: 10.1086/209331

Arnould, E. J., Price, L. L., & Otnes, C. (1999). Making magic consumption: A study of white-water river rafting. *Journal of Contemporary Ethnography, 28*(1), 33–68. doi: 10.1177/089124199129023361

Bennett, G., Henson, R. K., & Zhang, J. (2003). Generation Y's perceptions of the action sports industry segment. *Journal of Sport Management, 17*(2), 95–115.

Bouchat, P., & Brymer, E. (2019). BASE jumping fatalities between 2007 and 2017: Main causes of accidents and recommendations for safety. *Wilderness & Environmental Medicine, 30*(4), 407–411.

Brymer, E. (2002, October 25). *Extreme sports: Theorising participation—a challenge for phenomenology.* Paper presented at the ORIC research symposium, University of Technology, Sydney.

Brymer, E. (2005). *Extreme dude: A phenomenological exploration into the extreme sport experience* (Doctoral Dissertation), University of Wollongong, Wollongong. Retrieved from http://ro.uow.edu.au/theses/379

Brymer, E. (2009). Extreme sports as a facilitator of ecocentricity and positive life changes. *World Leisure Journal, 51*(1), 47–53.

Brymer, E. (2010). Risk and extreme sports: A phenomenological perspective. *Annals of Leisure Research, 13*(1&2), 218–239.

Brymer, E., Downey, G., & Gray, T. (2009). Extreme sports as a precursor to environmental sustainability. *Journal of Sport and Tourism, 14*(2–3), 1–12.

Brymer, E., & Feletti, F. (in press). Beyond risk: The importance of adventure in the life of young people. *Annals of Leisure Research.* Retrieved from www.tandfonline.com/doi/abs/10.1080/11745398.2019.1659837?journalCode=ranz20

Brymer, E., Feletti, F., Monasterio, E., & Schweitzer, R. (2020). Editorial: Understanding extreme sports: A psychological perspective. *Frontiers in Psychology, 10*, 3029. doi:10.3389/fpsyg.2019.03029

Brymer, E., & Gray, T. (2010a). Dancing with nature: Rhythm and harmony in extreme sport participation. *Adventure Education & Outdoor Learning, 9*(2), 135–149.

Brymer, E., & Gray, T. (2010b). Developing an intimate "relationship" with nature through extreme sports participation. *Loisir, 34*(4), 361–374.

Brymer, E., & Houge Mackenzie, S. (2017). Psychology and the extreme sport experience. In F. Feletti (Ed.), *Extreme sports medicine.* Cham: Springer

Brymer, E., & Oades, L. (2009). Extreme sports: A positive transformation in courage and humility. *Journal of Humanistic Psychology, 49*(1), 114–126.

Brymer, E., & Schwietzer, R. (2012). Fear is good for your health: A phenomenological exploration of extreme sports. *Journal of Health Psychology, 18*(4), 477–487.

Brymer, E., & Schweitzer, R. (2013a). Extreme sports are good for your health: A phenomenological understanding of fear and anxiety in extreme sport. *Journal of Health Psychology, 18*(4), 477–487. doi:10.1177/1359105312446770

Brymer, E., & Schweitzer, R. (2013b). The search for freedom in extreme sports: A phenomenological exploration. *Psychology of Sport and Exercise, 14*(6), 865–873. doi:10.1016/j.psychsport.2013.07.004

Brymer, E., & Schweitzer, R. (2017). *Phenomenology and the extreme sports experience.* New York: Routledge.

Buckley, R. C. (2018). To analyze thrill, define extreme sports. *Frontiers in Psychology, 9*, 1216. doi:10.3389/fpsyg.2018.01216

Celsi, R. L., Rose, R. L., & Leigh, T. W. (1993). An exploration of high-risk leisure consumption through skydiving. *Journal of Consumer Research, 20*, 1–23.

Cohen, R., Baluch, B., & Duffy, L. J. (2018). Defining extreme sport: Conceptions and misconceptions. *Frontiers in Psychology, 9*, 1974. doi:10.3389/fpsyg.2018.01974

Csikszentmihalyi, M. (1975). *Beyond boredom and anxiety.* Washington, DC: Jossey-Bass Publishers.

Csikszentmihalyi, M. (1988). The flow experience and its significance for human psychology. In M. Csikszentmihalyi & I. Csikszentmihalyi (Eds.), *Optimal experience: Psychological studies of flow in consciousness* (pp. 15–35). New York: Cambridge University Press.

Csikszentmihalyi, M. (2000). *Beyond boredom and anxiety: Experiencing flow in work and play* (25th Anniversary ed.). San Francisco: Jossey-Bass.

Davidson, L. (2008). Tragedy in the adventure playground: Media representations of mountaineering accidents in New Zealand. *Leisure Studies, 27*(1), 3–19.

Delle Fave, A., Bassi, M., & Massimini, F. (2003). Quality of experience and risk perception in high-altitude climbing. *Journal of Applied Sport Psychology, 15*, 82–98.

Farley, F. (1991). The type-T personality. In L. Lipsitt & L. Mitnick (Eds.), *Self-regulatory behavior and risk taking: Causes and consequences*. Norwood, NJ: Ablex Publishers.

Fletcher, D. (2018). Psychological resilience and adversarial growth in sport and performance. In *Oxford Research Encyclopedia of Psychology*.

Frenkel, M. O., Laborde, S., Rummel, J., Giessing, L., Kasperk, C., Plessner, H., . . . Strahler, J. (2019). Heidelberg risk sport-specific stress test: A paradigm to investigate the risk sport-specific psycho-physiological arousal. *Frontiers in Psychology, 10*, 2249. doi:10.3389/fpsyg.2019.02249

Frey, K. P. (1999). Reversal theory: Basic concepts. In J. H. Kerr (Ed.), *Experiencing sport: Reversal Theory* (pp. 3–17). New York: John Wiley & Sons.

Goma, M. (1991). Personality profiles of subjects engaged in high physical risk sports. *Personality and Individual Differences, 12*(10), 1087–1093.

Heirene, R. M., Shearer, D., Roderique-Davies, G., & Mellalieu, S. D. (2016). Addiction in extreme sports: An exploration of withdrawal states in rock climbers. *Journal of Behavioral Addictions, 5*(2), 332–341. doi:10.1556/2006.5.2016.039

Holmbom, M., Brymer, E., & Schweitzer, E. (2017). Transformations through proximity flying: A phenomenological investigation. *Frontiers in Psychology, 8*, 1831. https://doi.org/10.3389/fpsyg.2017.01831

Houge Mackenzie, S., Hodge, K., & Boyes, M. (2011). Expanding the flow model in adventure activities: A reversal theory perspective. *Journal of Leisure Research, 43*(4), 519–544, doi:0.1080/00222216.2011.11950248

Houge Mackenzie, S., Hodge, K., & Boyes, M. (2013). The multiphasic and dynamic nature of flow in adventure experiences. *Journal of Leisure Research, 45*(2), 214–232. doi:10.18666/jlr-2013-v45-i2-3012

Hunt, J. C. (1993). Straightening out the bends. *AquaCorps, 5*, 16–23.

Hunt, J. C. (1995a). Divers' accounts of normal risk. *Symbolic Interaction, 18*(4), 439–462.

Hunt, J. C. (1995b). *The neutralisation of risk among deep scuba divers*. Unpublished manuscript.

Hunt, J. C. (1996). Diving the wreck: Risk and injury in sport scuba diving. *Psychoanalytic Quarterly, LXV*, 591–622.

Hunt, J. C. (1998, November 16). Why some of us don't mind being in deep water. *The Star-Ledger*.

Jackson, S. A. (1995). Factors influencing the occurrence of flow states in elite athletes. *Journal of Applied Sport Psychology, 7*, 138–166.

Jackson, S. A. (1996). Toward a conceptual understanding of the flow experience in elite athletes. *Research Quarterly for Exercise and Sport, 67*(1), 76–90.

Kerr, J. H. (1991). Arousal-seeking in risk sport participants. *Personality and Individual Differences, 12*(6), 613–616.

Kerr, J. H., & Houge Mackenzie, S. (2012). Multiple motives for participating in adventure sports. *Psychology of Sport and Exercise, 13*(5), 649–657.

Kimiecik, J. C., & Harris, A. T. (1996). What is enjoyment: A concepetual/definitional analysis with implications for sport and exercise psychology. *Journal of Sport and Exercise Psychology*, *18*(3), 247–263.

Kruglanski, A. W., Szumowska, E., Kopetz, C. H., Vallerand, R. J., & Pierro, A. (2020). On the psychology of extremism: How motivational imbalance breeds intemperance. *Psychological Review*. Advance online publication.

Lambton, D. (2000, September). Extreme sports flex their muscles. *Allsport* (SB49), 19–22.

Lebeau, J. C., & Sides, R. (2015). Beyond the mainstream versus extreme dichotomy: A cyclical perspective on extreme sports. *Sport in Society*, *18*(6), 627–635. doi:10.1080/17430437.2014.982540

Lipscombe, N. (1999). The relevance of the peak experience to continued skydiving participation: A qualitative approach to assessing motivations. *Leisure Studies*, *18*(4), 267–288. doi:10.1080/026143699374853

Lois, J. (2001). Peaks and valleys: The gendered emotional culture of edgework. *Gender & Society*, *15*(3), 381–406. doi:10.1177/089124301015003004

Lyng, S. (1990). Edgework: A social psychological analysis of voluntary risk taking. *American Journal of Sociology*, *95*(4), 851–886. doi:10.1086/229379

Lyng, S. (2005). Sociology at the edge: Social theory and voluntary risk taking. In S. Lyng (Ed.), *Edgework: The sociology of risk-taking* (pp. 17–49). Abingdon: Routledge.

Lyng, S. (2008). Risk-taking in sport: Edgework and the reflexive community. In M. Atkinson & K. Young (Eds.), *Tribal play: Subcultural journeys through sport* (pp. 83–109). Bingley: Emerald.

Maslow, A. H. (1971). *The farther reaches of human nature*. Arkana/Penguin Books.

Maslow, A. H. (1977). *Religions, values and peak-experiences*. Harmondsworth: Penguin Books.

Milovanovic, D. (2005). Edgework: A subjective and structural model of negotiating boundaries. In S. Lyng (Ed.), *Edgework: The sociology of risk-taking* (pp. 51–88). Abingdon: Routledge.

Monasterio, E., & Brymer, E. (in press). Feeding time at the zoo: Psychological aspects of a serious rock climbing accident. *Journal of Adventure Education & Outdoor Learning*. doi: 10.1080/14729679.2020.1829494

Monasterio, E., Mulder, R., Frampton, C., & Mei-Dan, O. (2012). Personality characteristics of BASE jumpers. *Journal of Applied Sport Psychology*, *24*(4), 391–400.

Muir, J. (2003). *Alone across Australia*. Camberwell: Penguin Books Australia.

Murphy, M., & White, R. (1978). *The psychic side of sport*. Reading, MA: Addison-Wesley.

Ogilvie, B. C. (1974). The sweet psychic jolt of danger. *Psychology Today*, *8*(5), 88–94.

Olivier, S. (2006). Moral dilemmas of participation in dangerous leisure activities. *Leisure Studies*, *25*(1), 95–109.

Pain, M., & Kerr, J. H. (2004). Extreme risk taker who wants to continue taking part in high risk sports after serious injury. *British Journal of Sports Medicine*, *38*(3), 337–339. doi:10.1136/bjsm.2002.003111

Pain, M. T. G., & Pain, M. A. (2005). Essay: Risk taking in sport. *The Lancet*, *366*(1), S33–S34. doi:10.1016/S0140-6736(05)67838-5

Perkovik, A., & Rata, G. (2008). Notes on the language of adventure tourism. *Journal of Linguistic Studies*, *1*(1), 71–78.

Pizam, A., Reichel, A., & Uriely, N. (2002). Sensation seeking and tourist behavior. *Journal of Hospitality & Leisure Marketing*, *9*(3 & 4), 17–33.

Pollay, R. W. (2001). Export "A" ads are extremely expert, eh? *Tobacco Control*, *10*, 71–74.

Privette, G. (1983). Peak experience, peak performance, and flow: A comparitive analysis of positive human experiences. *Journal of Personality and Social Psychology*, *45*(6), 1361–1368.

Puchan, H. (2004). Living 'extreme': Adventure sports, media and commercialisation. *Journal of Communication Management*, *9*(2), 171–178.

Ravizza, K. (1977). Peak experiences in sport. *Journal of Humanistic Psychology*, *17*(4), 35–40. doi:10.1177/002216787701700404

Rinehart, R. (2000). Emerging arriving sports: Alternatives to formal sports. In J. Coakley & E. Dunning (Eds.), *Handbook of sports studies* (pp. 501–520). London: Sage.

Rinehart, R. (2005). "BABES" & BOARDS: Opportunities in new millennium sport? *Journal of Sport & Social Issues*, *29*, 232–255.

Rossi, B., & Cereatti, L. (1993). The sensation seeking in mountain athletes as assessed by Zuckerman's sensation seeking scale. *International Journal of Sport Psychology*, *24*, 417–431.

Schneider, M. (2002). Time at high altitude: Experiencing time on the roof of the world. *Time & Society*, *11*(1), 141–146.

Schrader, M. P., & Wann, D. L. (1999). High-risk recreation: The relationship between participant characteristics and degree of involvement. *Journal of Sport Behaviour*, *22*(3), 426–431.

Schroth, M. L. (1995). A comparison of sensation seeking among different groups of athletes and nonathletes. *Personality and Individual Differences*, *18*(2), 219–222.

Self, D. R., Henry, E. D., Findley, C. S., & Reilly, E. (2007). Thrill seeking: The type T personality and extreme sports. *International Journal of Sport Management and Marketing*, *2*(1–2), 175–190.

Shoham, A., Rose, G. M., & Kahle, L. R. (2000). Practitioners of risky sports: A quantitative examination. *Journal of Business Research*, *47*(3), 237–251.

Simon, J. (2002). Taking risks: Extreme sports and the embrace of risk in advanced liberal societies. In T. Baker & J. Simon (Eds.), *Embracing risk: The changing culture of insurance and responsibility* (pp. 177–208). Chicago: University of Chicago Press.

Slanger, E., & Rudestam, K. E. (1997). Motivation and disinhibition in high risk sports: Sensation seeking and self-efficacy. *Journal of Research in Personality*, *31*(3), 355–374. doi:10.1006/jrpe.1997.2193

Soreide, K., Ellingsen, C., & Knutson, V. (2007). How dangerous is BASE jumping? An analysis of adverse events in 20,850 jumps from the Kjerag Massif, Norway. *Journal of Trauma-Injury Infection & Critical Care*, *62*(5), 1113–1117.

Stookesberry, B. (2009). New waterfall world record. *Canoe & Kayak*. Retrieved from http://canoekayak.com/whitewater/kayaking/new-waterfall-record/

Storry, T. (2003). The games outdoor adventurers play. In B. Humberstone, H. Brown, & K. Richards (Eds.), *Whose journeys? The outdoors and adventure as social and cultural phenomena* (pp. 201–228). Penrith: The Institute for Outdoor Learning.

Stranger, M. (1999). The aesthetics of risk: A study of surfing. *International Review for the Sociology of Sport*, *34*(3), 265–276. doi:10.1177/101269099034003003

Swann, H., & Singleman, G. (2007). *Baseclimb 3 — The full story*. Retrieved from www.baseclimb.com/About_BASEClimb3.htm

Terwilliger, C. (1998). Type 'T' personality. *Denver Post*, *1*(4), 5–52.

Tofler, I. R., Hyatt, B. M., & Tofler, D. S. (2018). Psychiatric aspects of extreme sports: Three case studies. *The Permanente Journal*, *22*(17), 071. doi:10.7812/TPP/17-071. PMID: 29401052; PMCID: PMC5798939.

Warshaw, M. (2000). *Maverick's: The story of big-wave surfing*. San Francisco: Chronicle Books.

White, R. A. (1993). Working classification of EHE's. *Exceptional Human Experience: Background Papers: I, 11*, 149–150.

Willig, C. (2008). A phenomenological investigation of the experience of taking part in 'Extreme Sports'. *Journal of Health Psychology, 13*(5), 690–702.

Young, J., & Pain, M. D. (1999). The Zone an empirical study. *The Online Journal of Sport Psychology, 1*(3), 1–8.

Zarevski, P., Marusic, I., Zolotic, S., Bunjevac, T., & Vukosav, Z. (1998). Contribution of Arnett's inventory of sensation seeking and Zuckerman's sensation seeking scale to the differentiation of athletes engaged in high and low risk sports. *Personality and Individual Differences, 25*, 763–768.

Zuckerman, M. (1979). *Sensation seeking: Beyond the Optimal Level of Arousal*. Hillsdale, NJ: Erlbaum.

Zuckerman, M. (1984). Experience and desire: A new format for sensation seeking scales. *Journal of Behavioural Assessment, 6*(2), 101–114.

Zuckerman, M. (2000, November–December). Are you a risk taker. *Psychology Today*, 130.

Zuckerman, M., Bone, R. N., Neary, R., Mangelsdorff, D., & Brustman, B. (1972). What is the sensation seeker? Personality trait and experience correlates of the sensation seeking scales. *Journal of Consulting and Clinical Psychology, 39*(2), 308–321.

Zuckerman, M., Eysenck, S. B., & Eysenck, H. J. (1978). Sensation seeking in England and America: Cross-cultural, age, and sex comparisons. *Journal of Consulting and Clinical Psychology, 46*(1), 139–149. doi:10.1037/0022-006X.46.1.139

8
THE PSYCHOLOGY OF GREED

Katalin Takacs Haynes

Introduction to Greed Narratives

Greed has been a popular, sometimes even central topic of public discourse for millennia. For over 6,000 years, numerous thinkers, philosophers, authors of treatises, politicians, economists, ethicists, and religious and civic leaders as well as the greater public from as diverse locales as Ancient Egypt, Greece and Rome, China, Europe and the Americas, opined on this important social construct. Today, a search for the term 'greed' in Nexis Uni® yields over 500,000 hits from the news alone since 1888, likely one of the first mentions of greed in the printed news. The majority of the hundreds of thousands of mentions of greed is from more recent times, spanning the decades between 1989 and 2020. The 2008 Great Financial Crisis seems to have spurred a frenzy of activity in the news and opinions published about greed, causing on average 38,000 articles, blogposts, opinion pieces to appear each year, in contrast to the approximately 10,000 similar pieces per year in the years 1989 to 2008. The significant bump in the numbers that began around 2008 and has lasted until the current day, indicating an increased interest in greed likely tied to the Great Financial Crisis. The popular press covers greed in a wide range of industries, such as entertainment and the arts, media and telecommunications, banking, finance, professional services, retail and wholesale, travel hospitality and tourism, as well as in the not-for-profit sector, such as education, the government, and public administration. Why is greed such an attractive topic for the general public?

Perhaps people are fascinated by the idealized version of the power inherent in greed, recognized the world over from Gordon Gecko's famous monologue in the 1987 movie 'Wall Street'. Gordon Gecko's words inform shareholders of Teldar Paper that greed is good and righteous because it drives forward humanity's evolutionary spirit. Oliver Stone's intent for the viewers was to realize the absurdity and excess in the 1980s corporate raider's values and behavior. After all, Gordon Gecko's restructuring efforts at Blue Star created no value for customers nor secured jobs for employees. Breaking up the company was necessary only to extract value. Instead of being seen universally as the embodiment of a corporate villain however, Gordon Gecko became Wall Street's folk hero (Winter, 2007). The break signaled a rift between the domain of financial economics and the rest of the world. Stanley Wiser and Oliver Stone who co-wrote the screenplay for the movie *Wall Street* (Stone, 1987) claim to have based the fictional Gordon Gecko on several real-life financiers and corporate raiders, among them Carl Icahn, Ivan Boetsky, and Michael Ovitz. Real-life Gordon Geckos do exist outside Hollywood works of fiction, some of them in jail, some retired and sitting on their billions, and others yet, still amassing extraordinary wealth at significant cost to investors, their companies, employees, and society at large. What do 21st-century real-life 'greedy bastards' (Seuntjens, Zeelenberg, van de Ven, & Breugelmans, 2019) look like?

One type is the convict, like Bernie Madoff whose Ponzi scheme evaporated $60Bn of investor funds while he enriched himself and his family, or Dennis Kozlowski, former CEO of Tyco, who stole $600M from his own company to enrich himself and his family; they certainly qualify as individuals in pursuit of excessive and undeserved material wealth with no regard to others. However, greed also occurs within the bounds of the law. If greed is defined as extreme self-interest, manifesting in excessive pursuit of undeserved and extreme material wealth, with no concern for harm done to others, top managers walking away with $3.3Bn in bonuses and other compensation from the 25 largest bankruptcies between 1999 and 2001, at a cost of 100,000 jobs and $210 billion of shareholder wealth also appear as a form of greed (Wang, Malhotra, & Murnighan, 2011). The $66 billion payout to executives and directors of 1,035 companies that lost 75% of the value of their shares, or the $50 million average payout from insider stock sales received by 466 executives during the same time period seem to be in the same category (Gimein, 2002; Wang et al., 2011). The systemic problem of runaway acquisitiveness and hoarding of wealth was noticed even by Alan Greenspan, former chair of the Federal Reserve, who testified in 2002 before the Senate Committee on Banking, Housing, and Urban Affairs that, 'an infectious greed seemed to grip much of our business community', explaining that 'it is not that humans have become any more greedy than in generations past. It is that the avenues to express greed had grown so enormously'. In 2009, President Obama noted, 'the attitude that's prevailed from Washington to Wall Street to Detroit for too long; an attitude that valued wealth over work, selfishness over sacrifice, and greed over responsibility'. (Wang et al., 2011).

Yet, in spite of the powerful words by economic and political leaders and the public's interest in greed, a curious and somewhat troubling phenomenon exists in our scientific community. Namely, academic research in disciplines intuitively most proximal and adequate to understanding greed is scarce. Few research articles have been published about greed in top academic journals in business, organizational behavior, social psychology, and anthropology. Searches on Psychinfo result in 69 academic articles on the subject of greed during the 1980–2020 time period from various branches of the psychological sciences, including some that are about greed in name only. The ABI Inform database that covers multiple business disciplines, including management, finance, economics and accounting, returns 36 studies, with fewer than ten published in the past decade in top peer-reviewed journals. As in the case of the psychology studies, many of the listed references use the term greed casually or cursorily.

What is at the source of the discrepancy between the popularity of greed among members of the public and the lack of interest among academics, especially in business fields? Greed has been around for a long time and has even been labeled the oldest social construct (Haynes, Campbell, & Hitt, 2017; Balot, 2001). It has garnered a lot of attention around the world and across time and disciplines. If one asks someone in the street whether they know what greed is, or if they could recognize it, most people would answer in the affirmative. In addition, most individuals have strong views about the morality of greed, many considering it bad, others good, yet others denying its existence.

In the social sciences, theoretical foundations of various disciplines provide a variety of explanations for our cognitions, behaviors, morals, ethics, and emotions. Different theoretical approaches might explain why different fields treat and evaluate greed differently. Oka and Kuijt (2014) note that the terms greed and excess are difficult to define for general application to human behaviors and carry 'significant moral, social, political and cultural weight' (p. 30), often as synonyms of sin and vice, or as the source of social destruction. In other contexts, however, the same terms are hailed as virtues that elicit rewards and are the primary driver of social good. Indeed, Oka and Kuijt (2014) point out that both greed and excess are context dependent, both as core concepts that have accompanied human existence in every major culture for millennia. Thus, perception of greed as good, bad, or neutral depends on the socio-cultural, psychological, moral educational makeup of the individual evaluating the greed phenomenon.

In the rest of this chapter, in an effort to find answer to some of the questions posed earlier, I review the definition and characteristics of greed, including its extreme nature, contextualized by a brief historical overview of the social construct. An interdisciplinary review of the literature on greed incorporating theories and findings from psychology, anthropology, economics, and management highlights reasons why different disciplines interpret and value greed in vastly different ways. In the last section, I forward a multilevel model of greed as extremism. This chapter concludes with offering possible solutions to mitigate the problems caused by greed.

What Is Greed and Why Is It Extreme?

Merriam Webster's (2020) definition of the term 'greed' as 'a selfish and excessive desire for more of something (such as money) than is needed' points out selfishness as a necessary component. Wikipedia (2020) defines the term more broadly, noting that greed, '(or avarice) is an uncontrolled longing for increase in the acquisition or use: of material gain, be it food, land or animals/inanimate possessions); or social value, such as status, or power'. Collins Dictionary (2020) writes that greed is an 'excessive desire for getting or having especially wealth' and that greed is a 'desire for more than one needs or deserves'. Other definitions include 'excessive or rapacious desire, especially for wealth or possessions' (Collins, 2020). Synonyms of greed include terms like avarice, voracity, insatiability, ravenousness, and cupidity.

In aggregate, the definitions point to greed being an extreme or intense desire for and pursuit of material wealth, which is self-centered, and often unearned, or undeserved. Some interpretations also include power, food, or lust as the object of greed; however, in its most common form, greed is motivated by money and material wealth. Greed is generally seen as an individual-level construct, in which the actor is motivated to act in their own self-interest in an extreme fashion. When successful, the extreme and often opportunistic acquisitiveness leads to accumulation of wealth that comes at a cost to the community or society. Social, economic, or environmental harm is the cost born by those from whom the greedy person takes their 'undeserved' wealth. As such, the discussion of greed must extend to deserved versus undeserved wealth. High income earned by an author, a scientist, an artist, a doctor, or even a CEO does not necessarily mean they are greedy. Indeed, in most cases, such highly talented individuals contribute to society, the arts, the economy, and to science and are fairly, maybe even generously rewarded for their contribution. Thus, their income and wealth is earned and likely deserved. They exchange their talent for compensation while contributing to the greater good, and this exchange is balanced.

Conversely, what a greedy individual takes for themselves exceeds the contributions they make to the greater good. Greedy pursuits and actions are misaligned with the needs of society, serve only the needs of the individual, and cost others. As such, greed represents a realized motivational imbalance where the extreme desire for material wealth (or other physical or immaterial items) is unconstrained or unmoderated by other needs. Greed is extreme self-interest, running rampant, superseding the greedy person's other needs, as well as the needs of other individuals and groups. Greed is a state of intrapersonal imbalance that creates a state of imbalance in the social environment of the greedy individual. Such imbalances have significant social consequences such as economic inequality, which both creates and perpetuates racial, ethnic, and educational inequality, which all contribute to a divided society.

Kruglanski and colleagues (2017) discuss motivated deviance from general behavioral norms as a form of extremism that stems from a shift from a 'balanced satisfaction of basic human needs afforded by moderation to a motivational imbalance wherein a given need dominates the others' (p. 217). The conceptualization of extremism fits the definition of greed, where one need, the desire for excessive material wealth overrides concern for the needs of others, and in some cases, even other physical and psychological needs of the individual. Under conditions of moderation, the need to fulfill one's own self-interest is counterbalanced by concern for others, in the form of altruism, as well as the recognition that one has had enough, that is, that one is satiated. In the case of greed, however, satiation never arrives, and the individual continues to take for themselves, without regard for harm to other needs of their own or the needs of others.

The reason for this might be that the rewards of greed for material wealth reinforce the greedy behavior, creating a perpetual cycle of greed. What are the rewards of greed for the greedy individual? First and foremost, greed leads to increased material wealth. With material wealth comes power. These are rewards for the individual, as experienced by the individual. Greed, however, is also something that members of society observe and, depending on societal values, judge favorably or condemn. In general, the United States, a highly individualistic country (Hofstede, 2011; Schwartz, 1992, 2007), acting on one's own self-interest through the accumulation of wealth has been accepted, normalized, and even desirable. Individuals who pursue material wealth at any cost, which they then turn into power, are often revered, seen as leaders. Adjectives associated with such individuals include, 'aggressive', 'go-getter', 'smart', 'unstoppable', etc. Thus, in a society that values materialism, those who are greedy are rewarded intrinsically by their increased wealth and power, and extrinsically, by society's admiration. In countries that value equality, however, an internal conflict might develop in the greedy individual, as the rewards of their greed clash with society's evaluation of their actions.

Two basic characteristics of extremism are intensity and infrequency. The conceptualization of greed has always reflected an intense, excessive desire and pursuit, underscoring the appropriateness of viewing it through the lens of extremism. However, greed's perception as extreme also carries a value judgment, because greed is generally seen as either extremely good or extremely bad. During the course of history, different cultures have attached different narratives to the desirability and the accumulation of wealth. In some cultures, philosophers, scholars, and authors have presented greed as an extreme endeavor benefitting only the individual, and an endeavor that comes at great cost to the individual's psychological well-being as well as the community they harm. Elsewhere, such as in the fields of finance and economics, parallel narratives in celebration of greed emerged, recognizing greed as the driving principle of economic growth (Wang & Murnighan, 2011). Greed, while always seen as extreme, appears to have swung like a pendulum, depending on whether a given society focused on its beneficial effect on wealth accumulation or its destructive effect on society and the environment. Corresponding to where the pendulum swung, greed has been infrequent at times and quite frequent, even institutionalized and normative at others. Table 8.1 presents highlights of the greed narrative across time.

History of Greed

The history of greed goes back to over 4,000 years ago, when the first known written record of it was created. In the Maxims of Phahhotep, the vizier introduced the concept of the 'greedy heart', based on the notion that the heart was the organ responsible for reason, understanding and good intentions. Greed was conceptualized as coveting more than one's fair share, and leading to conflict (van den Dungen, 2016; Loprieno, 1996). A plethora of sources from Greek antiquity highlight that the foundations of our current thinking about greed were established during the 6th–4th centuries BCE. Solon's writings reveal that the Greek language in the 6th century BCE differentiated between multiple types of greed such as pleonexia (greediness), koros (greed or satiety), philochrematia (love of money), aischrokerdeia (base covetousness), and epithumia chrematon (desire for money) (Lewis, 2006), and also point out that greed exists in both upper and lower classes. Thus, there appears a key recognition that greed is not the same as wealth, and in fact, one does not need to be successful at the extreme pursuit of money or wealth to be greedy. In Solon's writings, greed is presented as unambiguously bad. In the 5th century BCE, the writings of Herodotus and Thuclydes (Balot, 2001) introduce the notion of elite greed. This is a different level of analysis than previously known: in elite greed, the Athenian city state directs its greed against other Greeks. Thus, greed unifies Athenians against others. In 4th–5th centuries BCE, in Plato's Gorgias, the pendulum swings further, through Callicles, who represents unbridled appetite for anything the stronger, more powerful ruling few desire. Callicles is an amoralist hedonist for whom there is no right or wrong. In the world of Callicles, the superior rule the inferior and are entitled to more without restraint. Plato through Callicles highlights a number of important issues that cut to the core of the definition of greed, justice, and inequality. For example, where does self-interest end and greed begin? Is being hedonistic the same as being greedy? What should be the basis on which we determine what is deserved and what is not deserved? How are superiority and inferiority decided? Do some deserve more than others, and if so how is that decided? In the 4th century BCE, Aristotle's Nicomachean Ethics provides some answers by tying greed (pleonexia) to injustice and attached a moral code to the pursuit and by proposing that the right and just thing is to aim for what benefits the greater good and the individual, rather than just the individual. Aristotle also differentiated between needs and wants that broadened the conceptualization of greed from material things to all desires and noted greed's violation of communal canons of distributive justice (Wang & Murnighan, 2011). Finally, Aristotle (2004) also noted that greed can be

TABLE 8.1 Selected Illustrations of the Treatment of Greed Between ca. 2300 BCE and the 20th Century CE.

Date	Place	Author or Source	Conceptualization	Observed Phenomenon	Consequences, Conclusions
ca. 2300 BCE	Egypt	Maxims of Ptahhotep: Maxim #20 (Van den Dungen, 2016)	The concept of the greedy heart is introduced, based on the belief that the heart is the organ of reason, understanding, and intention. Both good and ill will originate in the heart, knowing good from evil is driven by the knowledge of Maat, a divine force of order (Loprieno, 1996).	Greed in the division of goods, greed is coveting more than one's share, greed toward kin.	Craving leads to conflict.
6th century BCE	Greece	Writings of Solon (Lewis, 2006)	Differentiates between *pleonexia* (greediness), *koros* (greed or satiety), *philochrematia* (love of money), *aischrokerdeia* (base covetousness), and *epithumia chrematon* (desire for money).	Greed and injustice exist among lower and upper classes. In contrast to greediness, individual-level self-restraint and distributive fairness are the core features of proper political participation.	First source from ancient Greece to cite *koros* as unambiguously bad; *koros* leads to *hubris* (Lewis, 2006).
5th c. BCE	Greece	Writings of Herodotus and Thucydides (Balot, 2001)	Discuss the concept of elite greed.	Athenian democracy allows Athenians to direct greediness against other Greeks, unifying the city of Athens.	Elite greed in action exemplified by two coups against the city-state, illustrating Athenians' discontentment with what they saw as greedy management of the empire (Balot, 2001).

(*Continued*)

TABLE 8.1 (Continued)

Date	Place	Author or Source	Conceptualization	Observed Phenomenon	Consequences, Conclusions
5th and 4th centuries Before Common Era (BCE)	Greece	Plato: Gorgias and Republic (Barney, 2011)	Callicles, represents unbridled appetite for anything the stronger, more powerful ruling few desire; represents hedonism, and is contrasted with a fictional polis ruled by self-controlled, fair-minded aristocrats.	The superior man must 'allow his appetites to get as large as possible and not restrain them' and 'the superior rule the inferior and have a greater share than they'. (Barney, 2011).	When appetites are as 'large as possible, he ought to be competent to devote himself to them by virtue of his courage and intelligence, and to fill him with whatever he may have an appetite for at the time'. The extent and content of the unrestrained appetite are subjective.
4th c. BCE	Greece	Aristotle: Nicomachean Ethics, Book 5	Greed (*pleonexia*) tied to injustice (Aristotle, Tredinnick, Johnson & Barnes, 2004).	The 'right' or just thing is to pursue that which is universally good and good for the individual, rather than what is good only for the individual.	Fairness, or distributive justice, must take into account the worth of the individuals.
1st c. BCE	Rome	Cicero: De Legibus, Book 3, The Constitution. (Yonge, 1853)	The notion of natural law, that certain values and rights are inherent and universally recognizable by virtue of human nature, is expounded upon.	Acquisitiveness and greed in the Senate must be severely punished.	Greed disrupts the natural law. Greed in the Senate leads to greed and dissent in the rest of the state.

4th c. CE	Western Europe	Evagrius Ponticus: Skemata; John Cassian: Institutions (Newhauser, 2000)	*Philargyria*, the love of silver (money), and *philokalia*, the love of self, or self-esteem are explicitly considered vices, or sins in Christianity.	Avarice and hubris appear on the list of eight evil thoughts, the precursor of the seven deadly sins in Christianity. List compiled first in Greek then in Latin translation.
4th–5th c. CE	Western Europe	Ambrose, Bishop of Milan (Newhauser, 2000)	Critique of private property and individual ownership. *Avaritia* derives from the Latin *aveo*, to desire.	Avarice is the desire to own material goods or money, rather than the perversion of this desire. Avarice leads to laws of possession (property rights), which in turn results in social injustice and destroys human commonwealth.
6th c.	Western Europe	Pope Gregory I (Newhauser, 2005b)	Compilation of the modern concept of the Seven Deadly Sins, including *avarice*.	Greed is one of seven categories of sin, others being pride, envy, wrath, gluttony lust, and sloth. The 'deadly sins' are deadly because they kill the spirit.
13th c.	Western Europe	Thomas Aquinas: Summa Theologica (Kreeft, 1993)	Covetousness, defined as 'immoderate love of possessing' is categorized as a mortal sin.	The object of a sin is the good toward which an inordinate appetite tends. Covetousness or avaritia is an immoderate love of a useful good, i.e., riches. The useful good is different from the delightful good. Covetousness has no opposite vice. The lack of desire to possess more is not a sin.

(*Continued*)

TABLE 8.1 (Continued)

Date	Place	Author or Source	Conceptualization	Observed Phenomenon	Consequences, Conclusions
15th–16th c.	Western Europe	Protestant Reformation: Martin Luther (1520, 2012)	Some types of work are superior to other types of works and command different compensation.	Each person is assigned a place in social hierarchy on the basis of the work. Individuals work dutifully in their own profession, rather than try to change classes. A person's vocation is their calling and all callings are spiritually equal.	Affirms manual labor and that differences exist in the value of work. Earning an income to cover basics is necessary, but accumulating wealth is sinful.
15th–16th c.	Western Europe	Protestant Reformation: John Calvin (2008)	Predestination paves the way for accumulation of wealth through hard work. Work is a calling and all jobs are spiritually equal. Predestination manifests through 'elects', who are people chosen by God for eternal life.	The only outward evidence of being an 'elect' are a person's actions, daily deeds, and work. All men should work and accumulate profits for reinvestment. It is one's religious duty to select the most profitable profession for oneself and build as great of wealth as possible.	The foundations of a profit-oriented economic system are laid; wealth accumulation is aligned with spirituality and serving God.

16th–18th c.	Western Europe	Mercantilism (Wallerstein, 2011)	Economic nationalism based on the principles of mercantilism.	According to mercantilism, the protectionist political economic theory of the 16th–18th c., trade is a zero-sum game and the goal of nations should be to create a positive trade balance.	Mercantilism encourages exports and discourages imports, the use of all available land for agriculture, mining, or manufacturing. Preference is given to finished goods as opposed to raw materials due to added value. The goal of trade is to accumulate as much gold and silver as possible.
18th c.	Western Europe	Industrial Revolution	Productivity increased as economies shifted from agrarian to industrial, and as production shifted from small-scale home manufacturing to concentrated factory production (Bolt & van Zanden, 2013).	Unprecedented accumulation of wealth due to increased productivity. Drivers include innovations such as the steam engine, the flying shuttle and the spinning jenny; abundance of raw materials from colonies; increased demand for goods due to population growth; and increased supply of labor.	Per capita GDP in 1870 was double that in 1700; estimates indicate that during the Industrial Revolution, as much GDP per capita was created in 170 years as there had been in the preceding 1,700 years (Bolt & van Zanden, 2013).

(Continued)

TABLE 8.1 (Continued)

Date	Place	Author or Source	Conceptualization	Observed Phenomenon	Consequences, Conclusions
18th c.	British Empire	Adam Smith: The Theory of Moral Sentiment (1759); The Wealth of Nations (1776)	Argues free trade based on rational self-interest. Self-interest of small business owners leads to economic progress while also benefiting the greater good (society). In contrast to rational self-interest, Smith considered avarice to be excessive attention to 'the objects of self-interest' (Smith, 1759: VII.II.16/ ii), which was to be avoided.	Smith criticized mercantilism arguing free trade benefited both transacting parties. He noted that governmental policies protecting industry interests harmed greater society, by benefiting only government, industrialists and merchants. In a free-market economy, economic welfare is distributed more evenly and enjoyed by all members of society (LaHaye, 2008). The actions of rationally self-interested individuals in a free-market	Smith, primarily a moral philosopher, like other philosophers who preceded him (e.g., Aristotle, Solon), studied economic excess from an ethical standpoint, considering the greater good of society (Silvermintz, 2010). Smith differentiated a healthy level of self-interest that he considered the driver of economic progress, from greed, which he regarded as an extreme, exaggerated form of acquisitiveness, detrimental to economic and social development.

| Late 19th–early 20th c. | Western Europe | Emile Durkheim (1893); Max Weber (1905) | Durkheim writes that those holding power are driven by the desire to profit, which is equated to greed. Weber introduces the concept of Protestant ethic (later relabeled work ethic) in his analysis of the relationship between religion and economic behavior. | economy lead to economic well-being for society at large. Self-interested actors benefit the greater good by competing in the market with products and services that provide value. Protestant ethic, or work ethic, drives the 'spirit of capitalism'. Durkheim's concept of forced division of labor is one of the causes of social breakdown. | The desire to profit and the division of labor (Durkheim) lead to individuals performing work unsuitable for them, leading to a disruption of the natural equilibrium, and anomie. With anomie, there are insufficient moral constraints and concepts of what is proper and acceptable are unclear. |

a problem of the individual or a destructive social behavior (Wang & Murnighan, 2011). Early materialists at around the same time in South Asia and China (e.g., Kesha Kambali, Cervaka, Mo Zi) argued that the accumulation of wealth and material goods was not sinful in spite of the deleterious effects on society (Barua, 1920; Oka & Kuijt, 2014). The philosopher Mo Zi promoted utility maximizing behavior opposing the Confucian legalist concept of 'li' (propriety and restraint) (Zhang, 2007) and argued that even excessive wealth accumulation can serve as a social good, and remain moral, as long individuals acted out of universal love and respect. (Oka & Kuijt, 2014).

In the 1st century BCE Cicero expounded the notion of natural law, and that certain values, and rights are universal, and are disrupted by greed. He posited that acquisitiveness, one of the key components of greed, must be punished (Yonge, 1853). In subsequent centuries, a number of scholars, philosophers, and religious leaders in Western Europe theorized about greed in the context of Christian values, which eventually lead to the list of the Seven Deadly Sins. While these sins do not appear in the Bible itself, they form the basis of centuries of Christian narrative, including discussions on property rights and social injustice. Much of this narrative intended to be consumed by the common person, decries greed positioning it as the 'mother of all sins' (Tickle, 2004), from which the other sins originate, is at odds with the wealth accumulating behavior of the church (Kreeft, 1993; Newhauser, 2000, 2005a). Then in the 15th and 16th centuries, with the emergence of Protestantism, appeared the realization that even though all jobs are spiritually equal, differences exist in the value of work, and that different types of work command different levels and types of compensation. Accumulation of wealth was sinful according to Martin Luther (Luther, 1520/2012) but aligned with spirituality and serving God by Calvin (2008). Similar ideas can be found in other religions and other parts of the world as well. Both Buddhism and Hinduism taught that greed is an impediment to spiritual development (Nath, 1998; Sundararajan, 1989; Seuntjens, 2016). Indeed, Buddhist teachings point out that greed is one of the three poisons that lead to bad karma (Nath, 1998). Judaism holds that greed prevents other people from achieving what they deserve (Bloch, 1984). Islam, being a religion whose followers depend on commercial activity, resolves potential problems stemming from greed by mandating charitable donations (Oka & Kuijt, 2014). In sum, the world's major religions all take a stance against greed, and some even go as far as providing solutions for it via charity toward the less fortunate.

The 16th–18th centuries saw the rise of mercantilism in Europe (Wallerstein, 2011), a protectionist political theory, in which trade was a zero-sum game and nations sought to maximize exports and minimize imports. The goal of trade became to accumulate as much silver and gold as possible. Productivity increased as manufactured goods became preferred to agricultural production, and small-scale manufacturing gave way to concentrated factory production (Bolt & van Zanden, 2013). The Industrial Revolution witnessed unprecedented accumulation of wealth: per capita GDP doubled between 1700 and 1870, and the wealth created in that time period equaled that created in the preceding 1,700 years (Bolt & van Zanden, 2013).

Scholars like Locke, Hume, and Adam Smith criticized mercantilism, demonstrating that free trade need not be a zero-sum game and indeed might benefit both transacting parties. Adam Smith noted that mercantile policies protecting industry interests harmed greater society, by benefiting only government, industrialists, and the merchant classes. In contrast, in a free-market economy, economic welfare is distributed more evenly and enjoyed by all members of society (LaHaye, 2008). The actions of rationally self-interested individuals in a free-market economy lead to economic well-being for society at large. Self-interested actors benefit the greater good by competing in the market with products and services that provide value. While Adam Smith considers self-interested actors the drivers of the economic progress and contributors to the greater good of society (Silvermintz, 2010), he regards greed as an extreme and exaggerated form of acquisitiveness that is detrimental to economic and social development. Thus, Adam Smith, considered the father of capitalism by many, but also a moral philosopher, differentiated between the idea of self-interest and greed, which he viewed excessive and extreme. Hume acknowledged the duality of greed in that he saw it both as a vice that destroys society and a motivator and virtue in commerce. The final pieces of the historical treatment of greed puzzle come from the works of Durkheim (1893) and Max Weber (1905). According to Durkheim, those holding power are driven by the desire to profit, which he equates to greed. Weber introduces the concept of Protestant Ethic, which according to him drives the 'spirit of capitalism'. Durkheim argues that social breakdown originates in the desire to profit and the division of labor, which leads to individuals performing work unsuitable for them. The result is anomie, the disruption of the natural equilibrium and the breakdown of moral values and social bonds.

A noteworthy influence that found little resonance with professional philosophers was Ayn Rand's nonfictional work. Bhadwar and Long (2020) write, 'Rand developed a conception of metaphysical realism, rationality, ethical egoism (rational self-interest), individual rights, laissez-faire capitalism, and art, and applied her philosophy to social issues'. In her discussions of rational self-interest, Rand legitimized greed, by arguing that altruism, or caring about others' needs is immoral, while acting in one's own self-interest, even to the extreme, is not only moral but also virtuous. The tenets of Rand's 'philosophy' received a boost in Milton Friedman's shareholder primacy doctrine (1970). Friedman codified Rand's ideas promoting free markets and laissez-faire capitalism, paving the way to unbridled greed at the corporate level. Friedman's famous quotes, 'the business of business is business' and 'the social responsibility of business is to increase its profits' mandate that businesses care only about the bottom line and disregard the needs of all stakeholders other than shareholders. The shareholder primacy doctrine has formally and informally ruled the world of business from 1970 until in August 2019, when the Business Roundtable, a powerful pro-business lobbying group composed of about 200 top CEOs, broke with their official position since 1997 of serving solely the needs of shareholders by declaring to create value for all stakeholders.

Greed in the Social Sciences

The study of greed in the social sciences, including in psychology, anthropology, and business disciplines, such as economics, finance, marketing, and management is still in its early stages. Scholars have only recently started to engage scientific methods to study the mechanisms, motivations, heuristics, cognitive biases, and behaviors and related concepts surrounding the construct. As noted by Oka and Kuijt (2014):

> the topic of greed and excess also evokes extreme passions among people, as it forces us to make judgments about their social benefit or harm that are inherently accusatory or defensive. In point of fact, we are both unable and unwilling to resolve the moral underpinnings of greed and excess.
>
> *(p. 32)*

While theories present contradictory views on what greed is and is not, and whether it's good bad or neutral (Oka & Kuijt, 2014), we have relatively little to go on as far as scientific foundations of greed, such as a greed taxonomy or other systematic categorization of the construct. At the time of this writing, greed has been studied almost exclusively at the individual level in social dilemma games examining whether participants' actions are primarily motivated by greed or fear (e.g., Kuwabara, 2005; Balliet, Li, Macfarlan, & Van Vugt, 2011; Rapoport, Chammah, & Orwant, 1965; Coombs, 1973; Dawes, Orbell, Simmons, & Van De Kragt, 1986; van de Kragt, Orbell, & Dawes, 1983; Rapoport & Eshed-Levy, 1989; Murnighan, Kim, & Metzger, 1993) as well as through surveys (Seuntjens, 2016) that indicate greed might be dispositional. Thus, at the time of writing this essay, the evidence is mixed regarding whether greed is a state or a trait.

Social dilemma studies reveal important aspects of the psychology of greed. In basic Prisoners' Dilemma games, each of the two participants has a choice to make whether to cooperate or to defect. Armed only with information about what each partner's payoff/punishment is in case of defection and cooperation and prohibited from collusion, the rational choice for participants is to act in their own interest. Such actions, however, while leading to better outcomes for the individual who defects first, always lead to worse group outcomes than collaboration would have. In addition, due to the defection, the participants also end up worse off individually than they would have had they cooperated. Thus, the overall message of Prisoners' Dilemma games is that two rational players acting in their own interest will end up with worse outcomes for themselves as well as the whole than they could have had they considered the benefit to the greater good. Such social dilemma experiments lend themselves readily to the examination of greed, since they rely on elements familiar from the definition of greed, namely, acting in one's self-interest, harm to others, and the greater good. The tendency for Prisoners' Dilemma experiments to result in defection has been documented in many studies. Most pertinent to psychology of greed are studies that examine whether defection is motivated by fear or greed (Rapoport et al., 1965; Coombs, 1973). Reviewing such studies, Wang and Murnighan (2011) report that several social dilemma games in the mid-to-late 1980s shed light on the incentives to defect and thereby revealed underlying tensions of groups and societies. In one experiment, the fear incentive was neutralized by a money-back guarantee for those who cooperated when others defected, while in another classic study, the opposing, greed incentive was removed preventing defections from harming cooperators (van de Kragt et al., 1983; Dawes et al., 1986). Findings of these experiments showed that while removing fear made no difference, 'removing greed increased cooperation', indicating that greed is a stronger incentive than fear. Poppe and Utens' (1986) and Rapoport and Eshed-Levy's (1989) studies provided similar findings regarding greed being a stronger motivator. Extensions to the consistent baseline finding of greed being a stronger incentive than fear, include the finding that greed increases with the size of the group in Tragedy of the Commons experiments (Murnighan et al., 1993) and as the potential payoff increases (Dawes, 1980)[1]. In aggregate, these studies show that greed can be triggered by situations and that different situational factors can increase greed in individuals.

Without disputing the situational or 'state' aspect of greed, more recently, Seuntjens (2016) has made inroads exploring greed as disposition. In a series of studies, Seuntjens conceptualizes greed using a prototype analysis, a methodology used to better understand 'fuzzy' psychological constructs. The most salient features of greed are to acquire more combined with dissatisfaction of never having enough (Seuntjens, 2016). By using results of the prototype analysis, Seuntjens (2016) develops and validates a Dispositional Greed Scale (DGS), while also establishing greed's distinctiveness from maximization, self-interest, envy, and materialism. This work corresponds with the previous findings on greed being independent from related constructs such as envy, narcissism, and hubris (Haynes et al., 2017). By using the DGS, Seutjens (2016) finds that greed is associated with individual differences with more income, more expenses, as well as less savings and more debt. Overearning, or earning beyond what is needed, based on the preference to earn more rather than the preference for working more, is also associated with a greedy disposition, and the behavior appears stable over time. Finally, greed is also related to unethical behavior: individuals with a greedy disposition have more positive attitudes toward transgression and are more likely to transgress (Seuntjens, 2016), although self-control might mitigate the relationship. Taken together, the findings of Seutjens' (2016) work and the studies on situational greed indicate that greed is likely both a state and a trait. Even individuals who score low on the DGS might find themselves in situations that trigger them to act in a greedy manner. Conversely, those who score high on the DGS might display greedy behaviors in many or all situations they find themselves in. Such situations might include ones in which someone with decision-making authority has discretion over resources that they are able to divert in full or in part to benefit themselves, or to extract favors from others. Greedy actions are relatively easy to perform in weak regulatory environments and contexts difficult to monitor. Consider, for example, a corrupt official in charge of awarding government contracts. The corrupt official will award the contract to a company that provides a kickback. The greedy act diverts funds to the corrupt official, directly engages the government contractor in an illegal act, which simultaneously also benefits them financially, while also challenging the regulatory context. Decision-making authority and discretion over funds exist at all levels of society, and in all settings, including private, public, and not-for-profit. Therefore, greed can manifest in any setting as well.

The fields of management and economics are likely to consider greed either neutral or morally acceptable, perhaps even desirable due to its focus on the accumulation of wealth. Inquiring into the possible mechanisms of greed acceptance, researchers investigated effects of economic and business education on individuals' acceptance of greed. In a Dictator Game, students are paired up with one student taking the role of dictator who decides how to split up a set amount of money between themselves and their partner, who act as passive recipients. Students with economic and business education acted in their own self-interest, assigning themselves more money and giving less to their partners than students without economic and business education (Wang et al., 2011). In another study, the authors asked participants to recall an instance in which they succumbed to and another in which they resisted greed. Participants then were asked to report their post hoc feelings for the stories in which they had succumbed to greed, their moral perspective of greed, and whether their definition included a consideration of greed's harm on others. The findings showed that economic education shaped participants' positive attitudes toward greed and toward the morality of greed (Wang et al., 2011). In a third study, Wang and colleagues (2011) provided participants with one of three reading samples and asked them to assess the morality of greed based on the sample. One sample contained positive statements about self-interest and greed by famous economists, the second one contained negative statements about the same by other famous economists, and the third sample was a control about a topic unrelated to self-interest and greed. Participants had to then summarize what they had read. The analysis of the open-ended survey reinforced the findings of the previous two studies on the connection between greed and economic and business education by showing that even a brief positively worded text on self-interest will result in more positive attitudes toward greed, even for nonbusiness and noneconomic students. Greed, therefore, in business disciplines based on rational decision-making (e.g., areas of economics, management, finance) is seen as a simple extension of self-interest and as such considered a universal trait of rational profit-maximizing managers. Scholars of agency theory, a central and fundamental tenet of which is that managers are self-interested rational profit maximizers, might not consider greed worthwhile topic to study, since it would not be considered significantly different from the universal axiomatic managerial trait of self-interest.

In management, Haynes and colleagues (2015a, 2015b, 2017) and Sajko and colleagues (2020) have been examining greedy behavior in leaders. These scholars theorize about and empirically examine effects of greedy CEOs, top managers and others occupying leadership positions on their organizations. Haynes and colleagues (2017) build their theory around the idea that there is a tipping point at which rational self-interested behavior in managers turns into extreme pursuit of wealth, that is, greed. According to Haynes and colleagues (2015a, 2017), the tipping point is where managerial actions stop benefiting the greater good and begin to harm it. Defining the 'greater good' narrowly as shareholder wealth and focusing attention to the harm to others inherent in greed, Haynes and colleagues (2017) hypothesize that managerial greed negatively affects shareholder wealth. Adding situational components to the mix, the authors further posit that managerial discretion, or the opportunity to be greedy, increases the negative effect of greed on shareholder wealth. On the brighter side, Haynes and colleagues (2017) find evidence of the importance of strong governance in reducing the deleterious effects of greed on shareholder wealth, and on that greed in CEOs decreased, as their tenure increased. The study broke new ground in the greed literature for three reasons. First, it was the first empirical study to develop the greed construct and its nomological net in the strategic management literature. Second, it was the first study to develop and test measures of managerial greed using secondary data. Prior studies in management were either theoretical, literature review-type articles, or used experimental design and relied on laboratory studies. Third, Haynes and colleagues (2017) were able to show empirically that while the agency relationship does exist, and managers do act opportunistically in their own self-interest, extreme self-interest varies across individuals, rather than being a universal characteristic of all managers. Haynes and colleagues' (2017) empirical study paved the way for other researchers in strategic management to examine the effects of CEO greed on their organizations, shareholders, and stakeholders. A follow-on study by Sajko, Boone, and Buyl (2020) expand the idea of 'the greater good' to stakeholders and explore how greedy CEOs investment into corporate social responsibility (CSR) prior to the great financial crisis in 2008 affects their companies' fate during and after the crisis. The authors posit that the myopic behavior of greedy CEOs prevents them from taking a stakeholder perspective and pursuing socially and environmentally sound practices. Rather, greedy CEOs pursue their own interests to the extreme. The CEO's unbridled pursuit of personal wealth coupled with lack of investing into CSR, however, harms the company because CSR investments have been shown to create a more resilient firm. Thus, normal times, greedy CEOs might get away with syphoning wealth from the company; however, in case of a crisis, such behaviors lead to slower and weaker recovery for the firm. Sajko and colleagues (2020) show that firms lead by greedy CEOs experience greater losses in the short term and take longer to recover from crisis than firms that are led by nongreedy CEOs and those who invest into CSR. A significant contribution made by Sajko and colleagues (2020) is showing that investment into corporate social responsibility is a sound strategic practice that creates resilience and can improve recovery from an external shock. As such it well complements Haynes and colleagues' (2017) work on how greedy CEOs harm shareholders by showing how greedy CEOs also harm stakeholders and their own company's recovery from crisis. Both sets of authors present ways to mitigate the greedy CEOs' behavior, by stronger governance mechanisms, reducing CEO discretion and increasing tenure, and by investing into corporate social responsibility.

Multilevel Model of Greed as Extremism

Earlier I discussed the notion of greed as a type of extremism, followed by a review of historical narratives and current research findings about greed. According to motivational theory, a balanced state is achieved when multiple needs are satisfied, some needs keeping others in check, by constraining, and providing boundaries. Motivational imbalance occurs when one need overrides all others and is pursued and expressed without constraint. Much of what ancient philosophers theorized and modern-day scholars found in their research about self-interest and greed fit well into this framework. At the individual level, satisfying multiple needs and achieving a balanced motivational state translates to balancing self-interest and the needs of the greater good. This is the type of self-interested behavior Adam Smith referred to writing, 'It is not from the benevolence of the butcher, the brewer, or the baker that we expect our dinner, but from their regard to their own self-interest', in The Wealth of Nations (1776/1904) and also:

> How selfish soever man may be supposed, there are evidently some principles in his nature, which interest him in the fortune of others, and render their happiness necessary to him, though he derives nothing from it, except the pleasure of seeing it.
>
> *(Theory of Moral Sentiments, 1759/1976)*

When self-interest is taken to an extreme, in the form of excessive pursuit of material wealth, one need overrides all others, including concern for a balanced life, deep meaningful relationships, and caring for and about others. The behavioral expression of the motivational imbalance is greed. A greedy person's actions will reflect their single-minded desire for excessive wealth accumulation at any cost. Extreme acquisitiveness that has been historically associated with greed may be accompanied by retention of wealth as well, evocative of the miserly Ebenezer Scrooge in Charles Dickens' 1843 novella, A Christmas Carol. A noteworthy characteristic of both the acquisitive and the miserly greed is that it is pursued at all cost. Kets de Vries, Distinguished Professor of Leadership Development and Organizationsl Change at INSEAD lists seven characteristics of uncontrolled greed: extreme self-centeredness, being envious of others, lacking empathy, never being satisfied, being manipulative, a short-term focus, and knowing no limits. In combination, these characteristics paint a picture of a person to whom the needs of family, friends, and those less fortunate are inconsequential, someone who is myopic, and will strive to get their way no matter what (Kets de Vries, 2016). For such a person, the fulfilment of the single need to accumulate comes at the cost of harming the self and others. The former, harm to the self is reflected in lost or shallow relationships and interpersonal problems due to their lack of empathy and engagement. In some cases, health issues might arise from the social isolation a greedy person's self-centeredness causes (Kets de Vries, 2016).

While greed is normally associated with the individual, it always occurs in context, affecting individuals, groups, and society. An emerging stream of literature examines greed in a broader, organizational context. Studies by Haynes and colleagues (2017) and Sajko and colleagues (2020) demonstrate the negative effects of greedy CEOs on shareholder wealth and corporate social responsibility in corporations. Haynes, Hitt, and Campbell (2015b) show the effects of greed in entrepreneurial settings. Building on the premise that greed is ubiquitous, but that young startups, family firms, and corporate ventures are different from large established public firms, Haynes and colleagues theorize that greedy entrepreneurs have deleterious effects on their firms' human and social capital. Specifically, building on studies that examine the dark side of leadership (e.g., Silvia, Kaufman, Reiter-Palmon, & Wigert, 2011; Veselka, Giammarco, & Vernon, 2014: Lee & Ashton, 2014; Linstead, Marechal, & Griffin, 2014), Haynes and colleagues (2015b) argue that entrepreneurial greed in each of the three settings stands in the way of the venture's success: in young startups, greed prevents attracting human and social capital; in family firms, it leads to the underutilization of human and social capital; and in corporate ventures, to wasting human and social capital. Time horizon and firm size also play a role in the effects of greed on human and social capital. Greed generally is associated with a myopic or short term-focus, since the greedy person's goal is to extract wealth as soon and quickly as possible. In addition, empirical research on the temporal aspect of greed indicates that greed is more often observed early in a CEO's tenure than later and that greed is more pronounced in younger than in older individuals (Haynes et al., 2017; Gillespie, 2016). Expanding the effects of greed in organizations, in a recent study, Haynes and Bragaw (2020) introduced the concept of greed at the corporate level, theorizing that corporate greed stems from a shared heuristic in the top management team that leads to a cognitive bias, which then translates to actions that are perceived as corporate greed. According to the shareholder primacy doctrine (Friedman, 1970), managers are mandated to focus on one goal, maximizing shareholder wealth, at any cost. Milton Friedman's (1970) doctrine is summed up in the title of his essay: The Social Responsibility of Business is to Increase Its Profits (Friedman, 1970). Business schools around the world have indoctrinated hundreds of thousands of students in its tenets. The result of the single-minded focus on profit generation at any cost is that companies stray from their strategic initiatives and manage the company by financial, rather than strategic controls by (a) opportunistically pursuing revenues and (b) relentlessly implementing cost saving measures (Haynes & Bragaw, 2020). Focus on these two types of actions, which are intended to benefit shareholders at the cost of harming other stakeholders, results in a short-term focus, which initially results in short-term financial success. However, in the long term, corporate greed is likely to show negative consequences in lessened organizational commitment and a hollowing out of the firm that materializes through loss of human capital (Haynes & Bragaw, 2020). Companies' very nature changes as they lay off employees and outsource their work to overseas contractors in the ruthless pursuit of efficiency and opportunistic revenue and profit-seeking. Apart from massive increases to shareholder wealth, one of the outcomes of the shareholder primacy heuristic is that the pursuit of profits at any cost has normalized. We no longer question excessive layoffs and other cost-cutting measures, or the opportunistic pursuit of profits.

Mitigating the Problem

What are possible solutions to the problem of greed? Before answering that question, it's important to note that, as demonstrated throughout this essay, the perception of greed as a problem is not shared universally. Consequently, anyone approaching greed as a problem might encounter significant obstacles trying to resolve it. Considering greed as something that causes harm to others frames greed as a phenomenon that needs to be mitigated. At the individual level, it might be reasonable to assume that if exposure to economic and business education primes individuals to develop more favorable attitudes toward greed (Wang et al., 2011), exposure of business and economic students to theories and practices that are more other-focused might reduce greed. Wang and colleagues (2011) note that emphasis on numerical analyses and calculations in economics and business in general 'may blunt decision-makers abilities to see the social consequences of their decisions' (Wang, Zhong, & Murnighan, 2014; Wang et al., 2011, p. 657). Wang and colleagues (2011) call for future research to investigate whether economics curricula might need to include courses that provide alternative perspectives of the consequences of economic and financial decisions. Specifically, should individuals and managers be more aware of and place more emphasis on understanding how others, including those who do not stand to benefit financially, are affected by their decisions?

Another potential approach relates to the intrapersonal aspect of greed that arises due to focusing on one goal at the expense of others. A greedy person is more focused on the pursuit of wealth than other goals, even if they can't use the resources they acquire later (Seuntjens, 2016). The focus on pursuing a goal based on greed allows less time to pursue other goals, or build interpersonal relationships, spend time with family and friends. Yet, money does not necessarily make people happy, especially if it comes at a cost of less time spent with loved ones (Whillans, Weidmen, & Dunn, 2016). A possible solution to this would be to encourage self-reflection, self-monitoring, and self-assessment of personal values.

In the case of greedy CEOs, less managerial discretion, and stronger, better boards whose power matches or exceeds that of the CEO might help mitigate the self-serving activities of the CEO (Haynes et al., 2017). In addition, since greed seems more prevalent during early years of a CEO's tenure than in later years of tenure, stronger monitoring in the early years might counteract effects of CEO greed (Haynes et al., 2017). The problem of corporate greed, which appears to be a systemic issue of global business in the 21st century, might be more difficult to resolve. Quarterly financial reporting mandates a short-term focus, and the shareholder primacy doctrine requires continuous revenue and profit growth to benefit shareholders. While the announcement at the 2019 Business Roundtable meeting to move away from their previously stated purpose to serve only shareholders and 'create value for all stakeholders' is a promising step in the right direction, it will only move past being considered empty rhetoric if it is followed by sustained action. Such actions must be supported by those in power to make the change, that is, the CEOs, and also those who own the corporation, namely, the shareholders. At the writing of this chapter, noticeable large-scale changes that demonstrate a stakeholder focus have yet to occur in US and global corporations.

Finally, another approach to solving the problem of greed is uncovering the origins of greed. What causes people to become greedy? There is a dearth of research on this topic. Preliminary theorizing indicates that greed, as a universal construct, is present across societies, cultures, and generations, but manifests in different forms. We know there is greed for money, for love, for food, and for power, all pointing to an underlying mechanism of wanting more of something. A question probing deeper is what causes greed to develop in individuals in the first place? Are some people born greedy, or do we learn to be greedy due to environmental influences? Does our social class of origin predispose us to be more or less greedy? Work on social class and risk-taking (e.g., Kish Gephart & Campbell, 2015) shows that executives' childhood experiences have lasting effects on their risk preferences as CEOs. Since acts of greed involve risk-taking, could childhood experiences also influence whether someone becomes greedy later in life? Or, does social class origin simply lead to a greedy person being evaluated differently by outsiders? Growing up privileged might endow one with a sense of deservingness and entitlement. Acting upon such entitlement later in life or at work could be interpreted as greedy by outsiders, even though for the greedy actor, fulfilling their material needs at any cost is not extreme, rather it might seem normal and even fair. Curiously, someone from a more modest background, after transitioning into a higher social class, might also be motivated to act in a greedy way, and pursue material wealth that goes beyond their needs. However, in such cases, greedy acts are driven by the need to compensate for the deprivation they experienced growing up. In a way, this is also an argument for deservingness and fairness. However, unlike in the case of privilege, lack of privilege in one's background might result in a more forgiving attitude from outsiders and even be interpreted as a 'rags to riches' success story. Such instances of greed might even inspire others to follow the example, which in turn points to greed contagion. Such effects have been found in recent experimental studies (Cardella, Kugler, Anderson, & Connolly, 2019).

Note

1. Tragedy of the Commons refers to an economic problem that simulates limited resources to which everyone is entitled to use without constraint. While the resources would be sufficient if everyone only took their fair share, once any member begins to take more than their fair share, others follow suit and the resource is quickly depleted.

References

Aristotle, H. (2004). *Nicomachean ethics* (H. Tredinnick, Ed., J. Barnes, Intro., J. A. K. Thomson, Trans.). Harmondsworth: Penguin.

Balliet, D., Li, N. P., Macfarlan, S. J., & Van Vugt, M. (2011). Sex differences in cooperation: A meta-analytic review of social dilemmas. *Psychological Bulletin, 137*(6), 881.

Balot, R. K. (2001). *Greed and injustice in classical Athens*. Princeton: Princeton University Press.

Barney, R. (2011). Callicles and thrasymachus. In Edward N. Zalta (Ed.), *The Stanford encyclopedia of philosophy* (Winter 2011 ed.). Retrieved June 2020, from https://plato.stanford.edu/archives/win2011/entries/callicles-thrasymachus/

Barua, B. M. (1920). *The Ajivikas*. Calcutta: University of Calcutta Press.

Bhadwar, N. K., & Long, R. T. (2020). Ayn Rand. In E. N. Zalta (Ed.), *Stanford encyclopedia of philosophy* (Summer 2020 ed.). Retrieved June 2020, from https://plato.stanford.edu/entries/ayn-rand/

Bloch, A. P. (1984). *A book of Jewish ethical concepts: Biblical and postbiblical*. New York, NY: KTAV Publishing House, Inc.

Bolt, J., & Van Zanden, J. L. (2013, January 5). *The first update of the Maddison project: Re-estimating growth before 1820* (Maddison-Project Working Paper WP-4). Groningen, The Netherlands: University of Groningen.

Calvin, J. (2008). *Institutes of the Christian religion* (Henry Beveridge, Trans.). Peabody, MA: Hendrickson Publishers.

Cardella, E., Kugler, T., Anderson, J., & Connolly, T. (2019). Is greed contagious? Four experimental studies. *Journal of Behavioral Decision Making, 32*(5), 579–598.

Collins. (2020). Retrieved from www.collinsdictionary.com/us/dictionary/english/greed
Coombs, C. H. (1973). A reparameterization of the prisoner's dilemma game. *Behavioral Science, 18*(6), 424–428.
Dawes, R. M. (1980). Social dilemmas. *Annual Review of Psychology, 31*(1), 169–193.
Dawes, R. M., Orbell, J. M., Simmons, R. T., & Van De Kragt, A. J. (1986). Organizing groups for collective action. *American Political Science Review, 80*(4), 1171–1185.
Durkheim, E. (1893/1984). *The division of labor in society.* New York: The Free Press.
Friedman, M. (1970, September 13). The social responsibility of business is to increase its profits. *The New York Times Magazine,* 122–126.
Gillespie, P. (2016). *Don't be greedy . . . it works against you.* CNN Business. Retrieved June 2020, from https://money.cnn.com/2016/08/16/investing/love-of-money-state-street-survey/index.html
Gimein, M. (2002, September 2). The greedy bunch. *Fortune.*
Haynes, K. T., & Bragaw, N. (2020). *Corporate greed: Formalizing the construct and explicating its organizational consequences.* Academy of Management Proceedings, 2020, Vancouver, CA.
Haynes, K. T., Campbell, J. T., & Hitt, M. A. (2017). When more is not enough: Executive greed and its influence on shareholder wealth. *Journal of Management, 43*(2), 555–584.
Haynes, K. T., Hitt, M. A., & Campbell, J. T. (2015b). The dark side of leadership: Towards a mid-range theory of hubris and greed in entrepreneurial contexts. *Journal of Management Studies, 52*(4), 479–505.
Haynes, K. T., Josefy, M., & Hitt, M. A. (2015a). Tipping point: Managers' self-interest, greed, and altruism. *Journal of Leadership & Organizational Studies, 22*(3), 265–279.
Hofstede, G. (2011). Dimensionalizing cultures: The Hofstede model in context. *Online Readings in Psychology and Culture, 2*(1).
Kets de Vries, M. (2016). *Seven signs of the greed syndrome.* Retrieved June 2020, from https://knowledge.insead.edu/blog/insead-blog/seven-signs-of-the-greed-syndrome-4624
Kish-Gephart, J. J., & Campbell, J. T. (2015). You don't forget your roots: The influence of CEO social class background on strategic risk taking. *Academy of Management Journal, 58*(6), 1614–1636.
Kreeft, P. (1993). *A Shorter Summa, the essential philosophical passages of Saint Thomas Aquinas' Summa Theologica.* San Francisco: St Ignatius Press.
Kruglanski, A. W., Jasko, K., Chernikova, M., Dugas, M., & Webber, D. (2017). To the fringe and back: Violent extremism and the psychology of deviance. *American Psychologist, 72*(3), 217–230. doi:10.1037/amp0000091
Kuwabara, K. (2005). Nothing to fear but fear itself: Fear of fear, fear of greed and gender effects in two-person asymmetric social dilemmas. *Social Forces, 84*(2), 1257–1272.
LaHaye, L. (2008). Mercantilism. *The concise encyclopedia of economics. Library of Economics and Liberty.* Retrieved from http://www.econlib.org/library/Enc/Mercantilism.html June 2020.
Lee, K., & Ashton, M. C. (2014). The dark triad, the big five and the HEXACO model. *Personality and Individual Differences, 67,* 2–5.
Lewis, J. (2006). Solon the thinker. *Political Thought in Archaic Athens,* 42–59.
Linstead, S., Marechal, G., & Griffin, R. W. (2014). Theorizing and researching the dark side of organization. *Organization Studies, 35,* 165–188.
Loprieno, A. (Ed.). (1996). *Ancient Egyptian literature: History and forms* (Vol. 10). Leiden, The Netherlands: Brill.
Luther, M. (1520/2012). *Treatise on good works.* Minneapolis, MN: Fortress Press.

Merriam Webster. (2020). Retrieved from www.merriam-webster.com/dictionary/greed

Murnighan, J. K., Kim, J. W., & Metzger, A. R. (1993). The volunteer dilemma. *Administrative Science Quarterly*, 515–538.

Nath, S. (1998). *Encyclopedic dictionary of Buddhism*. New Delhi: Sarup & Sons.

Newhauser, R. G. (2000). *The early history of greed: The sin of avarice in early medieval thought and literature* (Vol. 41). Cambridge: Cambridge University Press.

Newhauser, R. G. (2005a). Justice and liberality: Opposition to avarice in the twelfth century. In I. Bejczy & R. Newhauser (Eds.), *Virtue and ethics in the twelfth century* (pp. 295–316). Leiden, The Netherlands: Kloninklijke Brill NV.

Newhauser, R. G. (2005b). *In the garden of evil: The vices and culture in the middle ages*. Volume 18 of Papers in Mediaeval Studies. Toronto, CA: PIMS.

Oka, R., & Kuijt, I. (2014). Greed is bad, neutral, and good: A historical perspective on excessive accumulation and consumption. *Economic Anthropology*, 1(1), 30–48.

Poppe, M., & Utens, L. (1986). Effects of greed and fear of being gypped in a social dilemma situation with changing pool size. *Journal of Economic Psychology*, 7(1), 61–73.

Rapoport, A., Chammah, A. M., & Orwant, C. J. (1965). *Prisoner's dilemma: A study in conflict and cooperation* (Vol. 165). Ann Arbor, MI: University of Michigan Press.

Rapoport, A., & Eshed-Levy, D. (1989). Provision of step-level public goods: Effects of greed and fear of being gypped. *Organizational Behavior and Human Decision Processes*, 44(3), 325–344.

Sajko, M., Boone, C., & Buyl, T. (2020). CEO greed, corporate social responsibility, and organizational resilience to systemic shocks. *Journal of Management*. doi:0149206320902528

Schwartz, S. H. (1992). Universals in the content and structure of values: Theory and empirical tests in 20 countries. In M Zanna (Ed.), *Advances in experimental social psychology* (Vol. 25, pp. 1–65). New York: Academic.

Schwartz, S. H. (2007). Cultural and individual value correlates of capitalism: A comparative analysis. *Psychological Inquiry*, 18, 52–57.

Seuntjens, T. G. (2016). *The psychology of greed*. Tilburg, The Netherlands: Tilburg University.

Seuntjens, T. G., Zeelenberg, M., van de Ven, N., & Breugelmans, S. M. (2019). Greedy bastards: Testing the relationship between wanting more and unethical behavior. *Personality and Individual Differences*, 138, 147–156.

Silvermintz, D. (2010). How much is too much? An Aristotelian view of irrational exuberance. *Yale Economic Review*, 6(1), 47-49.

Silvia, P. J., Kaufman, J. C., Reiter-Palmon, R., & Wigert, B. (2011). Cantankerous creativity: Honesty-humility, agreeableness, and the HEXACO structure of creative achievement. *Personality and Individual Differences*, 51, 687–689.

Smith, A. (1776/1904). *An inquiry into the nature and causes of the wealth of nations* (5th ed.). London: Methuen and Co., Ltd.

Smith, A. (1759/1976). *The theory of moral sentiments*. Oxford: Oxford University Press.

Stone, O. (1987). *Wall Street*. Twentieth Century Fox.

Sundararajan, K. R. (1989). *Hindu spirituality: Vedas through Vedenata*. New York: The Crossroads Publishing Company.

Tickle, P. (2004). *Greed: The seven deadly sins*. Oxford: Oxford University Press.

Van de Kragt, A. J., Orbell, J. M., & Dawes, R. M. (1983). The minimal contributing set as a solution to public goods problems. *American Political Science Review*, 77(1), 112–122.

Van den Dungen, W. (2016). *Ancient Egyptian readings*. Brasschaat: Taurus Press.

Veselka, L., Giammarco, E. A., & Vernon, P. A. (2014). The Dark Triad and the seven deadly sins. *Personality and Individual Differences, 67,* 75–80.

Wallerstein, I. (2011). *The modern world-system I: Capitalist agriculture and the origins of the European world-economy in the sixteenth century* (Vol. 1). Berkeley, CA: University of California Press.

Wang, L., Malhotra, D., & Murnighan, J. K. (2011). Economics education and greed. *Academy of Management Learning & Education, 10*(4), 643–660.

Wang, L., & Murnighan, J. K. (2011). On greed. *Academy of Management Annals, 5*(1), 279–316.

Wang, L., Zhong, C. B., & Murnighan, J. K. (2014). The social and ethical consequences of a calculative mindset. *Organizational Behavior and Human Decision Processes, 125*(1), 39–49.

Weber, M. (1905/1930). *The protestant ethic and the spirit of capitalism* (T. Parsons & A. Giddens, Trans.). London: Unwin Hyman.

Whillans, A. V., Weidman, A. C., & Dunn, E. W. (2016). Valuing time over money is associated with greater happiness. *Social Psychological and Personality Science, 7*(3), 213–222.

Wikipedia. (2020). Retrieved from https://en.wikipedia.org/wiki/Greed

Winter, J. (2007, September 25). Greed is bad. Bad! 2007. *Slate.* Retrieved June 20, 2021, from https://slate.com/culture/2007/09/how-wall-street-s-gordon-gekko-inspired-a-generation-of-imitators.html

Yonge, C. D. (1853). *The treatises of MT cicero.* London: Henry G. Bohn.

Zhang, Q. (2007). Human dignity in classical Chinese philosophy: Reinterpreting Mohism. *Journal of Chinese Philosophy, 43*(2), 239–255.

9
MORAL, EXTREME, AND POSITIVE

What Are the Key Issues for the Study of the Morally Exceptional?

William Fleeson, Christian Miller, R. Michael Furr, Angela Knobel, and Eranda Jayawickreme

Piotr Wrona was a Polish locksmith in Warsaw during World War II. Every day, when he went to his workshop, he encountered starving children:

> I could not move backwards or forwards, I was all surrounded and they all begged me with cries, imploring me for food. . . . The hands that were turned toward me were not hands, they were bones, simply bones covered with skin. These children were kept in a church, they looked inhuman, they looked like ghosts . . . You know what it is when a human being is starved? One is capable of doing anything! These were helpless children who were swarming me in the hope of receiving something to eat. So I would bring five rolls. I could not pass through the gate with more. The Germans would not let me.

(Tec, 1986, p. 21)

The goal of this chapter is to introduce some issues that might be of interest to scholars considering conducting empirical or conceptual scholarship on the morally exceptional. We define the morally exceptional as *the morally best or morally exemplary individuals among us*. Such people are the exception rather than the rule, representing the moral heights that people are capable of achieving. We mean people like Piotr Wrona, who accepted a measured risk to his own life in order to prevent children from starving to death. We hypothesize that between just a few people and 10% of people in a given society will be morally exceptional.

We wish to (a) describe the promise of studying the morally exceptional; (b) review existing knowledge about the morally exceptional, with special attention to their motivation and motivational balance; (c) suggest possible avenues of research on the morally exceptional; (d) raise some key difficulties for scholarship on the morally exceptional; and (e) propose some possible solutions to these difficulties. This chapter is similar to the other chapters in the book in that it considers extreme behavior; it is different from the other chapters in the book in that this chapter considers positive rather than negative extreme behavior. It also shares the goals of the book in considering the motivations behind this behavior and considering in particular whether positive extreme behavior results from a motivational imbalance the way that negative extreme behaviors might.

The Promise of Studying the Morally Exceptional

One reason for studying the morally exceptional is the fascination these individuals evoke. What is the best that humans can do? Are the morally exceptional happy, from doing good works, or do they live bleak and dreary lives full of constant sacrifice and deprivation? Do others like them out of admiration for who they are and what they do, or do others find them intolerable because of their moralizing and grandstanding? Are the morally exceptional necessarily boring individuals who live life in one note, or can they be rich, interesting individuals with texture and liveliness? Are they uncompromising idealists or are they able to negotiate with the realities of common sense? Do they always have a dark side or are they moral in all respects? As humans, we often wonder whether we would be so good if we were placed in similar situations—do we have what it takes to be morally great?

Another reason to study the morally exceptional is their practical importance. The morally exceptional can prevent atrocity, crime, and bullying. Whistleblowers can save societies from fraud. Organ donors can save lives. Active bystanders can slow genocide and stand up to bullying. Activists can create more humane societies. These consequences are enormous, and the more we understand the morally exceptional, the more we might magnify their effects.

Another reason is that the study of the morally exceptional provides insights into the development of moral character. By studying what people regularly can do at their best, researchers can identify the factors that improve moral character. Such factors may prompt new interventions. The morally exceptional can also serve as inspiration and guidance to normal individuals, by acting as exemplars.

The morally exceptional may provide information about moral psychology that would not be obtained or obtained only slowly by studying those who are less extreme. Because their behavior is highly moral, their psychological systems may offer particularly clear insights into the factors that lead to moral behavior. We take as analogies the study of exceptional cognitive talent or of highly effective businesses. By analyzing the successful, we may learn processes that lead to that success.

Do Morally Exceptional People Exist?

We take an ecumenical approach to the definitions of moral, morally good, and morally exceptional because we believe the morally exceptional can be fruitfully studied even with differing definitions of these terms. We explain this approach and other strategies later in the chapter. For now, we mean 'morally good' to refer roughly to those actions that have to do with right and wrong, such as compassionately caring for another in need, telling the truth in even trying situations, refraining from excessively wasteful consumption, and treating others with respect. We mean 'exceptionally moral' to refer to unusual high levels of morally good behavior, such as donating a kidney to save another's life, setting up an effective charity organization over the course of years, or blowing the whistle on a powerful organization. Note that the existing research has not yet put us in a position to commit to particular dimensions of morality, particular virtues, or particular features of the morally exceptional at this point. Furthermore, we do not see a benefit in limiting future discoveries by limiting such dimensions this early in the research.

The study of the morally exceptional hinges on two assumptions: (a) some individuals are stably more moral than others and (b) some individuals are exceptionally more moral than others. However, across both philosophy and psychology, some doubt has arisen about the existence of meaningful differences. Such doubt, if valid, implies that there cannot be morally exceptional individuals. This doubt is based on the following ideas:

1. Slight changes in situational characteristics appear to lead to large changes in moral behavior.
2. Therefore, situational characteristics are more powerful in determining moral behavior than are broad character traits such as honesty, fairness, and care for others.
3. Thus, broad character traits do not exist or at least are weak in their determinative power.

In response to this argument, several of us have countered (Fleeson & Furr, 2016; Jayawickreme, Meindl, Helzer, Furr, & Fleeson, 2014) that the effects of situational characteristics on behavior do not constitute the critical evidence bearing on the existence of broad character traits. Rather, we argued that the critical evidence bearing on the existence of broad character traits consists of correlations of an individual's behaviors across multiple contexts. Briefly, if some people engage in virtuous actions more often than others do and they do so repeatedly, there are sufficient reasons to regard character traits as real and important.

When turning our attention to such evidence, we see no room for doubt that the scientific literature supports the existence and importance of broad character traits (e.g., Bleidorn & Denissen, 2015; Helzer et al., 2014; Meindl, Jayawickreme, Furr, & Fleeson, 2015; Fleeson & Jayawickreme, 2017; Miller, 2013, 2014). Just as with other traits, the empirical evidence indicates that people differ from each other in the frequency of virtuous action and that these differences are due to internal psychological traits. However, the research also suggests that most people are 'mixed' in their degree of virtue—neither highly virtuous nor highly vicious. Thus, most people are not exceptionally moral.

People differ in their degree of moral character, which means that some people might be very morally good. If there are morally exceptional individuals then they would be those people. We believe that previous research has identified enough such examples, as we will review later in the chapter, to conclude that morally exceptional people do exist. Just how high along the scale of moral goodness people can reach is uncertain.

Theoretical, Philosophical, and Theological Accounts of the Morally Exceptional

Given that some people are more moral than others and given that it is possible that some people may be exceptionally moral, philosophers and theologians have contemplated what such people would be like, what would motivate them, and what their existence would mean for theories of morality and ethics.

What Features Are Relevant to Qualifying as Morally Exceptional?

What is it that makes someone morally exceptional? One approach is to focus on the behaviors enacted by the moral person. People may qualify as morally exceptional because they reliably behave in a morally good way. They may have performed one or a small number of extraordinary acts, such as rescuing Jews during the Holocaust. They may have often enacted a variety of highly moral acts, such as involved in the complex task of founding charities. If behavior can qualify people to be morally exceptional, behavior may also be capable of disqualifying people, if it is morally reprehensible (Doris, 2002; Fleeson & Jayawickreme, 2017).

But one might wonder whether more is involved, such as being reliably responsible to the appropriate moral reasons or having morally worthy motives for action. Indeed, perhaps the leading answer among philosophers would be to say that a morally exceptional person is someone who has a certain kind of character, namely a virtuous character (Zagzebski, 2017). This in turn invites questions about what the virtues are and what the most promising account of them should be. Virtues are often thought to bring in motivations and beliefs as key components or qualifying features, in addition to behavior (Wright, Warren, & Snow, in press).

The Desirability and Motivational Balance of the Morally Exceptional

Historically, there has been skepticism, and even antagonism, toward purportedly morally exceptional people (MacFarquhar, 2015). They are seen as necessarily flawed in some important regard. For example, Williams (1985) argued that there are values such as aesthetic values and relationship values other than moral values; extreme morality necessarily crowds those values out, creating imbalance. Market economies and evolutionary theory appear to suggest that selfishness is both natural and morally—or at least practically—good (e.g., Joyce, 2007). The psychoanalytic theory argues that morally extreme people are masking aggression toward another with whom they have identified (Sandler & Freud, 1983).

Wolf (1982) has famously argued that, at least on leading models in contemporary ethics, it would not be rational or desirable for human beings to aspire to become morally perfect. There are other values in life, and these other values arise in authentic desires and concerns that are unique to each person. Denying these values for the sake of morality inherently imposes an external and abstract motivation that is not true to the person, creating an imbalance. According to this line of argument, highly moral people are likely to be drab, boring, and stark, which is not ideal.

Wolf's arguments have met with criticism. Adams (1984) noted that actual highly moral people (e.g., historical saints) do not appear to be drab, uninteresting, or to have forsaken other values. Their motivation appears to have arisen in wanting to share goodness with others. Carbonell (2009) argued that morally exceptional people actually use hobbies, wit, and negativity, etc., in order to be effective. Morally exceptional individuals do not seem to be motivated by extreme and stark morality, but rather at least partly by the good they will accomplish.

A related issue is whether morally exceptional actions are required of all people or are merely *supererogatory*, such that failing to live up to a standard of excellence is not wrong and not blameworthy for the rest of us (Urmson, 1958). Pybus (1982), McGinn (1992), and others, though, have argued that being morally exceptional (indeed, being morally perfect) is a moral requirement. The morally exceptional could also have a 'ratcheting-up effect', whereby their morally relevant actions serve to increase the moral expectations that the rest of us are under (see Carbonell, 2012).

The Role of the Morally Exceptional in Defining and Founding the Virtues

It has been common in recent thinking about virtue ethics to place the morally exceptional person into a foundational, definitional role in the theory. Morally virtuous people's behavior is taken to define what it is for an action to be right or obligatory. For example, if morally virtuous people tend to read extensively about history, then virtue would include reading about history. But problems have arisen as well, such as cases involving situations which a virtuous person would never get herself into in the first place (Johnson, 2003). More work needs to be done in sorting out the relationship between the morally exceptional person and the central ethical concepts of the most defensible versions of virtue ethics.

If the morally exceptional define what actions are obligatory, then obligatory actions cannot also be the means of identifying the morally exceptional. This would be circular. Rather, an alternative means of identifying the morally exceptional is required if the morally exceptional define the obligatory. One kind of possible alternative consists of the emotional and intuitive responses to the morally exceptional. Zagzebski (2017) argues that admiration can be a reliable indicator of virtuous individuals, who can then be used as guides to understand the concepts of the good, right action, and virtue.

For example, for centuries Christian scholars have treated Jesus as the exemplar of the moral life. Since the Christian tradition maintains that Christ was both fully human and fully divine, he is the paradigmatic instance of moral perfection in that tradition. An individual's life will be morally perfect to the degree he or she manages to imitate Christ. Similarly, a large body of theological literature appeals to the saint as an explanatory paradigm (Clark, 2019; Zagzebski, 2017). What can the lives of saints tell us about our own lives and our own struggles, about what it means to practice a specific virtue, and about the moral life in general in Christianity and in other religious traditions?

Similarly, the Buddhist Jātaka stories depict the Boddhisatva (earlier incarnations of Siddhartha Gautama, the future Buddha) in a series of more than 500 stories. Each story highlights a virtue displayed by the Boddhisatva that results in a positive resolution of that story. The Jātaka tales have been widely used as an educational tool by Buddhists to promote the cultivation of key virtues (McRae & Cozort, 2018).

Empirical Findings: Why Are Some People Extremely Moral?

The fact that some people are more moral than others does not necessarily imply that any one individual is extremely moral. Perhaps even the most moral of us, in fact, fall well short of extreme moral praiseworthiness. However, realizing the promise of the morally exceptional requires that some such people actually exist.

The first step in the study of the morally exceptional is thus to develop a method for finding and identifying the morally exceptional. Two types of method have been employed (Floyd, 2016). The first type is to identify people who have demonstrated an exemplary action or set of actions. For example, people who have rescued Jews during the Holocaust received awards for bravery or humanitarian actions, donated organs to strangers, been whistleblowers, or fought against bullies. These actions are widely regarded as morally extreme and as morally good actions.

The second type of method is to obtain nominations of morally exceptional people who fit theoretically developed criteria. Colby and Damon (1994) pioneered this method by working with diverse experts to develop the following five criteria for selecting nominees:

- A sustained commitment to moral ideals
- Action consistent with moral ideals
- Willingness to risk self-interest
- The tendency to inspire others
- A realistic humility about one's own importance

The resulting nominees included creators of charities, crusaders for social justice, and a university president. Similar criteria were developed for nominations of young adults (Hart & Fegley, 1995; Matsuba & Walker, 2004; Reimer & Wade-Stein, 2004), historical figures (Frimer, Walker, Lee, Riches, & Dunlop, 2012), and individuals from specific professions, such as coaches (Hamilton & Lavoi, 2019) and business leaders (Den Hartog, 2015).

Researchers have been able to find many morally exceptional people with these methods. For example, *Yad Vashem* (2019) has now recognized over 25,000 rescuers of Jews during the Holocaust, even using strict criteria. However, it is clear that these two methods only begin to scratch the surface of potential methods for identifying people, and there is fertile ground for innovation of additional methods.

The second step is to investigate the individuals so identified and compare them to nonexemplars. The goal is to discover the characteristics of the morally exceptional people, the causes of their moral exceptionality, and the consequences that follow from it. In line with a central question of this book, we will pay particular attention to the question of whether the motivations for the morally exceptional appear to be balanced or imbalanced.

To date, the dominant method for such investigation has been the interview and qualitative analysis, most frequently with Holocaust rescuers. Sometimes researchers have coded speeches or other texts, employed quantitative questionnaires, or obtained ratings and impressions from knowledgeable others. We have not discovered any experimental approaches.

Across the types of methods and the types of morally exceptional individuals, several categories of possible explanations for their moral exceptionality have emerged. Because the methods were empirical, rather than theoretical, the findings of these studies generate a reasonable degree of confidence that they accurately describe the way morally exceptional people actually are. However, because the methods were qualitative and nonexperimental, it is important to acknowledge that the findings are not definitive.

The Role of Social Structures

One category of possible explanations concerns the social structures surrounding the individual. These social structures are important because they can enhance the incentives for extreme moral behavior and reduce the risks attached to extreme moral behavior. One social structure supporting Jewish rescue was a history of respect for, interactions with, and prewar friendships with Jews among rescuers, enhancing the incentive for rescue (Oliner & Oliner, 1988). Rescuers lived in societies where churches protested against anti-Semitism and there was less prewar victimization and segregation of Jews. Their countries' citizens already had a strong national identification, so they did not require scapegoating to bolster the identification (Midlarsky, 1985). Local leaders tapped into social structures, norms, and infrastructures to elicit help from potential rescuers. Leaders provided emotional, physical, and financial support and reduced the risk. Leaders often had prior legitimacy in their communities and rose in communities with few informers (Gross, 1994). Indeed, in a regression model, the effect of perceived risk on rescue behavior was explained by the availability of social and material support, suggesting that having such support may have nearly eliminated the perception of risk as a barrier to rescue (Gross, 1994). Similarly, defenders of bully victims in school tended to have high social status (Sainio, Veenstra, Huitsing, & Salmivalli, 2011) and high regard in the community (Salmivalli, Lagerspetz, Björkqvist, Österman, & Kaukiainen, 1996). Whistleblowers tended to draw on their social capital for support (Glazer & Glazer, 1999), and ethical computing reformers tended to come from organizations supportive of ethical choices (Huff & Barnard, 2009).

Gross (1994) concluded that a surrounding infrastructure—safety, material enablers, and leadership—played a vital role in galvanizing individuals to act. These factors suggest that rescuers, whistleblowers, and defenders of bully victims—despite engaging in extremely rare and risky behaviors—did not appear to have an imbalanced set of motives. Rather, strong motives to help others were bolstered by supportive social structures, and risks were mitigated to a more reasonable level by protective factors in the social structure.

Proximate Situational Causes

A similar category of explanations zooms into more proximate situational causes of rescue behavior. These factors also served to enhance incentives and to reduce risk. More rescuers lived on farms or in small villages and fewer lived in small- to medium-sized cities, compared to matched controls (Oliner & Oliner, 1988). The number of rooms in a person's house was a positive predictor of helping (Varese & Yaish, 2000). Being directly asked to help was the strongest discriminator between those who helped and those who did not in a multivariate model (Varese & Yaish, 2000). Altruistic kidney donations were often incentivized by priority for future needs for self and family (Roth, Sönmez, & Ünver, 2004). A nanny for a Jewish boy and girl, Antonina Gordey, visited the children after their relocation to a ghetto. When she learned the boy had been killed by repeated blood drawing, she immediately removed the girl from the ghetto; later, when the nanny married, she let her husband believe that the girl was her own illegitimate child (Yad Vashem, 2019).

Whistleblowers tended to have had expertise in the domain of their whistleblowing and had strong and clear evidence for the wrongdoing (Glazer & Glazer, 1999). They depended on others to support them in the subsequent conflict (Glazer & Glazer, 1999). Thus, the proximate situation served to enhance the incentive for and reduce the risk of acting, allowing strong—but apparently reasonably balanced—motivations to be realized in behavior.

Developmental History

The third category of explanations focuses on developmental causes. Growing up, rescuers received less discipline, and when they did, it was less punitive and involved more reasoning. They had closer families with fathers who were more religious. Their families had valued universal ethics more than had control participant's families (Oliner & Oliner, 1988). Rescuers reported identification with a very moral parent (London, 1970). These kinds of upbringing experiences have been associated with later development of healthy, autonomous motivation for prosocial action (Assor, 2012).

Similarly, individuals who received awards for bravery or humanitarian actions reported having had positive childhood attachments (Walker & Frimer, 2009). They had life stories that featured themes of redemption, agency, and moral motivation (Matsuba & Walker, 2005; Walker & Frimer, 2007). They appeared to develop terminal communal motivation at an earlier age than did controls. Young activists had a serving orientation that was triggered by an event and supported by parents and other adults (Michaelson, 2002).

These findings do not paint a picture of a challenging environment producing imbalanced motivations, but rather a picture of a supportive environment in which the child was cared for and the desire to care for others was nurtured.

Personality Traits

Another category of explanations focuses on characteristics of the morally exceptional themselves. Across types, morally exceptional individuals tended to be motivated by high empathy, social responsibility, connection to other people, integrity, honesty, and universal moral values (Carlo, PytlikZillig, Roesch, & Dienstbier, 2009; Fagin-Jones & Midlarsky, 2007; Monroe, 2002; Oliner, 2002; Oliner & Oliner, 1988). Social responsibility and empathy in particular strongly discriminated rescuers from nonrescuers (Midlarsky, Jones, & Corley, 2005).

Kidney donors had larger amygdala and showed greater amygdala reactivity to fearful faces, revealing distinctive emotional characteristics of the morally exceptional in physiology (Marsh et al., 2014). Rescuers also had greater independence, self-reliance, risk-taking, sense of adventure, and internal sense of control (London, 1970; Midlarsky et al., 2005; Oliner & Oliner, 1988; Tec, 1986). Exemplars did not struggle with themselves, work in isolation, or live grim, joyless lives (Colby & Damon, 1994). Rescuers had relatively negative views of Nazis and positive views of Jews (Oliner & Oliner, 1988). They had a lifelong commitment to helping the needy, such that rescue seemed automatic and compulsory (Tec, 1986). Kidney donors felt very connected to others (Clarke, Mitchell, & Abraham, 2014). Exemplars, rescuers, and award winners tended to be relatively religious (Colby & Damon, 1994; Matsuba & Walker, 2004; Monroe, 2002; Varese & Yaish, 2000). Rescuers also had characteristics that were identifiable enough to victims to lead to victims asking them for help (Varese & Yaish, 2000).

Whistleblowers used anger to sustain their motivation against retaliation (Glazer & Glazer, 1999). Defenders of bully victims expected their defenses to be successful and valued that success (Pöyhönen, Juvonen, & Salmivalli, 2012). Hospice volunteers reported that being true to oneself and respecting others were among the most important things in life (Oliner, 2002). People who helped in a car accident had a relatively greater internal locus of control, belief in a just world, social responsibility, and empathy (Bierhoff, Klein, & Kramp, 1991).

Moral award winners and highly influential moral individuals from the 20th century had a motivational profile in which they employed agency in the service of communion (Frimer, Walker, Dunlop, Lee, & Riches, 2011; Frimer et al., 2012). For example, the suffragette Emmeline Pankhurst spoke of fighting for rights (agency) in the service of the betterment of women (communion). This pattern appeared to be the end point of normal adult development, such that many individuals reached this point by late life (Walker & Frimer, 2015). Morally exceptional public administrators similarly may be distinguished by their use of power for the public good and being humane and compassionate in the application of power (Cooper & Wright, 1992).

These characteristics of morally exceptional individuals do not present a clear imbalance in motivations. There is a motive to care for others that appears to be missing in the normal case, plus the capacity to carry out the actions. Opp (1997) argued that rescue was a rational choice based on a strong internalized intrinsic value of others' welfare. These highly altruistic actions were completed by ordinary people who cared deeply about helping others and who encountered a situation where help was needed.

Moral Reasoning

Another set of explanations focuses on the qualities of moral reasoning that led to the decision to help. It is a long-held hypothesis that moral and immoral behaviors are natural consequences of how one thinks about morality. Morally exceptional individuals tended to extend the natural concern for others to all others (Geras, 1995; Crimston, Bain, Hornsey, & Bastian, 2016; Monroe, 2002; Oliner, 2002; Tec, 1986). Because of this extension of common humanity to others, they could see the need for help in others and felt a direct connection to others (Monroe, 2002). Others were seen as fellow humans in need (Geras, 1995).

The reasons rescuers offered for their rescue behavior were mostly ethical, humanitarian, and universal (Oliner, 2002; Oliner & Oliner, 1988). Gross (1994) found one group of rescuers to be characterized by sophisticated, high-stage, universal, but inconsistent moral reasoning. For example, such rescuers believed that 'Our country is founded on the basis of equal rights and liberty. Each citizen has the duty to protect these rights when they are threatened even which such action is dangerous and against the law' (Gross, 1994). A second group was characterized by simple, low-stage, conventional, but consistent and strict moral reasoning. Such rescuers endorsed motivations such as 'A local authority such as the Church or Resistance was aware of how important it was to save the Jews and one should listen to the authorities especially in time of crisis' (Gross, 1994). Midlarsky et al. (2005), however, found more consistently high-level moral reasoning among rescuers. Fagin-Jones and Midlarsky (2007) also found that high-level abstract moral reasoning was the strongest discriminator between rescuers and bystanders. Such abstract reasoning led to the conclusion that rescue was a required behavior (Oliner, 2002; Tec, 1986).

A focus on abstract, high-level moral reasoning, and an extension of shared humanity to all humans do not appear on the surface to reflect an imbalance in motivation. Rather, they may represent an ideal toward which humans (should) strive. The one exception to this picture is that Varese and Yaish (2000) argued that rescuers considered only the gain of the action rather than the cost and risk. A lack of consideration of risk would be an imbalance. However, many other theorists did not agree with this conclusion (Fagin-Jones, 2017; Gross, 1994; Opp, 1997). Additionally, kidney donors appeared to have about the same appreciation for risks as did nondonors (Maple, Hadjianastassiou, Jones, & Mamode, 2010; McGrath, Pun, & Holewa, 2012).

Identity as Motivation

A final category emerged repeatedly and spontaneously from independent investigations and concerned the self-definition of morally exceptional individuals. Morally exceptional individuals included connection to others in their self-definitions, saw all humanity as the same as them, and defined themselves in terms of being moral and helping others.

Sergeant Edmonds was a POW in Germany. When the Germans ordered all Jews to stand separately, Sgt. Edmonds convinced all POWs to stand with the Jews. The German officer said they can't all be Jews, and Sgt. Edmonds replied,

'We are all Jews'. The officer threatened to shoot Edmonds, who replied that the officer would have to shoot all the POWs, which would result in being tried for war crimes. The officer relented (Yad Vashem, 2019).

Many morally exceptional individuals saw themselves as having common humanity with others, such that others were basically oneself (Brethel-Haurwitz et al., 2018; Dunlop, Walker, & Matsuba, 2012; McFarland, Brown, & Webb, 2013; Monroe, 2002). Rescuers saw Jews as the same as themselves, as equally needy as themselves (Oliner & Oliner, 1988). Seeing others as sharing a common humanity invoked an attachment to those others and a responsibility for those others (Shepela et al., 1999). People who had donated organs to a stranger showed significantly enhanced neural connectivity between self and others when observing others' pain (Brethel-Haurwitz et al., 2018). Seeing oneself as overlapping with others made others' pain and needs salient so the morally exceptional could see the magnitude of the suffering and the intensity of the need for help.

All three of philanthropists, Carnegie Hero Award winners, and Holocaust rescuers saw themselves as strongly linked to others through a shared humanity (Monroe, 2002). Others needed their help so they acted spontaneously to provide that help.

For morally exceptional people, morality is not outside the self or a constraint on the self, but rather part of the self-definition (Colby & Damon, 1993; Huff & Barnard, 2009; Matsuba & Walker, 2005; Monroe, 2012; Walker & Frimer, 2007, 2009). They see their actual self as integrated with their ideal and their parental selves (Hart & Fegley, 1995). Morally, exceptional individuals' goals for their own lives were integrated with their central moral values (Colby & Damon, 1995).

Although infrequent and unusual, such identity motivations similarly do not appear obviously to be imbalanced or overriding. Rather, connection to others, connection to all humanity, and inclusion of morality in one's self-definition seem reasonable and balanced aspects of identity. Such integrated motivation is associated with well-being and growth (Assor, 2012). Indeed, the case could be made that the normal situation—in which people define themselves as separate from others, as not sharing a common humanity with those who are different, or in terms not closely related to morality—may be the imbalanced case. Including morality and shared humanity in one's identity may be a healthy self-regulatory step for bringing one's goals in line with one's values, reducing fragmentation and alienation of the self.

Summary of Empirical Findings: Ordinary People With a Deep Concern for Others

The fairly consistent message across rescuers, award winners, philanthropists, organ donors, political heroes, defenders of bullies' victims, and whistleblowers is that although these people engage in extreme acts, they appear to be rather ordinary people who have a very deep and self-defining concern for others. This strong motivation to care for others is strong enough to compete with the motive for self-preservation.

Imbalance, Different Balance, or Balance? The behavior of morally exceptional people is often intense and rare. Rescuing victims of the holocaust, donating a kidney, or blowing the whistle on wrongdoings at a corporation are rare actions that could be labeled as intense, at least in magnitude of effects. Additionally, these actions involve sacrifice and risk. Performers of any of these actions can experience retaliation or silencing; sometimes, the action itself directly causes harm to the individual who performs it. Costs can be financial, physical, or relational.

Kruglanski et al. (in press) have developed a theory to explain why some people are willing to conduct intense and rare actions. They call these actions 'extremism'. Although extremism is usually applied to actions viewed negatively (e.g., terrorism), Kruglanski et al. (in press) intend the theory to explain both positive and negative forms of extremism. They explain extremism as the result of a motivational imbalance. Thus, the reason some people conduct extreme acts is because they have imbalanced motivations.

Kruglanski et al. (in press) argue that humans have basic needs, and when all of the needs are met within a reasonable and comparable range of satisfaction, then the individual has motivational balance. But when one of the needs becomes predominant, it crowds out the other needs to create an imbalance. In extremism, one need becomes extremely predominant, substantially interfering with the satisfaction of the other needs. In addictions, for example, the need for the drug or activity becomes extremely predominant, interfering with the satisfaction of hunger, security, and relational needs.

Even extremism that might be viewed positively, such as extreme athletic performance, artistic creation, or moral extremism, is argued to be caused by such an imbalance. Although Kruglanski et al. (in press) argue that imbalance is not necessarily a bad thing, but is rather merely a descriptive account of extremists' motivational profiles, they also propose that balance is essential to well-being and that imbalance is a source of great cost to individuals and societies.

Kruglanski et al. (in press) might argue that the explanation for people conducting morally exceptional acts is that such people have a strong drive to do good, which is greater than their motivation for other, more selfish needs. The basic human need to be morally good (Prentice et al., 2019) is predominant. They might answer the question 'Why are some people morally exceptional?' by responding that such people are the way they are because they care much more about doing good than they care about their self-serving needs.

Although we concur that morally exceptional people have a deep and abiding concern for others, we are not able to agree that the morally exceptional are imbalanced. Partly, this is because we consider the term 'motivational imbalance' to suggest inappropriate, self-defeating, or detrimental motivational profiles. We do not believe that morally exceptional behavior is inappropriate, self-defeating, or detrimental to life quality. We believe that a deep concern for others is often or usually an appropriate and self-affirming motivation that enhances life quality.

Nonetheless, if we temporarily accept their definitions of motivational balance and imbalance, we are not sure it applies to the morally exceptional. The limited existing evidence is not enough to be certain one way or the other, but the evidence that does exist suggests that morally exceptional people may have raised their concern for others to a strong enough level that it can be brought into dialogue with their concern for themselves. Concern for others can be weighed against self-serving needs, rather than always being secondary to self-serving needs. Such people are appropriately humble, in the sense that they see themselves as no more nor no less important than others (Wright et al., 2018).

For example, Holocaust rescuers typically did not take reckless risks. They were more likely to help victims when the rescuers had relatively undetectable hiding places, when in supportive communities, and/or when they had access to food or funds to mitigate the costs (Fagin-Jones, 2017; Gross, 1994; Opp, 1997; although see Varese & Yaish, 2000). Piotr Wrona limited the bread he brought to starving children to five rolls to reduce the risk of angering the Germans.

However, Holocaust rescuers did take some risks and did suffer costs, sometimes quite high ones. They did so because of their deep and abiding concern for others. The idea is that they weighed the need of others against their own needs. Morally exceptional people were willing to accept a modest risk to their own safety in order to prevent almost certain death for another. That is, they compared the severity of cost to the other against the severity of cost to the self, in a way that might represent a balancing. Piotr Wrona definitely took a mortal risk by feeding starving Jewish children, but he was able to keep the risk to a low level in order to prevent a much greater risk to the lives of the children.

Similarly, defenders of bully victims and whistleblowers often intervened only when they had existing social support to mitigate the risk of the intervention (Glazer & Glazer, 1999; Huff & Barnard, 2009; Sainio et al., 2011; Salmivalli et al., 1996). Risk mitigation puts the trade-off of large risk to others versus small risk to self into a reasonable balance.

Another relevant point from the evidence is that morally exceptional individuals tended to come from healthy families with close attachments and less discipline (Assor, 2012; Oliner & Oliner, 1988; Walker & Frimer, 2007). Healthy families would not be the expected breeding ground for imbalanced motivations.

Finally, the morally exceptional had more sophisticated moral reasoning (Midlarsky et al., 2005). They were found to have a relatively equal pursuit and integration of agency and communion motivation (Frimer, Walker, Lee, Riches, & Dunlop, 2012).

The evidence—as limited as it is—may suggest a counter-possibility. Namely, it is possible that in the Kruglanski et al. (in press) sense, the morally exceptional are the motivationally balanced people and the nonmorally exceptional are the ones who are imbalanced. The nonmorally exceptional (average people) strongly privilege self-serving needs over the needs of others and the need to be a good person. This heavy privileging means that the need to do good and to help others has become unsatisfied and diminished. For example, normal citizens in Germany during the Holocaust may have been too frightened, angry, or dazzled to take on small risks in order to prevent certain death for others. One Polish citizen guarded the children of a Holocaust rescuer while the Germans searched for hidden Jews. The children tried to jump out a window to escape, but the citizen interfered, shortly after which the children were executed. A witness stated 'How his conscience allowed him to do this I don't know. . . . In those days, at the Baranek's farm, we were terrorized and paralyzed into inaction' (Tec, 1986, p. 66).

Closer to home, many people are unable to make small sacrifices in order to achieve great impacts on other people's lives. The average American is estimated to donate only 2–5% of income annually even though just a few dollars could save the lives of children (Singer, 2015).

In contrast, the morally exceptional are willing to bring others' needs into conversation with their own needs, so reasonable and balanced trade-offs can be made between the needs of others and the needs of oneself. The need to be a morally good person (Prentice et al., 2019) is as strong as needs for the self are. In the normal case, it is self-preservation and self-pleasure that may be too prominent, leading to the neglect of caring for others, especially of caring for those not close to the individual. This imbalance may lead to values and goals being misaligned, and the self becoming fragmented. However, the evidence is somewhat limited on these points, and future research is needed.

Another possibility is that the morally exceptional and the nonmorally exceptional are both balanced, but the morally exceptional appear to be imbalanced because they are balanced in an unusual way. They appear to be imbalanced because they satisfy multiple motivations at once when they engage in moral actions, and this is unusual. Such unusual integration of motives leads to the misapprehension that the morally exceptional are imbalanced. Specifically, because their moral motivation is incorporated into their self-concept, they satisfy self-serving needs when they act in moral ways. But because many people have not incorporated morality into their self-concept, they satisfy their need to be moral (Prentice et al., 2019) in separated actions. And such people see all moral actions this way, so the integrated actions of the morally exceptional appear to neglect self-serving motives when they in fact are satisfying them. The evidence regarding this possibility is also unfortunately minimal.

What Is Not Known

As this brief review makes clear, there is still much to learn about the morally exceptional. Current research is only at the beginning stages. Given what is known and what remains unknown, we see the following set of questions as essential to the study of the morally exceptional.

1. Can the morally exceptional be reliably identified? This question is critical because the field cannot have a firm basis without some level of agreement about who counts as morally exceptional. This question is difficult because it is hard to satisfactorily define morality and to satisfactorily define exceptionality. Nonetheless, we believe it is possible to do so, with at least some degree of reliability.
2. What are the statistical and methodological problems and solutions, in the study of a select group of individuals such as the morally exceptional? Studying extremes creates unique statistical and methodological problems. For example, it is not clear whether self-reports of extremely good status can be trusted. There are also issues of generalizability from extreme groups to the population, and the nonnormal distributions involved in exceptionality. One challenge to the morally exceptional frequently raised is whether it would be better to study the whole distribution in order to use continuous scales rather than truncate the scale at one end.
3. Which general psychological faculties associated with moral thought and behavior are involved in the morally exceptional? How do these psychological faculties operate? What are the mechanisms underlying morally exceptional behavior? Although some correlates of moral exceptionality have been identified in the literature, there are many more to test. Furthermore, very little research has tried to discover the processes that underlie moral exceptionality and how those processes produce morally exceptional behavior. These questions seem urgent to us.
4. If it can be shown that a morally exceptional individual has deep-seated flaws or even vices, what conclusions should that lead us to draw about the virtues we have traditionally attributed to him or her? Can exemplary virtues exist alongside deep-seated vices? Is Thomas Jefferson's contribution to the creation of democracy on the planet disqualified by this ownership of slaves?
5. What are the lay conceptions of the morally exceptional (what do people believe about moral exceptionality, how are the morally exceptional seen by others)? Lay theories about the morally exceptional provide one way to ground the study of the morally exceptional in a common view of the morally exceptional. It is important to discover who people think of as being morally exceptional, what criteria people normally use to designate someone as morally exceptional, and what qualities people ascribe to the morally exceptional. This also reveals common perceptions of what is good, and what guides people are using in their lives.
6. Should we aspire to become morally exceptional people? Are they worthy of admiration, and are they living lives that should be deemed flourishing lives? As elaborated in the section on balance and in the philosophical review, there

is a long history of debating whether morally exceptional individuals are to be taken as models of how everyone should strive to be or rather are admirable but unbalanced persons, and not taken to be models. This question gets to the very heart of one of the most basic questions, which is how we are to live. If the morally exceptional are worth aspiring to, does it become mandatory to be like them or only optional? Is being morally exceptional required by morality or is it merely supererogatory?

7. What role should the concept of an exceptional person play in the foundation of virtue ethics? Should it play a role in an account of right action, for instance? Is such a theory already implicit in theological writing on moral exceptionalism? Although familiar perspectives on ethics and morality are based in rules and principles, a long-standing perspective on ethics is based on virtues, as embodied in real people. Seeing what virtuous people do is the way to learn what are the ethical and moral ways of living. In this case, the morally exceptional could become the basis for all morality.

8. Are there social consequences (costs or benefits) of being morally exceptional? It is also important to understand how the morally exceptional interact with others. The predominance of moral concerns could interfere with normal social interaction, it could elevate normal social interaction or it could be irrelevant to normal social interaction. Morally exceptional people could be off-putting or they could be elevating.

9. What are the developmental origins of moral exceptionality and its life course? If scholars can discover the developmental etiology of moral exceptionality, scholars will be able to develop and test programs for increasing the moral goodness of the rest of us. Of course, this requires to determine whether being morally exceptional is even psychologically realistic for most people.

Three Difficulties for the Study of the Morally Exceptional and Strategies for Addressing Them

Answering the earlier questions will help to achieve the promise of studying the morally exceptional. It will answer questions presented by the mere existence of the morally exceptional, such as revealing the best humans can achieve, whether morally exceptional individuals are deprived and imbalanced, whether they are integrated in society or shunned because of their serving as a reminder of our own failings, and whether even morally exceptional individuals might have a dark side. It will help societies learn how to enhance the positive effects of the morally exceptional and to reduce the negative effects of immorality. It will reveal practical clues about how to aid in the development of moral character. And it might accelerate the growth of knowledge about morality in general.

However, the study of the morally exceptional has unique difficulties due to its subject matter. These difficulties can hinder effective research on the morally exceptional and often intimidate researchers away from this topic. Thus, we wish to acknowledge these difficulties and offer a menu of possible solutions to those difficulties. We believe each of these solutions is viable—adopting any of them will likely overcome the difficulties and lead to effective research. As a result, the promise of studying the morally exceptional will be realized.

One particularly exciting feature of these difficulties and solutions is that they consist of interdisciplinary mixtures of insights from both empirical and theoretical disciplines. The insights of philosophers and theologians have been integrated with the findings of psychologists to both realize important obstacles and suggest realistic solutions for overcoming those obstacles.

Difficulty 1: What Counts as 'Moral' and What Counts as 'Morally Good'?[1]

Julian Assange founded WikiLeaks in 2006. WikiLeaks has revealed significant confidential information about corruption, seedy political behavior, and national security issues. Some claim he is a hero of accountability and freedom of information; others claim he is a sensationalist spy and traitor to his nation.

Any research on topics relevant to morality is confronted simultaneously with the lack of agreement on the definitions of 'moral' or 'morally good', and with the intensity of concern about the correct definition. People disagree whether a given action is morally relevant or not. For example, some might think that arriving at work late or taking office supplies home are morally relevant actions, while others may think they are morally neutral actions. People also disagree about whether a given action is morally good or morally bad. For example, some might judge whistleblowing about government spying to be morally good whereas others may judge it as morally bad.

The need to make a commitment about the domain of morality and about what is morally good versus bad threatens the value of studying the morally exceptional altogether. Widespread disagreement and the inability of empirical research to resolve such disagreement may lead to complete disagreement about who qualifies as morally exceptional. Complete disagreement about who qualifies may reduce communication between scholars or accumulation of findings. For example, if researchers disagree about whether hospice nurses count as morally exceptional, then it is not clear whether to include findings about hospice workers in knowledge about the morally exceptional. This difficulty becomes especially severe if findings differ across contested populations.

We do not believe this difficulty is insurmountable. We have developed several strategies that we illuminate in the following for dealing with this difficulty. We believe at least some of these strategies will make research on the morally exceptional possible despite disagreements. Additionally, we do not believe there is as much widespread disagreement about the domains of morality or about the morally good as is feared. For most actions, we expect that most people agree on whether the action is morally relevant, whether it is morally good, and whether it is morally exceptional. Most people agree that saving a drowning child, anonymously donating organs to those in need, and rescuing Jews in the Holocaust are morally relevant, morally good, and even morally exceptional.

Strategy 1. Select a single moral domain and study individuals exceptional in that domain without regard for the ultimate morality of the domain. In this strategy, each researcher selects one of the various domains, dimensions, or virtues within morality and studies people who are exceptional in this particular domain or dimension. From this point forward, the relevance to morality is disregarded. For example, one might pick honesty and study people who are exceptionally honest. The researcher would not subsequently attend to whether honesty is moral or not. If the selected domain is interesting enough, then its ultimate relevance to morality would not be significant. For example, if honesty were ever deemed not relevant to morality, research on exceptionally honest individuals would nonetheless be interesting because the domain of honesty is inherently interesting.

Moreover, with time, convergences and divergences across domains in the findings about moral exceptionality will accumulate, creating an interlocking network of findings. Findings that are common across domains will be as revealing as findings that differ across domains.

Strategy 2. Compare morally exceptional individuals across moral domains. A second strategy is to take moral disagreement as a scientific opportunity rather than a threat. Suppose that moral exceptionality in different moral domains results from differing underlying processes. That is, the mechanisms that produce and sustain morally exceptional people in one moral domain differ from those mechanisms producing morally exceptional people in another domain. For example, it may be penetrating hope that sustains morally exceptional behavior in the domain of compassion but unflinching self-control that sustains morally exceptional behavior in the domain of honesty.

Even within one domain, such as fairness, individuals who are regarded as morally exceptional for very different types of actions (perhaps even actions that reflect opposing values) may be so regarded for very different reasons. For example, activists in favor of the Affordable Care Act may show different psychological processes than activists opposed to the Affordable Care Act. People on different sides of the abortion issue may follow different processes because of the different values they pursue.

Revealing different processes for becoming exceptional in different domains of morality would be an important finding. It would suggest that the process of becoming morally exceptional is not separable from the domain within morality in which one is becoming exceptional or from the values one is expressing with one's morally relevant actions.

Conversely, it may be true that the processes underlying moral exceptionality turn out to be common across domains of morality. For example, exceptionally honest people, exceptionally compassionate people, and exceptional activists on opposing sides of the gun control debate may all have gotten there via the same psychological mechanisms. Revealing the same processes for becoming exceptional across different domains of morality would likewise be an important finding. It would suggest that the processes of becoming morally exceptional are domain- and value-neutral.

Either discovery would be important for revealing the psychological processes involved in morality. Thus, research has an opportunity to determine which of these possibilities is correct.

Strategy 3. Start with a selection of exceptional individuals and discover the domains in which they are exceptional. A third strategy is a bottom-up approach, in which the researchers take a value-neutral position, and let the domains within morality emerge from the iterative process of the research. The identification of exceptional people comes first; the definition of morality is discovered subsequently during the process of research. For example, researchers could select people whose action or pattern of action is clearly morally good. This is the approach taken by those who study Holocaust rescuers and those who study award winners. Or people could simply be asked to nominate the most moral people they know, without specifying what is meant by 'moral'.

Once individuals are identified, researchers can employ venerated construct validity procedures to validate the identifications and refine the criteria iteratively. Toward this end, identified individuals would be assessed in a variety of ways, including self-report questionnaires, narratives, experience-sampling, conceptual analysis, and behavioral tests in order to iteratively narrow in on a group of people who count as morally exceptional. Whatever emerges as characteristics of the morally exceptional will comprise the domains within morality and the kinds of actions that are considered morally exceptional. For example, if rescuers of Jews during the Holocaust tend to have high levels of compassion, then compassion may be a relevant moral domain.

Difficulty 2: What Counts as Exceptional?

Mother Teresa readily comes to mind as a morally exceptional individual. Her limitless energy in creating a massive organization to help those dying of AIDS/HIV, leprosy, and tuberculosis is heralded as one of the clearest examples of exceptionally moral behavior. But even Mother Teresa has been criticized as less than saintly. Accusations include arbitrary medical treatment, baptisms regardless of the baptized individual's religion, and being primarily a missionary and political organization (Morgan, 1994).

In addition to wrestling with the definition of morally good, the study of the morally exceptional needs to wrestle with the criteria required for the elevation from ordinary morality to exceptional morality. The criteria will be inherently normative because the criteria have to define some actions or features as morally better than other actions or features. Despite the need for these criteria and their normative nature, there is no obvious choice, the decision may be somewhat arbitrary, and there may be disagreement about whether the criteria are correct. For example, most treatments of Holocaust rescuers, including the official World Holocaust Remembrance Center of Israel, do not include as rescuers those who fought in the resistance against Nazi Germany.

For all these reasons, justification for the criteria is necessary. When justifying the criteria, scholars should keep in mind the reason for studying specifically exceptional individuals. One of the main reasons for studying the morally exceptional is that studying those individuals will produce insights otherwise not reachable. The criteria for the morally exceptional thus should identify individuals who will provide those insights.

While considering the qualifying criteria, scholars should also consider the disqualifying criteria. Do any moral flaws disqualify an individual or only heinous flaws? Individuals may be morally exceptional in one respect and morally average or worse in other respects.

Strategy 1: Use a set of ideal criteria for exceptionality without regard for whether any existing individuals meet those criteria. In this approach, criteria of exceptionality are developed theoretically on the basis of considerations of what is ideally required to count as exceptional. This proceeds without regard for whether actual humans meet these criteria, based on the assumption that the accidental nature of humans should not constrain the definition of what is ideal. For example, religious sainthood involves a set of ideal criteria for moral exceptionality based on theological criteria and without regard to the number of humans who meet these criteria.

One source of such criteria could be individuals' life priorities and motivations. Wolf (1982) defines the morally exceptional as those whose lives are dominated by a commitment to improving the welfare of others or society. Similarly, the morally exceptional may be those who devote a significant amount of resources (their career, free time, or finances) to improving the welfare of others. They may be individuals who are almost always motivated to do the right thing or who have other morally appropriate motivations. The criteria may select individuals who are reliably responsible for the appropriate moral reasons or select individuals who notice moral issues more often. The clear and evident possibility of self-sacrifice as a result of those commitments could also be added.

Alternatively, criteria may be based on actions. Behavioral criteria may be focused on the frequency of actions, selecting those who always (or mostly) do the moral action. For example, criteria may select individuals who are gentle, helpful, considerate, and compassionate in their everyday lives. Behavioral criteria could also consist of a single or small number of extraordinary actions on the behalf of others, such as rescuing Jews in the Holocaust.

Another source of criteria could be a panel of experts on morality, who assemble a consensual set of criteria (e.g., Colby & Damon, 1994). Of course, requiring consensus may affect the types of criteria able to be agreed upon commonly, and the make-up of the expert panel is important to the results of the strategy.

Strategy 2: Let the criteria for exceptionality emerge from a nomination procedure. One strategy is to take a bottom-up approach, in which the scholars take a value-neutral position, and let the definition of moral exceptionality emerge from an iterative process of scholarship. In this strategy, a nomination procedure—rather than a set of criteria—is specified, and then those who emerge from this nomination procedure are taken as the exceptional. Further research will determine how these people differ from others and thereby will reveal the criteria for being exceptional.

The umbrella of nomination procedures is quite broad. It is broad enough to include procedures such as canonization, admiration, Internet voting, general public acceptance, and direct nominations. One can give loose guidance to the nominators, with instructions ranging from 'Use your own definition of morality' to 'Nominate people who have made a sustained commitment to a moral principle at substantial cost to themselves'. Or one can include more details in the guidance, recognizing that this builds ideal criteria into the nomination. Interestingly, such procedures can produce examples that may seem very morally good, but perhaps not exceptional. For example, a person might nominate the youth minister at their home church who was always available with a listening ear and wise advice. This person might have been willing to help a person deal with a crisis such as navigating a court date or moving out of the house. This level of compassion and availability might lead to a nomination from one of the beneficiaries of the help but may not qualify in some evaluations as exceptional.

Identified individuals are evaluated on several relevant criteria and compared to nonexceptional individuals. The results of those comparisons then feed into revising the original criteria. In a second round, a new set of criteria leads to selection of a new set of individuals, who are compared. This process continues iteratively until a satisfactory set of criteria is settled upon.

Strategy 3: Use Ordinary Conceptions ('Lay Theories') of Exceptionality to Set the Criteria. In this strategy, scholars first obtain lay theories of moral exceptionality and then use the resulting theories to define the criteria (Walker & Hennig, 2004). For example, scholars might read popular treatises on moral exceptionality and derive criteria from those treatises. Scholars might have subjects evaluate several possible criteria for exceptionality and adopt the ones endorsed most strongly by the subjects. At the very least, this strategy acknowledges that individuals who meet the exceptionality criteria of the public are of some interest, even if they may not meet philosophically rigorous criteria for exceptionality.

Strategy 4: Develop an Assessment Tool to Identify the Morally Exceptional. Learning from the intelligence domain, researchers could develop an assessment tool for moral standing and then select the highest scorers as the morally exceptional. Creating the assessment tool would require some a priori definitions of moral and morally good. However, usual construct validation techniques would refine those criteria based on the results of the assessments. Eventually, a set of coherent and valid criteria may emerge.

In this strategy, the criteria for being exceptional would likely be proportional. Some top percentage of individuals would qualify as morally exceptional. For example, scholars may determine that the morally best 10% of individuals will be designated as morally exceptional or maybe only the top 2%.

Difficulty 3: What Will Be Added to the Study of Morality by Studying the Morally Exceptional in Particular?

This chapter is based on the assumption that studying the morally exceptional will provide unique, important information about issues pertaining to morality and moral psychology that would not be obtained or obtained only slowly, by studying those at all levels of the moral spectrum. However, studying the extreme goes against the standard approach to studying dimensions, which is to study the whole spectrum of that dimension. Thus, researchers will need to justify focusing on the extreme end. Justifications may include factors such as the illumination of the best humans can do, the clear functioning of moral processes in the morally exceptional, the practical impact of the morally exceptional, and the fact that a good way to learn about a topic is often to study that topic at peak performance.

One justification is that the study of the morally exceptional may increase the power of studies investigating morality. This is for two reasons. First, contrasting the morally exceptional with individuals low in morality covers the whole dimension of morality and thus has the greatest chance of producing large effect sizes. A study of the effects of caffeine would most effectively contrast no caffeine against a large dose of caffeine then contrast a medium dose of caffeine to a slightly higher dose of caffeine. Similarly, a study of the effects of morality would better contrast people with very little morality to those with a large amount of morality than contrast people with average morality to those with slightly more morality.

A second way in which the study of the morally exceptional may increase power is by increasing the number of observations in a study. The morally exceptional may have more morally relevant encounters, life choices, and events than those who are less moral. Particularly in studies of the psychological processes underlying morality, the number of such encounters, choices, and events can directly affect the likelihood that analyzes will reveal detectable effects. In sum, samples of morally exceptional individuals may increase the power of studies investigating morality processes, because sample size and effect size are the two determinants of statistical power. Of course, later research will be needed to test the generality of the findings to the whole spectrum of morality.

Connecting findings about the morally exceptional to the rest of people will also require some scholarly effort. It is not known whether the morally exceptional differ from those less morally good in a qualitative or in a quantitative manner. Differing in a qualitative manner would mean that the morally exceptional function differently from others, that they have different psychological processes, that different features are needed to describe them, or that there is a large gap between them and the next most morally good individuals. Differing in a quantitative manner would mean that the morally exceptional differ from others in amount but not in kind. They would then be valuable to study because they have high levels of morality processes, making research more efficient, because they reveal the best humans can be, and because of their practical impact on the world.

Conclusion

Just as literature can be understood and comprehended by studying the best works in the canon, we believe that a complete understanding of morality and virtue can only be arrived at through examining the morally excellent. Just as much attention has been paid to how geniuses and high-performing teams in business function and thrive, we believe the morally excellent represent another form of 'genius' that is equally deserving of such attention. This paper is intended to provide guidance for those beginning work on the morally exceptional. It describes the reasons the morally exceptional are a particularly fruitful population to study and starts with an account of the existing knowledge about the morally exceptional. After reviewing the existing empirical research, it concludes that the morally exceptional do not have an immediately obvious motivational imbalance. Indeed, a case could be made that they are more balanced than others are because they care about others as well as about themselves. The chapter then raises some questions that might be worth addressing when studying the morally exceptional. The bulk of the paper takes on two major difficulties facing the study of the morally exceptional. Both difficulties arise largely from the normative nature of the construct and the resulting disagreement about what is morally good and about what is morally exceptional. We suggest several alternative strategies for addressing these difficulties. The chapter finishes with some smaller difficulties. We do not believe these difficulties are particularly insurmountable, and with the right strategy, we believe they are resolvable.

Acknowledgments

We thank the generous members of the White Paper Expert Group, Lawrence Walker, Candace Vogler, Jennifer Herdt, and Linda Zagzebski, for their consultation in the creation of an earlier version of this document. We also thank the members of the WFU Beacon Project Campus Reading Group for their thoughts about the morally exceptional.

This paper was made possible through the support of a grant from the Templeton Religion Trust. The opinions expressed in this publication are those of the authors and do not necessarily reflect the views of the Templeton Religion Trust.

Note

1. 'Moral' here is meant to refer to the domain of morality in general, without specification of whether an action is morally good or morally bad. 'Morally good' refers to actions that are evaluated positively on the moral dimension.

References

Adams, R. M. (1984). Saints. *The Journal of Philosophy*, 392–401.

Assor, A. (2012). Autonomous moral motivation: Consequences, socializing antecedents, and the unique role of integrated moral principles. In M. Mikulincer & P. R. Shaver (Eds.), *The social psychology of morality: Exploring the causes of good and evil* (2011–09275–013, pp. 239–255). Washington, DC: American Psychological Association.

Bierhoff, H. W., Klein, R., & Kramp, P. (1991). Evidence for the altruistic personality from data on accident research. *Journal of Personality*, *59*(2), 263–280.

Bleidorn, W., & Denissen, J. J. A. (2015). Virtues in action—The new look of character traits. *British Journal of Psychology*, *106*(4), 700–723.

Brethel-Haurwitz, K. M., Cardinale, E. M., Vekaria, K. M., Robertson, E. L., Walitt, B., VanMeter, J. W., & Marsh, A. A. (2018). Extraordinary altruists exhibit enhanced self—other overlap in neural responses to distress. *Psychological Science*, *29*(10), 1631–1641.

Carbonell, V. (2009). What moral saints look like. *Canadian Journal of Philosophy*, *39*(3), 371–398.

Carbonell, V. (2012). The ratcheting-up effect. *Pacific Philosophical Quarterly*, *93*(2), 228–254.

Carlo, G., PytlikZillig, L. M., Roesch, S. C., & Dienstbier, R. A. (2009). The elusive altruist: The psychological study of the altruistic personality. In *Personality, identity, and character: Explorations in moral psychology* (pp. 271–294). Cambridge: Cambridge University Press.

Clark, P. (2019). The particularity of sanctity: Why paradigms of exemplarity matter for Christian virtue ethics. *Journal of the Society of Christian Ethics. 39*, 111–127.

Clarke, A., Mitchell, A., & Abraham, C. (2014). Understanding donation experiences of unspecified (altruistic) kidney donors. *British Journal of Health Psychology*, *19*(2), 393–408.

Colby, A., & Damon, W. (1993). The uniting of self and morality in the development of extraordinary moral commitment. In G. G. Noam, T. E. Wren, G. Nunner-Winkler, & W. Edelstein (Eds.), *The moral self* (1993-98685-008, pp. 149–174). Cambridge, MA: The MIT Press.

Colby, A., & Damon, W. (1995). The development of extraordinary moral commitment. In M. Killen & D. Hart (Eds.), *Morality in everyday life: Developmental perspectives* (1995-98840-011, pp. 342–370). Cambridge: Cambridge University Press.

Colby, A., & Damon, W. (1994). *Some do care*. New York: Free Press.

Cooper, T. L., & Wright, N. D. (1992). *Exemplary public administrators: Character and leadership in government*. Hoboken, NJ: Jossey-Bass.

Crimston, D., Bain, P. G., Hornsey, M. J., & Bastian, B. (2016). Moral expansiveness: Examining variability in the extension of the moral world. *Journal of Personality and Social Psychology*, *111*, 636–653.

Damon, W., & Colby, A. (2015). *The power of ideals: The real story of moral choice*. New York: Oxford University Press.

Den Hartog, D. N. (2015). Ethical leadership. *Annual Review of Organizational Psychology and Organizational Behavior, 2*, 409–434.

Doris, J. (2002). *Lack of character: Personality and moral behavior*. Cambridge: Cambridge University Press.

Dunlop, W. L., Walker, L. J., & Matsuba, M. K. (2012). The distinctive moral personality of care exemplars. *The Journal of Positive Psychology, 7*(2), 131–143.

Fagin-Jones, S. (2017). Holocaust heroes: Heroic altruism of non-Jewish moral exemplars in Nazi Europe. In S. T. Allison, G. R. Goethals, & R. M. Kramer (Eds.), *Handbook of heroism and heroic leadership* (2017–00230–011, pp. 203–228). New York: Routledge and Taylor & Francis Group.

Fagin-Jones, S., & Midlarsky, E. (2007). Courageous altruism: Personal and situational correlates of rescue during the Holocaust. *The Journal of Positive Psychology, 2*(2), 136–147.

Fleeson, W., & Furr, R. M. (2016). Do broad character traits exist? Repeated assessments of individuals, not group summaries from classic experiments, provide the relevant evidence. In I. Fileva (Ed.), *Questions of character* (pp. 231–248). New York: Oxford University Press.

Fleeson, W., & Jayawickreme, E. (2017). Challenging Doris' attack on aggregation: Why we are not left "completely in the dark" about global virtues. *Ethical Theory and Moral Practice, 20*, 519–536.

Floyd, A. L. (2016). *Mapping moral exemplarity: How honors students construct meaning of the collegiate environment* (2016–31152–195). Birmingham: ProQuest Information & Learning, University of Alabama.

Frimer, J. A., Walker, L. J., Dunlop, W. L., Lee, B. H., & Riches, A. (2011). The integration of agency and communion in moral personality: Evidence of enlightened self-interest. *Journal of Personality and Social Psychology, 101*(1), 149.

Frimer, J. A., Walker, L. J., Lee, B. H., Riches, A., & Dunlop, W. L. (2012). Hierarchical integration of agency and communion: A study of influential moral figures. *Journal of Personality, 80*(4), 1117–1145.

Geras, N. (1995). Richard Rorty and the righteous among the nations. *Journal of Applied Philosophy, 12*(2).

Glazer, M. P., & Glazer, P. M. (1999). On the trail of courageous behavior. *Sociological Inquiry, 69*(2), 276–295.

Gross, M. L. (1994). Jewish rescue in Holland and France during the Second World War: Moral cognition and collective action. *Social Forces, 73*(2), 463–496.

Hamilton, M. G. B., & LaVoi, N. M. (2019). Coaches who care: Moral exemplars in collegiate athletics. *Journal of Applied Sport Psychology, 32*, 81–103.

Hart, D., & Fegley, S. (1995). Prosocial behavior and caring in adolescence: Relations to self-understanding and social judgment. *Child Development, 66*(5), 1346–1359.

Helzer, E. G., Furr, R. M., Hawkins, A., Barranti, M., Blackie, L. E. R., & Fleeson, W. (2014). Agreement on the perception of moral character. *Personality and Social Psychology Bulletin, 40*(12), 1698–1710.

Huff, C., & Barnard, L. (2009, Fall). Moral exemplars in the computing profession. *IEEE Technology and Society Magazine*, 47–54.

Jayawickreme, E., Meindl, P., Helzer, E. G., Furr, R. M., & Fleeson, W. (2014). Virtuous states and virtuous traits: How the empirical evidence regarding the existence of broad traits saves virtue ethics from the situationist critique. *Theory and Research in Education, 12*, 283–308.

Johnson, R. (2003). Virtue and right. *Ethics, 113*, 810–834.

Joyce, R. (2007). *The evolution of morality*. Cambridge, MA: MIT Press.
London, P. (1970). The rescuers: Motivational hypotheses about Christians who saved Jews from the Nazis. In J. Macaulay & L. Berkowitz (Eds.), *Altruism and helping behavior: Social psychological studies of some antecedents and consequences* (pp. 241–250). Cambridge: Academic Press.
MacFarquhar, L. (2015). *Strangers drowning: Grappling with impossible idealism, drastic choices, and the overpowering urge to help*. New York: Penguin Press.
Maple, N. H., Hadjianastassiou, V., Jones, R., & Mamode, N. (2010). Understanding risk in living donor nephrectomy. *Journal of Medical Ethics: Journal of the Institute of Medical Ethics, 36*(3), 142–147.
Marsh, A. A., Stoycos, S. A., Brethel-Haurwitz, K. M., Robinson, P., VanMeter, J. W., & Cardinale, E. M. (2014). Neural and cognitive characteristics of extraordinary altruists. *PNAS Proceedings of the National Academy of Sciences of the United States of America, 111*(42), 15036–15041.
Matsuba, M. K., & Walker, L. J. (2004). Extraordinary moral commitment: Young adults involved in social organizations. *Journal of Personality, 72*(2), 413–436.
Matsuba, M. K., & Walker, L. J. (2005). Young adult moral exemplars: The making of self through stories. *Journal of Research on Adolescence, 15*(3), 275–297.
McFarland, S., Brown, D., & Webb, M. (2013). Identification with all humanity as a moral concept and psychological construct. *Current Directions in Psychological Science, 22*(3), 194–198.
McGinn, C. (1992). Must I be morally perfect? *Analysis*, 32–34.
McGrath, P., Pun, P., & Holewa, H. (2012). Decision-making for living kidney donors: An instinctual response to suffering and death. *Mortality, 17*(3), 201–220.
McRae, E., & Cozort, D. (2018). The psychology of moral judgment and perception in Indo-Tibetan Buddhist ethics. In D. Cozort & J. M. Shields (Eds.), *The Oxford handbook of Buddhist ethics* (pp. 335–358). Oxford: Oxford University Press.
Meindl, P., Jayawickreme, E., Furr, R. M., & Fleeson, W. (2015). A foundation beam for studying morality from a personological point of view: Are individual differences in moral behaviors and thoughts consistent? *Journal of Research in Personality, 59*, 81–92.
Michaelson, M. D. (2002). *Poised to act: Profiles of fifteen young activists* (2002–95024–026). Cambridge: ProQuest Information & Learning.
Midlarsky, E., Jones, S. F., & Corley, R. P. (2005). Personality correlates of heroic rescue during the holocaust. *Journal of Personality, 73*(4), 907–934.
Midlarsky, M. I. (1985). Helping during the Holocaust: The role of political, theological, and socioeconomic identifications. *Humboldt Journal of Social Relations, 13*(1–2), 285–305.
Miller, C. B. (2013). *Moral character: An empirical theory* (p. 368). Oxford: Oxford University Press.
Miller, C. B. (2014). *Character and moral psychology* (p. 288). Oxford: Oxford University Press.
Monroe, K. R. (2002). Explicating altruism. In S. G. Post, L. G. Underwood, J. P. Schloss, & W. B. Hurlbut (Eds.), *Altruism & altruistic love: Science, philosophy, & religion in dialogue*. (pp. 106–122). New York: Oxford University Press.
Monroe, K. R. (2012). Ethics in an age of terror and genocide: Identity and moral choice. *PS: Political Science & Politics, 44*(3), 503–507.
Morgan, J. (Director). (1994). *Hell's Angel*. Bandung Productions.

Oliner, S. P. (2002). Extraordinary acts of ordinary people: Faces of heroism and altruism. *Altruism & altruistic love: Science, philosophy, & religion in dialogue* (pp. 123–139). Oxford: Oxford University Press.

Oliner, S. P., & Oliner, P. M. (1988). *The altruistic personality: Rescuers of Jews in Nazi Europe* (1988–97707–000). New York: Free Press.

Opp, K. D. (1997). Can identity theory bettter explain the rescue of Jews in Nazi Europe than rational actor theory? *Research in Social Movements, Conflict and Change, 20,* 223–253.

Pöyhönen, V., Juvonen, J., & Salmivalli, C. (2012). Standing up for the victim, siding with the bully or standing by? Bystander responses in bullying situations. *Social Development, 21*(4), 722–741.

Prentice, M., Jayawickreme, E., Hawkins, A., Hartley, A., Furr, R. M., & Fleeson, W. (2019). Morality as a basic psychological need. *Social Psychological and Personality Science, 10*(4), 449–460.

Pybus, E. M. (1982). Saints and heroes. *Philosophy, 57*(220), 193–199.

Reimer, K., & Wade-Stein, D. (2004). Moral identity in adolescence: Self and other in semantic space. *Identity: An International Journal of Theory and Research, 4,* 229–249.

Roth, A. E., Sönmez, T., & Ünver, M. U. (2004). Kidney exchange. *Quarterly Journal of Economics, 119*(2), 457–488. Scopus.

Sainio, M., Veenstra, R., Huitsing, G., & Salmivalli, C. (2011). Victims and their defenders: A dyadic approach. *International Journal of Behavioral Development, 35*(2), 144–151.

Salmivalli, C., Lagerspetz, K., Björkqvist, K., Österman, K., & Kaukiainen, A. (1996). Bullying as a group process: Participant roles and their relations to social status within the group. *Aggressive Behavior, 22*(1), 1–15.

Sandler, J., & Freud, A. (1983). Discussions in the Hampstead index on "The ego and the mechanisms of defence": XI A form of altruism. *Bulletin of the Hampstead Clinic, 6*(4), 329–349.

Shepela, S. T., Cook, J., Horlitz, E., Leal, R., Luciano, S., Lutfy, E., . . . Worden, E. (1999). Courageous resistance: A special case of altruism. *Theory & Psychology, 9*(6), 787–805.

Singer, P. (2015). *The most good you can do: How effective altruism is changing ideas about living ethically* (2015-31407-000). New Haven: Yale University Press.

Tec, N. (1986). *When light pierced the darkness: Christian rescue of Jews in Nazi-occupied Poland.* Oxford: Oxford University Press.

Urmson, J. (1958). Saints and heroes. In *Essays in moral philosophy* (pp. 196–216). Seattle: University of Washington Press.

Varese, F., & Yaish, M. (2000). The importance of being asked: The rescue of Jews in Nazi Europe. *Rationality and Society, 12*(3), 307–334.

Walker, L. J., & Frimer, J. A. (2007). Moral personality of brave and caring exemplars. *Journal of Personality and Social Psychology, 93*(5), 845–860.

Walker, L. J., & Frimer, J. A. (2009). Moral personality exemplified. In *Personality, identity, and character: Explorations in moral psychology* (pp. 232–255). Cambridge: Cambridge University Press.

Walker, L. J., & Frimer, J. A. (2015). Developmental trajectories of agency and communion in moral motivation. *Merrill-Palmer Quarterly, 61*(3), 412–439.

Walker, L. J., & Hennig, K. H. (2004). Differing conceptions of moral exemplarity: Just, brave, and caring. *Journal of Personality and Social Psychology, 86*(4), 629–647.

Williams, B. (1985). *Ethics and the Limits of Philosophy.* New York: Routledge.

Wolf, S. (1982). Moral saints. *The Journal of Philosophy*, 419–439.
Wright, J. C., Nadelhoffer, T., Ross, L., & Sinnott-Armstrong, W. (2018). Be it ever so humble: Proposing a dual-dimension account and measurement for humility. *Self & Identity*, *17*(1), 92–125.
Wright, J. C., Warren, M. T., & Snow, N. E. (in press). *Understanding virtue: Theory and Measurement*. Oxford University Press.
Yad Vashem. (2019) *The World Holocaust Remembrance Center | www.yadvashem.org |*. Retrieved December 1, 2019, from index.html
Zagzebski, L. T. (2017). *Exemplarist moral theory*. Oxford: Oxford University Press.

10
THE SOCIAL PSYCHOLOGY OF VIOLENT EXTREMISM

Erica Molinario, Katarzyna Jasko, David Webber, and Arie W. Kruglanski

Introduction

Actions aimed to achieve political change can take on a number of forms and range from voting and taking part in demonstrations to civil disobedience, violence, and terrorism. The collective action literature distinguishes specifically between actions that conform to the norms of an existing social system and nonnormative actions that violate these rules (e.g., Wright, Taylor, & Moghaddam, 1990a). Importantly, this distinction focuses on the individual intentions; an action is defined as nonnormative if the actor is aware of violating social expectations and conventions (Wright, 2001). Moreover, this definition highlights the contextual character of political actions since the perceived extremity of actions depends on the norms of a given group. Thus, actions that can be defined as normative for one group can be perceived as radical and nonnormative for another group and vice versa. In this chapter, we focus on the latter forms of political actions, and particularly on individuals who violate strong norms against violence by pursuing a political cause through violent means.

We consider violent extremism as a specific form of extreme behavior, in that it is 'exceeding the ordinary, usual or expected' (Merriam-Webster, 1986, p. 441). Violent political extremism involves inflicting a high severity aggression or violence (e.g., serious injury or death; Allen & Anderson, 2017) against people or properties as a means to achieve a political, ideological, or religious goal. An individual can support or endorse the use of violence in this manner, and/or they can act on this belief through actual engagement in violence.

Violent extremism is rampant these days, and its recent proliferation represents a threat to national and international security. Whereas the rise of extremism has been evident over the large part of the last decade last year has witnessed an acceleration of this process and a mounting volume of calls for violent acts among hate-based far-right groups in response to the global COVID-19 pandemic and the resulting restrictions imposed by governments (Kruglanski, Gunaratna, Ellenberg, & Speckhard, 2020). One instance among many is the thwarted attack against a hospital struggling with the coronavirus pandemic in Belton, Missouri, and the foiled kidnapping of the Michigan governor by anti-government extremists.

From a rational perspective, the costs involved in the use of violence to achieve political or ideological goals seem to outweigh the benefits. Often, individuals who pursue violent extremism pay a high price for their actions. At an individual level, this includes arrest and possible incarceration, and at a group level, societal opprobrium and repressive actions taken by states to quell a violent movement. More importantly, past research suggests that violence is not efficacious in serving the perpetrators' objectives. In this vein, Thomas and Louis (2014) found that violent collective actions are ineffective as a tool of social influence (i.e., influencing a sympathetic bystander audience). Studies have shown that in comparison to peaceful activism, violent protests reduced identification with activists (Feinberg, Willer, & Kovacheff, 2017) and reduced their perceived legitimacy (Wang & Piazza, 2016). A systematic analysis by Chenoweth and Stephan (2011) comparing nonviolent and violent campaigns found that violent campaigns achieved their political goals almost half as often as nonviolent campaigns. And finally, research shows that peaceful tactics such as mass peaceful protests and civil disobedience have been critical in the success of social movements (Karatnycky & Ackerman, 2005; Wasow, 2020).

Given the costs and relative ineffectiveness of political violence, it is imperative to understand why actors continue to pursue violence. In this chapter, we review the abundant psychological literature that has examined the social and psychological causes of the use of violence in political and social movements. Our aim is to better understand what makes violence so appealing. We end by discussing a theoretical framework that proposes a comprehensive understanding of violent extremism.

Perceived Efficacy of Violence

Although the objective evidence shows that violence is less efficacious than nonviolence at achieving political goals, individual actors or groups are more likely to engage in such acts if they *perceive* them to be instrumental and effective. Accordingly, Saab, Tausch, Spears, and Cheung (2015) found that the more efficacious the aggressive actions are perceived, the greater their appeal and the less they are reduced by the perceived efficacy of peaceful action. Violent means are perceived as highly strategic since they can be seen as fulfilling a number of short-term goals, such as influencing a wider public opinion, building a movement, and winning third parties for the cause (see Hornsey at al., 2006). These smaller victories (or process goals) help to sustain terrorist organizations, for instance, by allowing them to increase recruitment, funding, or media attention (Abrahms, 2012). Process goals are the stepping stones for the actors' journey toward (hopefully) achieving larger political outcome(s) for which they strive. For instance, using violence in a situation can help garner support for the cause by provoking the opponent into extreme counter-action (Sedgwick, 2004). In this vein, Sageman (2004) described how Egyptian Islamic Jihad used violence to provoke more repressive measures by the government which would then alienate the general population and mobilize them against the regime. This was also observed post-9/11 when in an effort to counter al Qaeda, the United States engaged in controversial tactics such as torture and curtailing of privacy rights of American citizens. Violence-based collective actions can help focus the attention on neglected issues or actors, raising stakes in setting goals or weakening the opponents (Cunningham, Bakke, & Seymour, 2012; Lawrence, 2010). This is because violent acts attract more attention and shine the spotlight on the political issue in question.

Perceptions of effectiveness are also likely to be higher when actors perceive violence as the only available means to their ends. Research supports this contention (Jackson, Huq, Bradford, & Tyler, 2013; Saab, Spears, Tausch, & Sasse, 2016; Schumpe, Bélanger, Giacomantonio, Nisa, & Brizi, 2018). Violence is likely to be viewed as the only available means when a social group is marginalized within the existing political system (Gurr, 1993; Schwarzmantel, 2010). Corrupt political systems undermine the viability of nonviolent alternatives, which may ultimately set up the conditions for the emergence of violence (Thomas & Louis, 2014; Worchel, Hester, & Kopala, 1974). Accordingly, Drury and Reicher (2005) found that the indiscriminate and illegitimate use of force by an authority plays a crucial role in radicalizing crowd dynamics. For instance, right-wing political violence is more likely to develop when political opportunities are closed off by the state than by sustained grievances (Koopmans, 2005). The goal behind resistance can thus change under overwhelming oppression and with it the definition of success (Drury & Reicher, 2009; Drury, Evripidou, & Van Zomeren, 2014). Hence, it may be that aggression comes to be seen as a moral response to injustice particularly when group members feel overwhelmed by the power of the adversary. For example, Scheepers, Spears, Doosje, and Manstead (2006) demonstrated that people with a stable low group status (i.e., perceiving their low position in society as unlikely to improve) engaged in more provocative forms of bias (i.e., outgroup derogation) than those who felt their low status was malleable. Similarly, Wright and colleagues (1990a) showed that nonnormative action was chosen when movement from a disadvantaged group to an advantaged group was impossible. These authors also demonstrated that those who chose nonnormative action (vs. normative) were best distinguished by their hope for an improvement of their position (Wright, Taylor, & Moghaddam, 1990b).

The conditions outlined earlier likely lead actors to perceive that they (or their social groups) are unable to advance their political goals. In fact, research has shown this form of group efficacy (or lack thereof) to be a robust explanation for collective action to address political or social disadvantage (Folger, 1987; Mummendey, Kessler, Klink, & Mielke, 1999; Smith, Cronin, & Kessler, 2008). In the context of violent extremism, Tausch et al. (2011) found that engagement in and endorsement of nonnormative collective action, such as violence and terrorism, was greater the lower the perceived efficacy of the ingroup to redress injustice by nonviolent means. Similarly, van Zomeren, Saguy, and Schellhaas (2013) found group efficacy to be negatively predictive of aggressive collective action tendencies.

Similarly, political violence is chosen when the group is too disorganized or unsupportive of the cause to bring about the mass action required to effect change with means within the system (see van Zomeren, Spears, Fischer, & Leach, 2004). Wang and Piazza (2016) found that activists tend to avoid the use of violence when their claims or goals are supported by a broad swath of the public. In fact, some researchers have reflected on how the use of violence has been interpreted as a signal of self-segregation, followed by their inability to develop into a social movement or to revitalize a social movement that has begun to decline (Melucci, 1982).

Feeling Noticed and Agentic

Feeling that one has no alternative means might activate a need for control (Fiske, 2003; Pittman & Zeigler, 2007) inducing the individual to search for ways to establish their sense of control and agency. Accordingly, Fritsche, Jonas, and Kessler (2011) found that threats to personal control increased relative attractiveness ratings of small groups when the relative agency of small groups was perceived to be high.

Violence may be particularly suited to restoring control and providing a sense of agency. Through violating all norms of human coexistence, radicalized groups can, in fact, provide a sense of collective agency that helps to restore their sense of control. In this vein, Ransford (1968) demonstrated in the context of the Watts Riots in the United States that feelings of powerlessness and lack of control over events were positively correlated with willingness to engage in violence.

Violence may provide a sense of agency through drawing attention to a cause in which name the group is launching its fight. In fact, nonviolent protests often go unnoticed, but ones that turn violent garner attention from the media and the authorities. Moreover, because violence stands in the way of achieving many other goals a person has (that require following norms and abiding by societal mandates) and because most individuals are at least somewhat reluctant to engage in violence (Grossman, 1996), the use of violence communicates a strong commitment to one's group values.

Thus, through violence, groups or individuals demonstrate not only immediate control but also power over other parties (at least in the short run). In fact, some studies have indicated that members of terrorist organizations are seen as courageous, honorable, and important in many communities and societies (Silke, 2008). For example, interviews with jailed terrorists have shown that they can gain high social status through their membership of terrorist groups such as Hamas or Fatah (Post, Sprinzak, & Denny, 2003). For terrorist organizations that are competing for the attention of recruits and public support, escalating violence serves as a means to demonstrate their commitment and capability relative to other organizations (Bloom, 2005; Kydd & Walter, 2006). Indeed, as Goldman, Giles, and Hogg (2014) framed it: 'Acting violently sends the message that a person is not to be messed with and is capable of anything' (p. 8).

Thus, it is likely that violence has a deep psychological function related to establishing a sense of power and worthiness. In fact, Haslam and Reicher (2012) have argued that in desperate circumstances (such as Jewish people facing inevitable death in Nazi concentration camps), violent revolts occurred to preserve honor and pride rather than to secure survival. Similarly, there is some evidence that extreme and sacrificial actions for a collective cause show one's commitment to socially important values whose protection provides a sense of meaning, esteem, or pride (Becker, Tausch, & Wagner, 2011; Olivola & Shafir, 2013; Tropp & Brown, 2004). In the context of violence, Franco, Blau, and Zimbardo (2011) showed that the martial hero who participates actively in combat is perceived as the typical 'Hero' figure.

Consistent with these notions, research has found that leaders of violent organizations, although more committed to the ideological goals of the group, were less likely to engage in violence (Jasko & LaFree, 2020). Leaders, because of their position of authority, already garner respect within the group and thus have little credibility and esteem to gain from engaging in violence. Followers, on the other hand, may still need to prove their commitment to the group, thereby increasing the potential rewards that would come from using violence to advance their political cause. Thus, although violence comes with its share of potentially costly consequences (physical harm, incarceration, death, etc.), it carries immense potential benefits that may outweigh those costs and promote engagement in violent actions.

Ingroup Identification

Research has consistently found that the more a person identifies with a group, the higher the chances they will take part in collective action on behalf of that group (de Weerd & Klandermans, 1999; Simon & Klandermans, 2001; Simon et al., 1998). One such line of research studies the role of group fusion, which represents a particularly strong form of attachment to one's group (Swann, Jetten, Gómez, Whitehouse, & Bastian, 2012). When people are fused with a group, they come to endorse the goals of the group as their own, while also coming to view their group members like brothers and sisters. As a result, strongly fused persons are more likely to support fighting and dying for their ingroup (e.g., Besta, Gómez, & Vázquez, 2014; Gómez, Brooks et al., 2011; Gómez, Morales, Hart, Vázquez, & Swann, 2011; Swann, Gómez, Seyle, Morales, & Huici, 2009). As an example, engagement in violence among ex-combatants in Liberia and Uganda was correlated with their degree of identification with the violent group (Littman, 2018).

Interestingly, the relationship can also operate in the opposite direction—that is, participation in violence may increase the sense of unity and group identification. The use of violence to pursue a cause may make the cause especially valuable and important, to the extent to which an individual or a group goes beyond what is socially accessible to achieve it. This emphasis on value is in line with expectancy-value models of motivation that posit that individuals will be motivated to engage in goal pursuit to the degree that the goal is perceived to have value (e.g., Shah & Higgins, 1997). In the same vein, sacrificing for a valuable cause makes the group itself valuable and morally superior. Identifying with such a group increases the positive view about oneself, to the extent that important social identities and the feeling of belonging to a valued group provide individuals with meaning (Cialdini et al., 1976; George & Park, 2016; Tajfel & Turner, 1979).

For example, research shows that participation in radical collective actions fosters disidentification with moderate groups but enhances political identification with radical subgroups (Becker, Tausch, Spears, & Christ, 2011). Likewise, Simon and Klandermans (2001) proposed that when people become involved in a power struggle their social identity is politicized. A politicized identity is one wherein an individual is acutely sensitive to power divisions in society, sharply demarcating those who possess it from those who are deprived of it. Existing research suggests that participation in collective actions can lead to an increased politicized identity among participants (e.g., Drury & Reicher, 1999, 2005; Simon & Klandermans, 2001), thereby increasing their willingness to engage in activism (Kelly & Breinlinger, 1995).

Culturally Approved Violence

The discussion thus far has relied on the notion that violence is universally viewed as immoral and is disapproved, discouraged, and sanctioned as such. It carries with it the possibility of immense repercussions that can be mitigated by the various psychological benefits of violence. Too, violence may be deemed necessary because of political and social structures that limit nonviolent routes toward goal achievement. However, violence isn't condemned universally, and there are instances in which cultural norms promote rather than obstruct violence.

Examples include masculinity norms and the culture of honor. Masculine norms in Western countries typically put a premium on being agentic, dominant, tough, and capable of aggression and violence (Nowak, Gelfand, Borkowski,

Cohen, & Hernandez, 2016; Wong, Tsai, Liu, Zhu, & Wei, 2014). In honor cultures, men often carry out interpersonal aggressive retaliation, even for minor insults, to protect their reputation. For example, White men in the southern United States express anger toward, and attack, those who insult their honor (Nisbett & Cohen, 1996). This is reflected in the high number of argument-related homicides in the southern United States (Nisbett, 1993), as well as violence in schools. High school students in states with honor cultures are more likely to bring guns to school, and the frequency of shootings per capita in these states is more than twice that in other states (Brown, Osterman, & Barnes, 2009). Not only is interpersonal aggression related to honor culture but also is national and intergroup aggression by individuals (e.g., Nawata, 2020). For example, masculine honor ideology has been found to have a positive relation to endorsement of aggressive responses to illegal immigration and terrorism (Barnes, Brown, & Osterman, 2012; Barnes, Brown, Lenes, Bosson, & Carvallo, 2014) and to a greater willingness to risk one's life during combat operations (Mandel & Litt, 2013).

The link between masculine norms and violence is highlighted in the criminology literature that has linked masculinity norms with homicides (Gastil, 1971; Hackney, 1969; Nisbett & Cohen, 1996; Wolfgang & Ferracuti, 1967). And Leander et al. (2020) found that, under frustrating conditions, the endorsement of violence was higher primarily among US adults of a lower educational background and/or men who endorse a masculine honor culture.

Other research has examined how normative tightness (vs. looseness) promotes political violence (Gelfand, LaFree, Fahey, & Feinberg, 2013). Using a data source with records of over 80,000 terror attacks that occurred across the world between 1970 and 2007, the authors found that cultural values and norms that promote rigid thinking—fatalistic beliefs, strict gender roles, and greater tightness—are related to a greater number of terrorist attacks or fatalities. They argue that low gender egalitarianism affords a culture that permits violent behavior in general due to the enhanced masculinity and toughness, potentially sparking violence even when minor provocations are present (or in the case of terrorism, grievances), resulting in a greater number of incidents and fatalities.

Violence as a Clear Response

Violence may be a preferred course of action because it sends an unambiguous message about the importance of a cause. In fact, extremist groups tend to hold clear-cut beliefs devoid of ambiguity (Hogg, Kruglanski, & Van den Bos, 2013). They tend to think about important political issues in dogmatic and simplistic terms (Conway et al., 2016), conceptualize political issues and stimuli in clearly defined and homogenous categories (Lammers, Koch, Conway, & Brandt, 2017), and have a firm conviction that their views are right and that alternative perspectives are wrong (van Prooijen & Krouwel, 2017).

Thus, violence can be a response to uncertainty and ambiguity. A recent review of the empirical evidence of the processes leading to political violence (Gøtzsche-Astrup, 2018) found that uncertainty related to an individual's place in the world and their future is a key antecedent of extremism, as proposed by uncertainty-identity theory (Hogg, 2014). Research has found that instilling a feeling of uncertainty within study participants increases the appeal of extreme groups (Hogg, Sherman, Dierselhuis, Maitner, & Moffitt, 2007), leads participants to react more extremely toward members of an outgroup (van den Bos, Euwema, Poortvliet, & Maas, 2007), and increases ideological extremism in the form of religious zeal (McGregor, Nash, Mann, & Phills, 2010). Feelings of uncertainty should activate the need for cognitive closure (Kruglanski, 2004; Kruglanski & Webster, 1996; Webster & Kruglanski, 1994), which has been linked to support for violence. In this vein, Federico, Golec, & Dial, (2005) found that the need for closure was associated with support for military action in Iraq among persons high in national attachment. Similarly, the need for closure was found to promote ingroup glorification and support for extreme measures against one's group's enemies (Dugas et al., 2018).

Relative Deprivation and Inequality

In examining the causes of political extremism and terrorism, a wide swath of scholars has focused on material and economic grievances (e.g., poverty, material deprivation, and unequal distribution of resources). Perhaps the most paradigmatic framework in which these studies are developed is relative deprivation theory (RTD, Gurr, 1970; Stouffer, Suchman, DeVinney, Star, & Williams, 1949; Pettigrew, 1967; Runciman, 1966; Walker & Smith, 2002), which links relative economic disparity to the propensity of individuals to resort to violent political action. For example, Piazza (2006, 2011) showed that economic discrimination of minorities was a significant predictor of terrorism at the state level. Similarly, in three experiments, Greitemeyer and Sagioglou (2017) showed that participants in a personal relative deprivation condition reported higher levels of aggressive affect and behaved more aggressively (toward targets that were the source for participants' experience of disadvantage) than participants in a personal relative gratification condition.

Relative deprivation can be personal (egoistic) or group-based (fraternal; Runciman, 1966). Group-based relative deprivation refers to feelings of discontent that occur when people perceive that members of their group have less than what they believe their group is entitled to (Smith, Pettigrew, Pippin, & Bialosiewicz, 2012). Research shows that political protests are a common response to group-based relative deprivation (e.g., Vanneman & Pettigrew, 1972). Likewise, Obaidi and colleagues (2019) found higher extremism among Western-born (vs. foreign-born) Muslims as a result of perceived relative deprivation; Western-born Muslims were particularly vulnerable to the impact of perceived relative deprivation because comparisons with peers in majority groups were more salient for them than for individuals born elsewhere.

Still, research is less clear as to whether economic disparity increases a preference for violence over nonviolent forms of collective action. Some scholars suggest that feelings of relative deprivation (or that one has been unfairly disadvantaged) evoke anger and resentment (Smith & Huo, 2014; Smith et al., 2012). Accordingly, within social identity research, perceptions of injustice, as accompanied by anger or contempt, are predictors of violent collective action. In this vein, research shows that Muslim immigrants who perceived themselves as victims of discrimination (Victoroff, Adelman, & Matthews, 2012) or who felt marginalized within society (Lyons-Padilla, Gelfand, Mirahmadi, Farooq, & Van Egmond, 2015) expressed increased support for radical actions and political views. Others link the occurrence of violent extremism to political grievances (e.g., marginalization by the political class; see, e.g., Koopmans, 2005; Schwarzmantel, 2010).

In the next section, we suggest that all of these instances that lead to violence have a common denominator. Economic grievances, political discrimination, and social marginalization activate the same motivational mechanism, which if inserted in a certain ideological and social milieu can guide one toward violence. We call this motivational force the *need for significance* (Kruglanski, Chen, Dechesne, Fishman, & Orehek, 2009; Kruglanski et al., 2013, 2014; Kruglanski, Jasko, Chernikova, Dugas, & Webber, 2017; Kruglanski, Bélanger, & Gunaratna, 2019), which together with the factors of the *narrative* and *network* constitute the 3N framework of violent extremism that we discuss next.

The 3N Framework: Need, Narrative, and Networks

As discussed in previous sections, the endorsement and perpetration of political violence is influenced by numerous factors. Here, we discuss a framework that incorporates many of these ideas into a comprehensive understanding of radicalization to violence. The framework proposes that political extremism is first determined by a motivational component (i.e., need for significance) that is activated by the opportunity of gaining considerable social recognition or by frustrations and humiliations that occur at the individual or collective level. In these circumstances, violence is likely to be perpetrated when it is embedded in an ideological narrative (e.g., based on one's culture, the norms of one's community and its worldviews) that justifies the use of violence and is endorsed by a social network that recognizes the individual's effort and rewards them for acting on behalf of the narrative. We discuss these three factors in details.

Needs

Our research suggests that the circumstances of relative deprivation and affront to one's social identity induce a feeling of insignificance and worthlessness that awakens the *need for significance*, which is the fundamental desire to matter, to merit respect, and to 'be someone' (Kruglanski et al., 2009, 2013, 2014, 2017, 2019).

As with any motivation, the quest for significance can be activated through the avenues of *deprivation* and *incentivization* (cf. Kruglanski et al., 2014). The deprivation route of activation involves the loss of one's sense of personal significance and mattering occasioned by, for instance, a personal failure or an insult to one's social identity (e.g., discrimination against, and/or humiliation of one's ethnic, religious, national, or political group). For instance, a person could fail at an important life pursuit or experience deep humiliation, and hence find extremism for a socially valued cause attractive as it lifts them from their experienced inferiority (Pedahzur, 2005). Individuals may also experience a loss of significance on behalf of a group with which they identify, such as Muslims in general, if they feel or are told that their Muslim brethren around the world are humiliated and discriminated against (Adib-Moghaddam, 2005; McCauley & Moskalenko, 2011; Zartman & Khan, 2011). In those latter circumstances, individuals experience an affront to their social identity (Tajfel & Turner, 1979), and hence a loss of significance as members of a disparaged group. Thwarting one's sense of significance arouses the motivation to regain significance (i.e., quest for significance) by whatever means are available including violence.

Incentivization occurs when one is concerned about achieving great levels of significance and respect. In this case, the discrepancy results from unfulfilled high aspirations. For example, individuals join terrorist groups because they wish to become heroes or martyrs (Kruglanski et al., 2013). Thus, engaging in violent extremism can be portrayed as an opportunity for a vast significance gain and a place in history (Post, 2006). For instance, one may seize the opportunity to elevate their sense of significance (e.g., the status of hero or martyr) by self-sacrificing for a hallowed cause.

Deprivation and incentivization induce need arousal; this creates a state of motivational imbalance in which the aroused need is dominant and alternative concerns are out of focus. Balance occurs when individuals strive to satisfy all of their basic needs, which thereby constrains behavior as actions that serve to satisfy one needs while frustrating another will be avoided. Motivational balance leads to moderation. Motivational imbalance, on the other hand, releases behavioral constraints as all behaviors that serve the dominant need (even those incompatible with other needs) become viable. The greater the imbalance, the weaker the constraints imposed by the suppressed needs, and the greater the perceived acceptability of extreme behaviors (Kruglanski, 2018). The motivational imbalance may underlie all kinds of extremism regardless of their specific nature. The looming dominance of the need for significance (due to the deprivation and/or incentivization factors) typically prompts sacrificial and extreme behaviors for an ideological cause that lends people significance (Kruglanski et al., 2009, 2013). Notably, these extreme behaviors are not necessarily violent, as they can include nonviolent, peaceful, and self-sacrificial actions. Whether they are violent or not depends on the narrative that spells out what actions lend one respect and significance (see subsequent section).

Instances of significance loss at the personal level (such as personal failures or relationship problems) were found related to the perpetration of violent (vs. nonviolent) politically motivated crimes (Jasko, LaFree, & Kruglanski, 2017). Similarly, former members of a terrorist organization revealed that higher experiences of personal humiliation and shame were related to their greater support for violence to achieve the political goals of the terrorist organization (Webber, Babush et al., 2018). Thus, engaging in violence is one of the ways people choose to obtain recognition and significance. Violence has the particular function of immediately establishing one's superiority and thus respect over the other. It is likely for that reason that engaging in violence is the primary way for gang members to gain respect from their peers (Melde, Taylor, & Esbensen, 2009; Stretesky & Pogrebin, 2007).

Narrative

As discussed previously, people usually avoid extreme and violent behaviors because they might be perceived as immoral and punishable, but under certain cultural contexts, violent actions are rewarded. Violence becomes more appealing when it is embedded in an ideological narrative that depicts violence as a route to significance (Kruglanski et al., 2013). An individual who subscribes to a violence-promoting narrative will be more likely to engage in violent extremism, because the narrative justifies violence as an effective way to attain significance and presents it as moral and acceptable. Accordingly, Fiske and Rai (2014) suggest that people engage in violence despite their reluctance to do so because most perpetrators of violence feel they are doing what they should do in this situation. In other words, political aggressors believe that the violence they use is a morally justified course of action under the circumstances. Ginges and Atran (2011) have argued that people endorse or pursue violent strategies not necessarily (only) for collective material gains but because violence seems like the right and moral thing to do. In their survey among Israeli settlers, the perceived righteousness of violence emerged as a predictor of support for violence whereas the general efficacy of violent actions for accomplishing a political objective did not.

Ideological narratives are shared systems of belief (Jost, Ledgerwood, & Hardin, 2008) that guide the individual to identify the actions required to achieve significance. Additionally, ideological narrative provides the moral justifications that transform extreme behaviors, violence included, into something that is acceptable and even desirable (Kruglanski et al., 2014; Bandura, 1996) in serving supreme values. A person who serves an important value and is willing to make sacrifices for its sake is thereby a person worthy of admiration and (self-)respect. That is the implication of an extremism justifying narrative.

Several studies on moral conviction and sacred values support the idea that high moralization of an issue is related to more extreme means chosen to pursue or defend it. For example, Ryan (2014) found that when people were convinced about the moral value of the cause they showed less respect for authorities (i.e., Supreme Court) when those authorities yielded decisions inconsistent with those convictions. Similarly, research on sacred values (e.g., Atran & Ginges, 2012; Ginges, Atran, Medin, & Shikaki, 2007; Tetlock, 2003) showed that when fundamental moral values were compromised, they caused increased support for violence (Ginges et al., 2007). In short, when people define a political issue in terms of moral values, they are more willing to approve of and use violence in their name. When one's actions serve these important values, they garner that individual admiration and respect, thus bestowing a sense of significance. However, research also shows that sacrifice for a moralized cause does not need to be violent. For example, a series of studies on peaceful activism showed that when activists perceived the cause in moral terms, they gained more personal significance from their commitment to the cause and as a consequence they were more willing to (peacefully) sacrifice for the cause in the future (Jasko, Szastok, Grzymala-Moszczynska, Maj, & Kruglanski, 2019).

Thus, people under quest for significance seek an ideological narrative that morally justifies the use of a certain course of actions to restore significance. However, to better serve the significance restoration goal, the ideological narrative cannot exist in a social vacuum but needs to be validated and cherished by a social network.

Network

To restore their significance, individuals will seek a social network that shares and endorses the ideology and that will value the individual's sacrifice for the cause. Social networks that promote violence as an effective (and preferred) means of significance attainment increase individuals' likelihood of engaging in violent extremism because they validate the narrative justifying violence as the means to significance. Such validation is typically performed by social networks that can range from informal assemblies of like-minded friends or family members to formal organizations devoted to a shared purpose (Sageman, 2004, 2008). For instance, Jasko and colleagues (2020) found evidence that radical social contexts strengthen the link between quest for significance—particularly collective significance—and support for political violence. Research also found that being connected to radicalized others is related to higher chances of individual radicalization. For example, a study among Swedish adolescents by Dahl and Van Zalk (2014) showed that peers' involvement in illegal political behavior predicted adolescents' engagement in similar actions. Similar effects were demonstrated in the context of real-life acts of violence (Holman, 2016; Jasko et al., 2017; Sageman, 2004; Webber, Chernikova et al., 2018).

Taken together, the research reviewed in this section is in line with the main assumptions of Significance Quest Theory, which states that the need to feel significance translates into specific (violent or nonviolent) means in part as a function of the accessible narratives and social networks.

Conclusion

In this chapter, we characterized violent extremism broadly in terms of a motivational imbalance, in which a given need assumes dominance over alternative concerns thus implying sacrifice as a defining feature of extremism. In the first part of this chapter, we examined why violence is appealing in the context of political engagement. After reviewing different perspectives that provide some answers to this query, we proposed an underlying framework that better explains the appeal of violence.

We proposed that violent extremism may be appealing to individuals seeking glory and significance. Thus, we discussed that an individual is more likely to choose a violent route toward significance restoration if he or she resides within a social context that both justifies violence as an appropriate means toward goal attainment, and bestows significance (respect, adulation) on people who act violently on behalf of a cherished cause. Commitment to extremism is additionally strengthened if such narrative is validated by the individual's network of significant others who are willing to reward it by their respect and admiration for the extremist's devotion and sacrifice.

The 3N framework allows the integration of prior research on violent extremism within its three parameters (of Needs, Narratives, and Networks). For instance, the relative deprivation notion relates to a loss of significance due to unfair treatment of one's group. Likewise feeling noticed and agentic pertains to one's sense of significance. Discussions of culturally appropriate violence, sacred values, and perceived efficacy of violence pertain to the narrative whereby violence is an effective means to glory and significance. Finally, discussions of group fusion, tightness of group norms, etc. pertains to the network aspect of the model. The advantage of the 3N framework is that it provides a functional explanation of a variety of different factors affecting violent extremism. Most human behavior is motivated; it has a goal that directs it. The quest for significance is the overarching goal underlying extremist behavior that serves as a means to the goal. The relation between the means and the goal is stated in the narrative to which individuals subscribe, and the validity of the narrative is established by the fact that respected members of the individuals' network believe it to be valid. In this manner, a wide variety of factors that thus far were discussed as isolated and unrelated to one another are now seen as functionally interdependent in their combining to produce violent extremism.

References

Abrahms, M. (2012). The political effectiveness of terrorism revisited. *Comparative Political Studies*, *45*(3), 366–393.

Adib-Moghaddam, A. (2005). Islamic utopian romanticism and the foreign policy culture of Iran. *Critique: Critical Middle Eastern Studies*, *14*(3), 265–292.

Allen, J. J., & Anderson, C. A. (2017). Aggression and violence: Definitions and distinctions. *The Wiley Handbook of Violence and Aggression*, 1–14.

Atran, S., & Ginges, J. (2012). Religious and sacred imperatives in human conflict. *Science*, *336*, 855–857.

Bandura, A. (1996). Failures in self-regulation: Energy depletion or selective disengagement? *Psychological Inquiry*, *7*(1), 20–24.

Barnes, C. D., Brown, R. P., Lenes, J., Bosson, J., & Carvallo, M. (2014). My country, my self: Honor, identity, and defensive responses to national threats. *Self and Identity*, *13*(6), 638–662.

Barnes, C. D., Brown, R. P., & Osterman, L. L. (2012). Don't tread on me: Masculine honor ideology in the US and militant responses to terrorism. *Personality and Social Psychology Bulletin*, *38*(8), 1018–1029.

Becker, J. C., Tausch, N., Spears, R., & Christ, O. (2011). Committed dis(s)idents: Participation in radical collective action fosters disidentification with the superordinate group but enhances political identification. *Personality and Social Psychology Bulletin*, *37*, 1104–1116.

Becker, J. C., Tausch, N., & Wagner, U. (2011). Emotional consequences of collective action participation: Differentiating self-directed and outgroup-directed emotions. *Personality and Social Psychology Bulletin*, *37*(12), 1587–1598.

Besta, T., Gómez, Á., & Vázquez, A. (2014). Readiness to deny group's wrongdoing and willingness to fight for its members: The role of poles' identity fusion with the country and religious group. *Current Issues in Personality Psychology*, *2*(1), 49–55.

Bloom, M. (2005). *Dying to kill: The allure of suicide terror*. New York: Columbia University Press.

Brown, R. P., Osterman, L. L., & Barnes, C. D. (2009). School violence and the culture of honor. *Psychological Science*, *20*(11), 1400–1405.

Chenoweth, E., & Stephan, M. J. (2011). *Why civil resistance works: The strategic logic of nonviolent conflict*. New York: Columbia University Press.

Cialdini, R. B., Borden, R. J., Thorne, A., Walker, M. R., Freeman, S., & Sloan, L. R. (1976). Basking in reflected glory: Three (football) field studies. *Journal of Personality and Social Psychology*, *34*(3), 366–375.

Conway III, L. G., Gornick, L. J., Houck, S. C., Anderson, C., Stockert, J., Sessoms, D., & McCue, K. (2016). Are conservatives really more simple-minded than liberals? The domain specificity of complex thinking. *Political Psychology*, *37*(6), 777–798.

Cunningham, K. G., Bakke, K. M., & Seymour, L. J. (2012). Shirts today, skins tomorrow: Dual contests and the effects of fragmentation in self-determination disputes. *Journal of Conflict Resolution*, *56*(1), 67–93.

Dahl, V., & Van Zalk, M. (2014). Peer networks and the development of illegal political behavior among adolescents. *Journal of Research on Adolescence*, *24*(2), 399–409.

De Weerd, M., & Klandermans, B. (1999). Group identification and political protest: Farmers' protests in the Netherlands. *European Journal of Social Psychology*, *29*, 1073–1095.

Drury, J., Evripidou, A., & Van Zomeren, M. (2014). The Intersection of identity and power in collective action. *Power and Identity*, 94–116.

Drury, J., & Reicher, S. (1999). The intergroup dynamics of collective empowerment: Substantiating the social identity model of crowd behavior. *Group Processes & Intergroup Relations, 2*(4), 381–402.

Drury, J., & Reicher, S. (2005). Explaining enduring empowerment: A comparative study of collective action and psychological outcomes. *European Journal of Social Psychology, 35*(1), 35–58.

Drury, J., & Reicher, S. (2009). Collective psychological empowerment as a model of social change: Researching crowds and power. *Journal of Social Issues, 65*(4), 707–725.

Dugas, M., Schori-Eyal, N., Kruglanski, A. W., Klar, Y., Touchton-Leonard, K., McNeill, A., . . . Roccas, S. (2018). Group-centric attitudes mediate the relationship between need for closure and intergroup hostility. *Group Processes & Intergroup Relations, 21*(8), 1155–1171.

Federico, C. M., Golec, A., & Dial, J. L. (2005). The relationship between the need for closure and support for military action against Iraq: Moderating effects of national attachment. *Personality and Social Psychology Bulletin, 31*(5), 621–632.

Feinberg, M., Willer, R., & Kovacheff, C. (2017). *Extreme protest tactics reduce popular support for social movements* (Rotman School of Management Working Paper No. 2911177). Retrieved from https://ssrn.com/abstract=2911177 or http://dx.doi.org/10.2139/ssrn.2911177

Fiske, A. P., & Rai, T. S. (2014). *Virtuous violence: Hurting and killing to create, sustain, end, and honor social relationships.* Cambridge: Cambridge University Press.

Fiske, S. T. (2003). Five core social motives plus or minus five. In S. J. Spencer & S. Fein (Eds.), *Motivated social perception: The Ontario symposium* (Vol. 9, pp. 233–246). Mahwah, NJ: Erlbaum.

Folger, R. (1987). Reformulating the conditions of resentment: A referent cognition model. In J. C. Masters & W. P. Smith (Eds.), *Social comparison, social justice, and relative deprivation* (pp. 183–215). London: Erlbaum.

Franco, Z. E., Blau, K., & Zimbardo, P. G. (2011). Heroism: A conceptual analysis and differentiation between heroic action and altruism. *Review of General Psychology, 15*(2), 99–113.

Fritsche, I., Jonas, E., & Kessler, T. (2011). Collective reactions to threat: Implications for intergroup conflict and for solving societal crises. *Social Issues and Policy Review, 5*(1), 101–136.

Gastil, R. D. (1971). Homicide and a regional culture of violence. *American Sociological Review*, 412–427.

Gelfand, M. J., LaFree, G., Fahey, S., & Feinberg, E. (2013). Culture and extremism. *Journal of Social Issues, 69*(3), 495–517.

George, L. S., & Park, C. L. (2016). Meaning in life as comprehension, purpose, and mattering: Toward integration and new research questions. *Review of General Psychology, 20*(3), 205–220.

Ginges, J., & Atran, S. (2011). War as a moral imperative (not just practical politics by other means). *Proceedings of the Royal Society B: Biological Sciences, 278*(1720), 2930–2938.

Ginges, J., Atran, S., Medin, D., & Shikaki, K. (2007). Sacred bounds on rational resolution of violent political conflict. *Proceedings of the National Academy of Sciences of the United States of America, 104*, 7357–7360.

Goldman, L., Giles, H., & Hogg, M. A. (2014). Going to extremes: Social identity and communication processes associated with gang membership. *Group Processes & Intergroup Relations, 17*(6), 813–832.

Gómez, A., Brooks, M. L., Buhrmester, M. D., Vázquez, A., Jetten, J., & Swann Jr, W. B. (2011). On the nature of identity fusion: Insights into the construct and a new measure. *Journal of Personality and Social Psychology, 100*, 918–933.

Gómez, Á., Morales, J. F., Hart, S., Vázquez, A., & Swann Jr, W. B. (2011). Rejected and excluded forevermore, but even more devoted: Irrevocable ostracism intensifies loyalty to the group among identity-fused persons. *Personality and Social Psychology Bulletin, 37*, 1574–1586.

Gøtzsche-Astrup, O. (2018). The time for causal designs: Review and evaluation of empirical support for mechanisms of political radicalisation. *Aggression and Violent Behavior, 39*, 90–99.

Greitemeyer, T., & Sagioglou, C. (2017). Increasing wealth inequality may increase interpersonal hostility: The relationship between personal relative deprivation and aggression. *The Journal of Social Psychology, 157*(6), 766–776.

Grossman, D. (1996). *On killing: The psychological cost of learning to kill in war and society.* Boston: Little Brown.

Gurr, T. R. (1970). *Why men rebel.* Princeton: Princeton University Press.

Gurr, T. R. (1993). Why minorities rebel: A global analysis of communal mobilization and conflict since 1945. *International Political Science Review, 14*(2), 161–201.

Hackney, S. (1969). Southern violence. *The American Historical Review, 74*(3), 906–925.

Haslam, S. A., & Reicher, S. D. (2012). When prisoners take over the prison: A social psychology of resistance. *Personality and Social Psychology Review, 16*(2), 154–179.

Hogg, M. A. (2014). From uncertainty to extremism: Social categorization and identity processes. *Current Directions in Psychological Science, 23*(5), 338–342.

Hogg, M. A., Kruglanski, A., & Van den Bos, K. (2013). Uncertainty and the roots of extremism. *Journal of Social Issues, 69*(3), 407–418.

Hogg, M. A., Sherman, D. K., Dierselhuis, J., Maitner, A. T., & Moffitt, G. (2007). Uncertainty, entitativity, and group identification. *Journal of Experimental Social Psychology, 43*, 135–142.

Holman, T. (2016). 'Gonna get myself connected' the role of facilitation in Foreign fighter mobilizations. *Perspectives on Terrorism, 10*(2), 2–23.

Hornsey, M. J., Blackwood, L., Louis, W., Fielding, K., Mavor, K., Morton, T., . . . White, K. M. (2006). Why do people engage in collective action? Revisiting the role of perceived effectiveness. *Journal of Applied Social Psychology, 36*(7), 1701–1722.

Jackson, J., Huq, A. Z., Bradford, B., & Tyler, T. R. (2013). Monopolizing force? Police legitimacy and public attitudes toward the acceptability of violence. *Psychology, Public Policy, and Law, 19*(4), 479–497.

Jasko, K., & LaFree, G. (2020). Who is more violent in extremist groups? A comparison of leaders and followers. *Aggressive Behavior, 46*(2), 141–150.

Jasko, K., LaFree, G., & Kruglanski, A. W. (2017). Quest for significance and violent extremism: The case of domestic radicalization. *Political Psychology, 38*, 815–831.

Jasko, K., Szastok, M., Grzymala-Moszczynska, J., Maj, M., & Kruglanski, A. W. (2019). Rebel with a cause: Personal significance from political activism predicts willingness to self-sacrifice. *Journal of Social Issues, 75*(1), 314–349.

Jasko, K., Webber, D., Kruglanski, A. W., Gelfand, M., Taufiqurrohman, M., Hettiarachchi, M., & Gunaratna, R. (2020). Social context moderates the effects of quest for significance on violent extremism. *Journal of Personality and Social Psychology, 118*(6), 1165–1187.

Jost, J. T., Ledgerwood, A., & Hardin, C. D. (2008). Shared reality, system justification, and the relational basis of ideological beliefs. *Social and Personality Psychology Compass, 2*(1), 171–186.

Karatnycky, A., & Ackerman, P. (2005). *How freedom is won: From civic resistance to durable democracy*. New York: Freedom House.

Kelly, C., & Breinlinger, S. (1995). Identity and injustice: Exploring women's participation in collective action. *Journal of Community & Applied Social Psychology, 5*(1), 41–57.

Koopmans, R. (2005). *Contested citizenship: Immigration and cultural diversity in Europe*. Minneapolis, MN: University of Minneapolis Press.

Kruglanski, A. W. (2004). *The psychology of closed mindedness*. New York: Psychology Press.

Kruglanski, A. W. (2018). Violent radicalism and the psychology of prepossession. *Social Psychological Bulletin, 13*(4), 1–18.

Kruglanski, A. W., Bélanger, J. J., Gelfand, M., Gunaratna, R., Hettiarachchi, M., Reinares, F., . . . Sharvit, K. (2013). Terrorism—A (self) love story: Redirecting the significance quest can end violence. *American Psychologist, 68*, 559–575.

Kruglanski, A. W., Bélanger, J. J., & Gunaratna, R. (2019). *The three pillars of radicalization: Needs, Narratives, and Networks*. Oxford: Oxford University Press.

Kruglanski, A. W., Chen, X., Dechesne, M., Fishman, S., & Orehek, E. (2009). Fully committed: Suicide bombers' motivation and the quest for personal significance. *Political Psychology, 30*, 331–557.

Kruglanski, A. W., Gelfand, M. J., Bélanger, J. J., Sheveland, A., Hetiarachchi, M., & Gunaratna, R. (2014). The psychology of radicalization and deradicalization: How significance quest impacts violent extremism. *Political Psychology, 35*, 69–93.

Kruglanski, A. W., Gunaratna, R., Ellenberg, M., & Speckhard, A. (2020). Terrorism in time of the pandemic: Exploiting mayhem. *Global Security: Health, Science and Policy, 5*(1), 121–132.

Kruglanski, A. W., Jasko, K., Chernikova, M., Dugas, M., & Webber, D. (2017). To the fringe and back: Violent extremism and the psychology of deviance. *American Psychologist, 72*(3), 217–230.

Kruglanski, A. W., & Webster, D. M. (1996). Motivated closing of the mind: "Seizing" and "freezing." *Psychological Review, 103*, 263–283.

Kydd, A. H., & Walter, B. F. (2006). The strategies of terrorism. *International Security, 31*(1), 49–80.

Lammers, J., Koch, A., Conway, P., & Brandt, M. J. (2017). The political domain appears simpler to the politically extreme than to political moderates. *Social Psychological and Personality Science, 8*(6), 612–622.

Lawrence, A. (2010). Triggering nationalist violence: Competition and conflict in uprisings against colonial rule. *International Security, 35*(2), 88–122.

Leander, N. P., Agostini, M., Stroebe, W., Kreienkamp, J., Spears, R., Kuppens, T., . . . Kruglanski, A. W. (2020). Frustration-affirmation? Thwarted goals motivate compliance with social norms for violence and nonviolence. *Journal of Personality and Social Psychology, 119*(2), 249–271.

Littman, R. (2018). Perpetrating violence increases identification with violent groups: Survey evidence from former combatants. *Personality and Social Psychology Bulletin, 44*(7), 1077–1089.

Lyons-Padilla, S., Gelfand, M. J., Mirahmadi, H., Farooq, M., & Van Egmond, M. (2015). Belonging nowhere: Marginalization & radicalization risk among Muslim immigrants. *Behavioral Science & Policy, 1*(2), 1–12.

Mandel, D. R., & Litt, A. (2013). The ultimate sacrifice: Perceived peer honor predicts troops' willingness to risk their lives. *Group Processes & Intergroup Relations, 16*(3), 375–388.

McCauley, C., & Moskalenko, S. (2011). *Friction: How radicalization happens to them and us*. Oxford: Oxford University Press.

McGregor, I., Nash, K., Mann, N., & Phills, C. E. (2010). Anxious uncertainty and reactive approach motivation (RAM). *Journal of Personality and Social Psychology, 99*(1), 133–147.

Melde, C., Taylor, T. J., & Esbensen, F. A. (2009). "I got your back": An examination of the protective function of gang membership in adolescence. *Criminology, 47*(2), 565–594.

Melucci, A. (1982). *L'invenzione del presente: movimenti, identità, bisogni individuali* (Vol. 146). Bologna: il Mulino.

Merriam-Webster, Inc. (1986). *Webster's ninth new collegiate dictionary*. Springfield, MA: Merriam-Webster.

Mummendey, A., Kessler, T., Klink, A., & Mielke, R. (1999). Strategies to cope with negative social identity: Predictions by social identity theory and relative deprivation theory. *Journal of Personality and Social Psychology, 76*(2), 229–245.

Nawata, K. (2020). A glorious warrior in war: Cross-cultural evidence of honor culture, social rewards for warriors, and intergroup conflict. *Group Processes & Intergroup Relations, 23*(4), 598–611.

Nisbett, R. E. (1993). Violence and U.S. Regional culture. *American Psychologist, 48*, 441–449.

Nisbett, R. E., & Cohen, D. (1996). *Culture of honour: The psychology of violence in the South*. Boulder, CO: Westview.

Nowak, A., Gelfand, M. J., Borkowski, W., Cohen, D., & Hernandez, I. (2016). The evolutionary basis of honor cultures. *Psychological Science, 27*(1), 12–24.

Obaidi, M., Bergh, R., Akrami, N., & Anjum, G. (2019). Group-based relative deprivation explains endorsement of extremism among western-born Muslims. *Psychological Science, 30*(4), 596–605.

Olivola, C. Y., & Shafir, E. (2013). The martyrdom effect: When pain and effort increase prosocial contributions. *Journal of Behavioral Decision Making, 26*(1), 91–105.

Pedahzur, A. (2005). *Suicide terrorism*. Cambridge: Polity Press.

Pettigrew, T. F. (1967). Social evaluation theory: Convergences and applications. In *Nebraska symposium on motivation*. Lincoln: University of Nebraska Press.

Piazza, J. A. (2006). Rooted in poverty? Terrorism, poor economic development, and social cleavages. *Terrorism and Political Violence, 18*(1), 159–177.

Piazza, J. A. (2011). Poverty, minority economic discrimination, and domestic terrorism. *Journal of Peace Research, 48*(3), 339–353.

Pittman, T. S., & Zeigler, K. R. (2007). Basic human needs. In A. W. Kruglanski & E. T. Higgins (Eds.), *Social psychology: Handbook of basic principles* (pp. 473–489). New York: Guilford Press.

Post, J. M. (2006). The psychological dynamics of terrorism. In L. Richardson (Ed.), *The roots of terrorism* (pp. 17–28). New York: Routledge and Taylor & Francis Group.

Post, J. M., Sprinzak, E., & Denny, L. (2003). The terrorists in their own words: Interviews with 35 incarcerated Middle Eastern terrorists** This research was conducted with the support of the Smith Richardson Foundation. *Terrorism and political Violence, 15*(1), 171–184.

Ransford, H. E. (1968). Isolation, powerlessness, and violence: A study of attitudes and participation in the Watts riot. *American Journal of Sociology, 73*(5), 581–591.

Runciman, W. G. (1966). *Relative deprivation and social justice*. London: Routledge and Kegan Paul.

Ryan, T. J. (2014). Reconsidering moral issues in politics. *The Journal of Politics*, 76(2), 380–397.

Saab, R., Spears, R., Tausch, N., & Sasse, J. (2016). Predicting aggressive collective action based on the efficacy of peaceful and aggressive actions. *European Journal of Social Psychology*, 46(5), 529–543.

Saab, R., Tausch, N., Spears, R., & Cheung, W. Y. (2015). Acting in solidarity: Testing an extended dual pathway model of collective action by bystander group members. *British Journal of Social Psychology*, 54(3), 539–560.

Sageman, M. (2004). *Understanding terror networks*. Philadelphia, PA: University of Pennsylvania Press.

Sageman, M. (2008). *Leaderless jihad: Terror networks in the twenty-first century*. Philadelphia, PA: University of Pennsylvania Press.

Scheepers, D., Spears, R., Doosje, B., & Manstead, A. S. (2006). Diversity in in-group bias: Structural factors, situational features, and social functions. *Journal of Personality and Social Psychology*, 90(6), 944–960.

Schumpe, B. M., Bélanger, J. J., Giacomantonio, M., Nisa, C. F., & Brizi, A. (2018). Weapons of peace: Providing alternative means for social change reduces political violence. *Journal of Applied Social Psychology*, 48(10), 549–558.

Schwarzmantel, J. (2010). Democracy and violence: A theoretical overview. *Democratization*, 17(2), 217–234.

Sedgwick, M. (2004). Al-Qaeda and the nature of religious terrorism. *Terrorism and Political Violence*, 16, 795–814.

Shah, J. Y., & Higgins, E. T. (1997). Expectancy × value effects: Regulatory focus as determinant of magnitude and direction. *Journal of Personality and Social Psychology*, 73(3), 447–458.

Silke, A. (2008). Holy warriors: Exploring the psychological processes of jihadi radicalization. *European Journal of Criminology*, 5(1), 99–123.

Simon, B., & Klandermans, B. (2001). Politicized collective identity: A social psychological analysis. *American Psychologist*, 56(4), 319–331.

Simon, B., Loewy, M., Stürmer, S., Weber, U., Freytag, P., Habig, C., . . . Spahlinger, P. (1998). Collective identification and social movement participation. *Journal of Personality and Social Psychology*, 74(3), 646–658.

Smith, H. J., Cronin, T., & Kessler, T. (2008). Anger, fear, or sadness: Faculty members' emotional reactions to collective pay disadvantage. *Political Psychology*, 29(2), 221–246.

Smith, H. J., & Huo, Y. J. (2014). Relative deprivation: How subjective experiences of inequality influence social behavior and health. *Policy Insights from the Behavioral and Brain Sciences*, 1(1), 231–238.

Smith, H. J., Pettigrew, T. F., Pippin, G. M., & Bialosiewicz, S. (2012). Relative deprivation: A theoretical and meta-analytic review. *Personality and Social Psychology Review*, 16(3), 203–232.

Stouffer, S. A., Suchman, E. A., DeVinney, L. C., Star, S. A., & Williams, R. A. (1949). *The American soldier: Adjustment during Army life* (Vols. 1 and 2). Princeton: Princeton University Press.

Stretesky, P. B., & Pogrebin, M. R. (2007). Gang-related gun violence: Socialization, identity, and self. *Journal of Contemporary Ethnography*, 36(1), 85–114.

Swann, W. B., Gómez, A. M., Seyle, D. C., Morales, J. F., & Huici, C. (2009). Identity fusion: The interplay of personal and social identities in extreme group behavior. *Journal of Personality and Social Psychology*, *96*(5), 995–1011.

Swann, W. B., Jr., Jetten, J., Gómez, Á., Whitehouse, H., & Bastian, B. (2012). When group membership gets personal: A theory of identity fusion. *Psychological Review*, *119*(3), 441–456.

Tajfel, H., & Turner, J. C. (1979). An integrative theory of inter- group conflict. In W. G. Austin & S. Worchel (Eds.), *The social psychology of intergroup relations* (pp. 33–47). Monterey, CA: Brooks-Cole.

Tausch, N., Becker, J. C., Spears, R., Christ, O., Saab, R., Singh, P., & Siddiqui, R. N. (2011). Explaining radical group behavior: Developing emotion and efficacy routes to normative and nonnormative collective action. *Journal of Personality and Social Psychology*, *101*(1), 129–148.

Tetlock, P. E. (2003). Thinking the unthinkable: Sacred values and taboo cognitions. *Trends in Cognitive Sciences*, *7*(7), 320–324.

Thomas, E. F., & Louis, W. R. (2014). When will collective action be effective? Violent and non-violent protests differentially influence perceptions of legitimacy and efficacy among sympathizers. *Personality and Social Psychology Bulletin*, *40*(2), 263–276.

Tropp, L. R., & Brown, A. C. (2004). What benefits the group can also benefit the individual: Group-enhancing and individual-enhancing motives for collective action. *Group Processes & Intergroup Relations*, *7*(3), 267–282.

Van Den Bos, K., Euwema, M. C., Poortvliet, P. M., & Maas, M. (2007). Uncertainty management and social issues: Uncertainty as an important determinant of reactions to socially deviating people 1. *Journal of Applied Social Psychology*, *37*(8), 1726–1756.

van Prooijen, J. W., & Krouwel, A. P. (2017). Extreme political beliefs predict dogmatic intolerance. *Social Psychological and Personality Science*, *8*(3), 292–300.

Van Zomeren, M., Saguy, T., & Schellhaas, F. M. (2013). Believing in "making a difference" to collective efforts: Participative efficacy beliefs as a unique predictor of collective action. *Group Processes & Intergroup Relations*, *16*(5), 618–634.

van Zomeren, M., Spears, R., Fischer, A. H., & Leach, C. W. (2004). Put your money where your mouth is! Explaining collective action tendencies through group-based anger and group efficacy. *Journal of Personality and Social Psychology*, *87*(5), 649–664.

Vanneman, R. D., & Pettigrew, T. F. (1972). Race and relative deprivation in the urban United States. *Race*, *13*(4), 461–486.

Victoroff, J., Adelman, J. R., & Matthews, M. (2012). Psychological factors associated with support for suicide bombing in the Muslim diaspora. *Political Psychology*, *33*(6), 791–809.

Walker, I., & Smith, H. J. (Eds.). (2002). *Relative deprivation: Specification, development, and integration.* Cambridge: Cambridge University Press.

Wang, D. J., & Piazza, A. (2016). The use of disruptive tactics in protest as a trade-off: The role of social movement claims. *Social Forces*, *94*(4), 1675–1710.

Wasow, O. (2020). Agenda seeding: How 1960s Black protests moved elites, public opinion and voting. *American Political Science Review*, 1–22.

Webber, D., Babush, M., Schori-Eyal, N., Vazeou-Niewenhuis, A., Hettiarachchi, M., Belanger, J. J., . . . Gelfand, M. J. (2018). The road to extremism: Field and experimental evidence that significance loss-induced need for closure fosters radicalization. *Journal of Personality and Social Psychology*, *114*(2), 270–285.

Webber, D., Chernikova, M., Kruglanski, A. W., Gelfand, M. J., Hettiarachchi, M., Gunaratna, R., . . . Belanger, J. J. (2018). Deradicalizing detained terrorists. *Political Psychology*, *39*(3), 539–556.

Webster, D. M., & Kruglanski, A. W. (1994). Individual differences in need for cognitive closure. *Journal of Personality and Social Psychology*, *67*, 1049–1062.

Wolfgang, M. E., Ferracuti, F., & Mannheim, H. (1967). *The subculture of violence: Towards an integrated theory in criminology* (Vol. 16). London: Tavistock Publications.

Wong, Y. J., Tsai, P. C., Liu, T., Zhu, Q., & Wei, M. (2014). Male Asian international students' perceived racial discrimination, masculine identity, and subjective masculinity stress: A moderated mediation model. *Journal of Counseling Psychology*, *61*(4), 560–569.

Worchel, P., Hester, P. G., & Kopala, P. S. (1974). Collective protest and legitimacy of authority: Theory and research. *Journal of Conflict Resolution*, *18*(1), 37–54.

Wright, S. C. (2001). Strategic collective action: Social psychology and social change. In R. Brown & S. L. Gaertner (Eds.), *Intergroup processes: Blackwell handbook of social psychology* (Vol. 4, pp. 409–430). Oxford: Blackwell.

Wright, S. C., Taylor, D. M., & Moghaddam, F. M. (1990a). Responding to membership in a disadvantaged group: From acceptance to collective protest. *Journal of Personality and Social Psychology*, *58*(6), 994–1003.

Wright, S. C., Taylor, D. M., & Moghaddam, F. M. (1990b). The relationship of perceptions and emotions to behavior in the face of collective inequality. *Social Justice Research*, *4*(3), 229–250.

Zartman, I. W., & Khan, M. (2011). Growing up in groups. In I. W. Zartman & G. O. Faure (Eds.), *Engaging extremists: Trade-offs, timing, and diplomacy* (pp. 27–56). Washington, DC: United States Institute of Peace.

11
MOTIVATIONAL IMBALANCE IN JIHADI ONLINE RECRUITMENT

Gabriel Weimann

Introduction

Without recruitment, terrorism cannot prevail, survive, and develop. Recruitment provides the killers, the suicide bombers, the kidnappers, the executioners, the engineers, the soldiers, and the armies of future terrorism. The Internet and online platforms have become a useful instrument for modern terrorists' recruitment. The success of the Islamic State (ISIS) in the massive recruitment of over 40,000 fighters from as many as 90 countries attracted global attention and scholarly interest. Once an offshoot of Al Qaeda in Iraq, ISIS has emerged into a global leader of the Sunni jihadist movement. Despite significant losses, ISIS continuously recruited thousands of enthusiastic fighters from many corners of the world. While the news media focused on Europe and the United States as recruiting centers, the terrorist group has drawn large numbers of fighters from South Asia, the Middle East, the former republics of the Soviet Union, sub-Saharan Africa, and the United States (Berger & Morgan, 2015). Within 3 years, thousands of foreign fighters and non-combatant supporters have travelled to join the Islamic State, including unprecedented numbers of Westerners. Followers who could not travel to the war zones in Syria and Iraq were urged to carry out attacks in their homelands.

Numerous studies explored the online recruitment of Jihadi groups and especially of ISIS. Yet, these studies, focusing mainly on the online platforms and the appeals used, lacked a theoretical framework that can serve the search for the factors explaining process and its success. We argue that the notion of motivational imbalance as a key persuasive appeal in terrorist online recruitment can provide the needed psychological theoretical concept to explain their success. Motivational imbalance is a state wherein a given need becomes dominant to the

DOI: 10.4324/9781003030898-14

point of inhibiting other needs. In the case of violent extremism, the dominant need may be the quest for significance, the desire to matter, to take revenge, and to get self- respect and others' respect (Kruglanski, 2018; Kruglanski, Fernandez, Factor, & Szumowska, 2019).

This chapter is organized as follows: First we examine terrorist online recruitment and its features (stages, type of messages, content of the messages, characteristics of the targets) and relate it to the notion of motivational imbalance. We then explore the trend of 'narrowcasting' (focusing persuasive communication on selected target audiences) and how well it combines with motivational imbalance in the process of recruitment by terrorist groups. Finally, we demonstrate how motivational imbalance has been used in recent Jihadi recruitment campaigns. The evidence presented comes from the contents of recruitment campaigns as well as from the testimonies of recruits. In conclusion, we suggest that the unique combination of medium (online platforms), persuasive content (motivational imbalance), and selected audiences (narrowcasting) is the key to the success of Jihadi recruitment campaigns.

Online Recruitment

Without recruitment, terrorism cannot prevail, survive, and develop. Recruitment provides the killers, the suicide bombers, the kidnappers, the executioners, the engineers, the soldiers, and the armies of modern terrorism. The Internet has become a useful instrument for modern terrorists' recruitment, combining several advantages for the recruiters: it makes information gathering easier for potential recruits by offering more information, more quickly, and in multimedia format; the global reach of the Net allows groups to publicize events to more people; and, by increasing the possibilities for interactive communication, new opportunities for assisting groups are offered, along with more chances for contacting the group directly.

How does online recruitment work? The rise and success of extreme Islamist online campaigns of radicalization and recruitment have been attributed to the combination of effective appeals (content) and platforms (e.g., social media) in a multistage process. Let us look first on the use of online platforms. As first coined by Ducol, the term *Jihadisphere* is understood simply as a loose network of online communities that support the Jihadi drive (Ducol, 2012, pp. 51–52). These online communities rely on online platforms that have evolved over time to support their interactions and contacts. The first of these platforms were official or 'top-down' websites (Weimann, 2006; Zelin, 2013, p. 5). These websites aimed to disseminate and promote the group's messages, ideology, and directives. However, these official websites have declined due to a combination of them being blocked or taken down and the growing need for more interactive platforms instead of the one-way websites. Thus, with the decline of websites came the development of forums that allowed members to communicate and bond directly with other activists and sympathizers while enjoying online anonymity. Given the scope these mediums provided for interactions, they soon began to outbid and replace static websites belonging to jihadist organizations as the main platforms from which to spread jihadist propaganda and create online networks (Ramsay, 2008; Zelin, 2013, p. 5).

While chat rooms and forums have become less dependent upon password protection (Weimann, 2010), Interned-savvy extremists were looking for more open, free, and public media outlets that will propagate their messages and recruit people (Weimann, 2015). This desire was answered with the emergence of social media, from Facebook, Twitter, and Instagram to Telegram and YouTube. These new platforms, very popular among younger people, have made online jihadist calls and directions far more accessible to the general public: 'The ability to exchange comments about videos and to send private messages to other users help jihadists identify each other rapidly, resulting in a vibrant jihadist virtual community' (Weimann, 2014, p. 10). Furthermore, these social networking sites have maximized accessibility; whether sympathetic or not, anyone can fall foul of material online that no longer exists in the periphery of the 'darkest corners of the Internet' (Neumann, 2012, p. 17).

These social media became very effective in establishing virtual networks through which 'a large number of foreign fighters receive their information about the conflict not from the official channels provided by their fighting group but through so-called disseminators' (Carter, Maher, & Neumann, 2014, p. 1). These online disseminators are sympathetic individuals who are able to contribute violent extremist narratives from the comfort and relative safety of their homes. They provide information, live updates, directions, and suggestions and thus become agents of persuasion, radicalization, and recruitment. One of the keys to ISIS' success in global recruiting is its use of Internet-based communications technology and especially social media platforms (Weimann, 2006, 2015, 2016). Western agencies monitoring ISIS online campaigns have described the extreme terrorist group's use of social media platforms as unprecedented in its sophistication and high quality. In what emerged as a three-prong social media strategy, ISIS used social media to raise its international profile, recruit members, and inspire lone actor attacks (Berger & Morgan, 2015).

The Internet and advanced online platforms provide powerful tools for recruiting and mobilizing group members through integrated communications (Ahmad, 2014; Weimann, 2016). In addition to seeking converts by using the full panoply of online contents (e.g., audio, digital video) to enhance the presentation of their message, terrorist organizations capture information about the users who browse their websites or access their postings. Users who seem most interested in the organization's cause or appear well suited to carrying out its work are then contacted. Recruiters may also use more interactive Internet technology to roam online chat rooms, Facebook, Twitter, and other platforms, looking for receptive members of the public, particularly young people (Just, 2015). The reach of these online outlets provides terrorist organizations and sympathizers with a global pool of potential recruits. The virtual forums offer an open venue for recruits to learn about and provide support to terrorist organizations, and they promote engagement in direct actions (GerDenning, 2010). Finally, through the use of interactive social media, it allows members or potential members of the group to engage in debate with one another to seduce, persuade, teach, radicalize, and provide training and instructions and even launch action.

Today, all terrorist groups are employing online recruitment campaigns. Of all these groups, ISIS invested the most effort on a vast spectrum of online platforms, with a special emphasis on foreign fighters. An indication of the group's online recruitment effort is the fact that ISIS operated a special media production unit for Western recruitment, the al-Hayat Media Center, launched in May 2014. This unit regularly produces and releases videos and other materials in many languages, including English, Turkish, Dutch, French, German, Indonesian, and Russian. Content released from al-Hayat included the English-language magazine called 'Dabiq' and recruitment videos from foreign fighters spanning from America, Britain, Australia, and Germany just to name a few. These online materials combine a rich variety of multimedia productions.

The online recruitment by terrorist groups has several distinctive characteristics. First, it is a multistep process, which requires a gradual transition and numerous phases (Berger, 2015). Second, unlike conventional or traditional recruitment, the process relies on online platforms. Applying the RAND Corporation's model of selection and recruitment, the first step is 'the Net': a target population may be engaged equitably by being exposed to an online message, video, taped lecture, or the like (Gerwehr & Daly, 2006). Some members will respond positively, others negatively; but in general, a vast group is viewed as primed for recruitment. More specifically, the target audience is viewed as homogeneous enough and receptive enough to be approached with a single undifferentiated pitch. The second stage is the 'funnel'. As the term implies, potential recruits start at one end of the process and are transformed, after some culling along the way, into dedicated members when they emerge at the other end. Here, the recruiter may use an incremental, or phased, approach when he or she believes a target individual is ripe for recruitment yet requires a significant transformation in identity and motivation. This stage capitalizes on a wealth of well-studied techniques in cognitive, social, and clinical psychology (Hegghammer, 2014). It involves online exchanges and exposure to religious, political, or ideological material. This stage relies on social bonding (albeit virtually) based on the target's alienation, social frustration, solitude, and personal pessimism. The next stage is the 'infection'. Selected target members who are dissatisfied with their social status or have a grudge against their political or religious system are directed to self-radicalization. The self-radicalization relies only on online sources and involves gradual advancement in the level of commitment and extremism. The final stage, the 'activation', involves the release of the recruit to carry out the terrorist action or to send directions for the actual joining of the fighters in Syria or Iraq.

ISIS recruitment campaigns combined a successful mix of platforms (i.e., social media) and contents (messages, appeals, visuals, etc.). As revealed by numerous studies, ISIS has been highly successful in the use of a sophisticated propaganda machine and slick media productive capacity to entice mostly young Muslim men and women to travel to Syria and Iraq and become members of the ISIS community by portraying a glorified and romanticized version of reality (Mahood & Rane, 2017). This initial online persuasion was supplemented by one-on-one personal interaction with professional recruiters, performed mostly on social media. In terms of contents, seven key factors emerged: (a) appeals to personal victimization and discrimination; (b) appeals to historical and political grievances; (c) appeals to personal grievances; (d) appeals to socio-economic disparities; (e) appeals to social alienation, solitude, and alienation; (f) appeals to inspirational motives; and (g) appeals to thrill and action.

The growing literature on radicalization is useful in revealing the myriad of circumstances, tactics, methods, and appeals that facilitate a person's transition to violence. It does not, however, illuminate the strategy behind groups such as ISIS does not provide an overall theoretical basis for understanding radicalization and recruitment by seductive political extremists. This is where the notion of motivational imbalance becomes a relevant and useful conceptual framework.

The Notion of Motivational Imbalance

Radicalization involves a human process wherein someone experiences a gradual change in mindset and behavior. How does it work, considering the fact that the process leads to violence, sacrifice, probable death, and victimization of others? Why would anyone in their right mind (and hardly anyone believes, nowadays, that all extremists are insane) commit acts that so clearly violate such basic concerns as health or survival (Kruglanski et al., 2014)? Blackwell (2016) considered radicalization with respect to human needs and Maslow's hierarchy. Blackwell argued that ISIS was successful in its radicalization effort because it responded to human needs and provided a means whereby individuals perceived their needs could be fulfilled. In other words, the radicalization process contains a mixture of human needs that people perceive may be satisfied through some type of affiliation with terrorism.

A wide range of human needs were and still are used by terrorist recruitment. The literature lists needs like honor, vengeance, religion, loyalty to the leader, perks in the afterlife, feminism, action, and thrill, meaning in life, social bonding, romance, and more. Terrorist organizations can use these needs to their advantage and offer recruits ways to meet their needs for significance or community, for instance, while framing their mission as an honorable one that will alleviate some of the hardships one might suffer. Kruglanski and his colleagues

(2014) highlighted the use of the 'quest for significance'. This quest share characteristics which 'constitute a major, universal, human motivation variously labeled as the need for esteem, achievement, meaning, competence, control, and so on' (Kruglanski et al., 2014, p. 73). While specific terrorists have their own specific goals, most include 'honor, vengeance, religion, loyalty to the leader, perks in the afterlife, even feminism' (Kruglanski et al., 2014, p. 73). These goals of terrorism are part of general human needs for 'esteem, achievement, meaning, competence, control, and so on' (Kruglanski et al., 2014, p. 73). It can be assumed, then, that people who commit radical behavior are seeking one of the previously outlined psychological human needs. They may adapt their morality to suit this behavior and may place these needs above any other, making radical behavior the means by which to achieve their main goal, whether personal or political.

This was further developed into a theory of motivational balance. Motivational balance as suggested by Kruglanski (2018; Kruglanski, Jasko, Chernikova, Dugas, & Webber, 2017; Kruglanski et al., 2019) is a state in which all the basic needs constitute active concerns whose fulfilment drives individuals' behavior. These needs constrain one another such that behavior that gratifies only some needs while undermining others tends to be avoided. For instance, one's need for intimacy and relatedness may temper one's need for achievement, thus promoting a work–family balance, etc. At times, however, a motivational imbalance may occur wherein a given need receives disproportionate emphasis, overriding the others. This state may be described as one of prepossession, in which one's mind is predominantly absorbed with a given motivational concern, and is oblivious to all else. Thus, there is a disproportionate commitment to ends served by the extreme behavior that prompts a devaluation or a forceful suppression of alternative considerations (Shah, Friedman, & Kruglanski, 2002). Illustrative of such exclusive commitment is a statement of a Black Tamil Tiger, member of the suicide cadre of the Liberation Tigers of Tamil Eelam:

> Family and relationships are forgotten in that place. There was no place for love . . . That means a passion and loyalty to that group, to those in charge, to those who sacrificed their lives for the group. Then I came to a stage where I had no love for myself. I had no value for my life. I was ready to give myself fully, even to destroy myself, in order to destroy another person.
>
> *(Quoted by Kruglanski et al., 2014)*

Is terrorist recruitment applying this over-emphasis on one need while undermining others and thus attract, seduce, and radicalize these prepossessed individuals? There is no doubt that extremist and terrorist groups are employing various persuasive appeals (Weimann, 2007, 2016). Moreover, Weimann (2015) highlighted the increasing use of 'narrowcasting' in terrorist online communication.

As we will demonstrate, 'narrowcasting' increases the effectiveness of the motivational imbalance tactic.

'Narrowcasting'

Terrorists rely on state-of-the-art techniques from the advertising industry to attract recruits. Rather than broadcast, or use one message to all, the extremists 'narrowcast' targeting small groups with specific messages that exploit their vulnerabilities. Narrowcasting aims messages at specific segments of the public defined by values, preferences, demographic attributes, or as this chapter argues—by needs and their prioritization. The use of 'narrowcasting', tailoring messages and platforms to specific sub-groups, appears to increase effectiveness. These focused campaigns involve also the use of specific human needs when addressing certain target audiences. Thus, instead of 'one-message-for-all', Internet-savvy terrorists target specific subpopulations, including children, women, 'lone wolves', overseas communities, or diasporas and imprisoned fans and followers. The unmistakable growth in the participation of women and youth in terrorist activity with the evident growth in persuasive online messages targeting these groups may provide alarming signals of the narrowcasting tactic's success

Social media allow terrorists to use this targeting strategy of narrowcasting more effectively. On social media, messages are tailored to match the profile of a particular social group or social category. Many social media users join interest groups, and these groups enable terrorists to target users whom they might be able to manipulate. These users often accept people as 'friends' on the social media site whether or not they know them, thereby giving strangers access to personal information and photos. Some people even communicate with strangers and establish virtual friendships. Terrorists can therefore apply the narrowcasting strategy used on the broader Internet to specific and more personal social networking. They can tailor their name, accompanying default image, and information on a group message board to fit the profile of a particular social group. Interest groups also provide terrorists with a list of predisposed recruits or sympathizers. In the same way that marketing groups can view a member's information to decide which products to target on their web pages, terrorist groups can view people's profiles to decide who they are going to target and how they should configure the message. The use of narrowcasting combined with social media platforms is evident in the case of recruiting foreigners to fight for ISIS in Syria and Iraq. While in the past, jihadi groups published most of their materials on traditional media outlets, such as websites, chatrooms, and forums, ISIS has pioneered the use of social media as the main means for recruitment of foreigners, especially from North America, Europe, and the Far East.

In 2001 Ayman al-Zawahiri's claim that 'with the available means, small groups could prove to be a frightening horror for the Americans and the Jews' completely fell on deaf ears. Lacking the power of the 'narrowcasting' capabilities of social media and encrypted apps like Viber, Telegram, and WhatsApp, al-Zawahiri could not hope to reach the vast audience or securely communicate with his would-be and actual followers that ISIS has achieved. In 2010, Osama bin Laden had also called on his followers to attack, relying on identically old-fashioned Internet communications. His orders accordingly gained little traction, much less the widespread attention that ISIS's effective use of cutting-edge social media has achieved. Weimann's studies of narrowcasting revealed the terrorists' focusing on social groups (especially children, women, 'diaspora' communities) by the combination of 'narrow' contents, appeals, and platforms (Weimann, 2015, 2016). This trend allowed the online recruiters to present tailored needs to specific subgroups, and especially apply motivational imbalance according to the recipients' characteristics.

Using Motivational Imbalance in Jihadi Campaigns

Kruglanski and his associates outlined a general psychological theory of extremism and applied it to the special case of violent extremism. Extremism is defined as motivated deviance from general behavioral norms and is assumed to stem from a shift from a balanced satisfaction of basic human needs afforded by moderation to a motivational imbalance wherein a given need dominates the others. Thus, while majorities of individuals practice moderation, extremism is the province of the few. This ability to deviate is partially determined by the activities of external agents who required motivational imbalance. Our continuous project on monitoring terrorist contents on the Internet (Weimann, 2006, 2016) provides a huge database for the study of extremists' persuasion tactics.[1] Moreover, it allows for testing the applicability of the motivational imbalance theorem to Jihadist recruitment. To illustrate the combination of motivational imbalance, narrowcasting, and online platforms, we will use two sources: (a) we will highlight the appeals used in the campaigns, as retrieved from our database on online terrorism and (b) we will demonstrate the use of motivational imbalance as revealed by testimonies of the recruits themselves. The narrowcasting of the persuasive communication promotes the effectiveness of the recruitment process by instilling the dominant motivation in a narrative that tells one how to satisfy it, a narrative that is supported by a trusted authority of the recruiter that became close to the intended recruited.

The Appeals Used

The Highest Demand

The best way to promote motivational imbalance is by presenting an order of needs or values, ordering them, and then highlighting the first as overriding all others. This has been done often by recognized and respected Jihadi online preachers, radicalizers, and recruiters. For example, in August 2019, Abu Obeida Yusuf al-Annabi, an official in al Qaeda in the Islamic Maghreb (AQIM), addressed Algerian audiences in an online speech, reiterating his call to them to make Shariah-based governance their highest demand. His audio speech, entitled, 'O Free People of Algeria, Do Not Leave Your Places', came in a 19.5-minute video produced by the group's al-Andalus Media Foundation. He presents his notion of the right order of demands:

> You have seen how the tyrants have sought since the beginning of the uprising to abort it before it achieves the most important of its demands . . . So be cautious O worshipers of Allah from giving up your legitimate demands and your ultimate goal in the rooted and comprehensive change . . . There is no value to an uprising if it does not aim to make these noble goals its highest demands . . . There is no value to the uprising of a Muslim people if it does not seek to return to the rational Islamic rule, and to resume a decent Islamic life . . . There is no value for the uprising to Allah the Almighty, if it does not call for the Tawhid [oneness] of Allah the Almighty, and to renounce polytheism, and to establish His Shariah and judgments in the land.

In September 2019, Abu Bakr al-Baghdadi, the leader of ISIS, published online a speech entitled, 'And Say: "Do Deeds!"' in which he urged Jihadists to be 'torches of guidance' and embrace Muslims and guide and lead them to 'pure monotheism'. He also called on them to remind the Muslims that there is 'no way to remove tawaghit [tyrant rulers] except through the way revealed by Allah in His book, that is, jihad for His sake'. Moreover, he promoted the devotion to Allah as the highest need:

> This is medicine for the heart when afflicted by calamities and intense dangers. It is freedom from the sins and ills of the soul and refinement of it against what slows it down from the march to its Lord. It is recognition of guilt, a request for forgiveness to the Great and All-Forgiving. It is closeness through acts of obedience and numerous sacrifices. The alienation of faith and its people requires those who seek to elevate it anew to hold on to what the predecessors held, so that, while they fight their enemy, they do not turn their faces towards what the people of kufr and falsehood and the callers to hell guide, where these seek to keep them idle and to confuse them about their faith so that they do not reach the objective for which they immigrated and fought.

Boko Haram, The Jihadi group based in Nigeria associated with ISIS and known also as Jama'at Ahl al-Sunnah Lil Dawa Wal Jihad distributed in October 2019 an

Arabic-language video entitled 'There is no Glory Except in Jihad'. The video presents the religious foundation of its fighters and their vow to fight until death because this is the call of Allah. The text includes the following statements:

> We are people whom Allah honored with Islam, so you O worshipers should accept the call of Allah and pursue jihad, and join the group of Allah, under the one banner, to please Your creator. Our call and jihad is for Allah. We do not fight except for the supremacy of the monotheism of Allah and for the establishment of His Shariah in every place on earth. We are people who fight polytheism and its people, Allah permitting, and we will displace them wherever they are. This is our faith in our creator, whom we worship alone and no other. The Almighty said: 'Jihad is ordained for you though you dislike it, and it may be that you dislike a thing which is good for you and that you like a thing which is bad for you. Allah knows but you do not know' [Al-Baqarah, 216] . . . My beloved, know that we do have two options and not a third: it is either a victory that we attain, or martyrdom, which is our highest goal.

In February 2020, a Palestinian music video was launched online, promoting death for Allah as Martyrs for Al-Aqsa and encourages suicide bombings. The song teaches Palestinians that 'life is insignificant' when the Al-Aqsa Mosque calls them, and that dying for Allah is preferable to living: 'God, grant us Martyrdom . . . A million grooms and brides . . . have written the marriage contract in blood'. The music video shows the photos of two female suicide bombers: Wafa Idris and Ayyat Al-Akhras who between them murdered 3 and wounded over 128.

The Power of Fatwa

There is clearly a tendency of online Jihadi recruiters to create such an imbalance of human needs. The 'new order' provided by them is highlighting and promoting a single need to a supreme dominant position, overruling and suppressing all other needs. One popular mechanism to promote such imbalance relies on issuing a fatwa. A fatwa is an Islamic religious ruling, a scholarly opinion on a matter of Islamic law. In 1989, the term 'fatwa' became globally known, following Ayatollah Khomeini's death-fatwa issued on Salman Rushdie for his novel, 'Satanic Verses'. Osama bin Laden issued two fatwas, in 1996 and then again in 1998. The 1996 fatwa is entitled 'Declaration of War against the Americans Occupying the Land of the Two Holy Places'. It is a long declaration, documenting American activities in numerous countries. The 1998 fatwa was signed by five people, four of whom represented specific Islamist groups: Osama bin Laden, Ayman al-Zawahiri, Ahmed Refai Taha, and alias Abu Yasser of al-Gama'a al-Islamiyya (in Egypt). Mir Hamzah, 'Secretary of the Jamiat Ulema-e-Pakistan', and Fazul Rahman, 'Emir of the Jihad Movement in Bangladesh'. The signatories were identified as the 'World Islamic Front for Jihad against Jews and Crusaders'. This fatwa declares:

> The ruling is to kill the Americans and their allies is an individual duty for every Muslim who can do it, in order to liberate the Al Aqsa mosque [Jerusalem] and the Holy Mosque [Mecca] . . . This is in accordance with the words of Almighty God . . . We, with God's help, call on every Muslim who believes in God and wishes to be rewarded to comply with God's order to kill the Americans and plunder their money wherever and whenever they find it.

Today, the Internet became a useful platform for posting of fatwas and interpretations of fatwas (Weimann, 2011).

Many of the new Jihadi fatwas are promoting motivational imbalance by endorsing a single motive, a dominant need, and use the power of fatwa to endorse it. Take for example the following fatwa issued in 2006 by Sheikh Yusuf Al-Qaradawi, the prominent Jihadi scholar and preacher:

> The martyr operation is the greatest of all sorts of Jihad in the Cause of Allah. A martyr operation is carried out by a person who sacrifices himself, deeming his life less value than striving in the Cause of Allah, in the cause of restoring the land and preserving the dignity . . . But a clear distinction has to be made here between martyrdom and suicide. Suicide is an act or instance of killing oneself intentionally out of despair and finding no outlet except putting an end to one's life. On the other hand, martyrdom is a heroic act of choosing to suffer death in the Cause of Allah, and that's why it's considered by most Muslim scholars as one of the greatest forms of Jihad.

Declaring martyrdom as the 'greatest of all sorts' and as 'a heroic act' chosen 'in the cause of Allah' is meant not only to promote suicide attacks but also to distinguish them from regular suicides that are forbidden by the Quran. Moreover, it states that the martyr is 'deeming his life less value than striving in the cause of Allah', a clear manifestation of motivational imbalance.

The Fatwas' impact was revealed by numerous testimonies of Jihadists. For example, Sa'd Ibrahim Al-Bidna, a young Saudi, who traveled to Afghanistan with the aim of joining the jihad. He was arrested 2 months later, and spent 4 years and 8 months at Guantanamo. In an interview with *Al-Riyadh*, he said that it was fatwas posted on the Internet that motivated him to wage jihad.

AL-RIYADH: 'Tell us about the beginning of your journey and the reasons [that motivated you] to set out for Afghanistan'.

AL-BIDNA: 'Many may find it difficult to believe, but I was not very devout, though I did pray regularly. But enthusiasm and zeal filled the hearts of many young people, and unfortunately, I followed certain fatwas that were posted on the Internet. [These fatwas] call upon young people to wage jihad in certain regions. They tempt them [by describing] the great reward [they will receive], the status of the martyrs in Paradise and the virgins that await them [there]. These fatwas have great influence on young people who have no awareness or knowledge [that enables them] to examine them and verify their validity'.[2]

Legitimizing and Promoting Suicide

Suicide is an undesirable act by most religions, including Islam (see the Quran's verse 'Spend yourself in the way of Allah; do not cast yourself into destruction'). Yet, Jihadi fatwas found creative ways to justify suicide attacks. In the 'forbidding verse' the words 'in the way of Allah' ('*fi sabil Allah*'in Arabic) are interpreted as 'for the sake of Allah'. Thus, some fatwas claim that this means you are supposed to commit suicide if it is done for Allah. The same verse that is interpreted traditionally in Islam as prohibiting suicide is interpreted in these fatwas as supporting suicidal actions if committed 'in the way of Allah'. In 2006, the Egyptian Al-Gama'a Al-Islamiyya group published a book titled *Islam and the Laws of War* (Al-Islam wa-tahdhib al-hurub). Discussing suicide, the text argues that:

> A Muslim is permitted to attack enemy ranks in order to inflict severe damage on them, even if he believes that he is likely to be killed. This is not the same as one who puts his own life in jeopardy, on whom many religious scholars of the righteous generations placed conditions that he must fulfill . . . One who ventures to blow himself up for the sake of elevating Allah's word and liberating his country from the imperialists cannot be called one who commits suicide, since one who commits suicide does not die for principles and for religion, in contrast with the one who carries out this attack.

Another tactic is to present suicide as a new conception of martyrdom, especially for certain followers. This shift challenges traditionally strong and Islamic prohibitions against suicide.[3] Freamon argues that:

> that this transformation of religious doctrine. . ., resulted in the appearance of a new norm of jihadist battlefield behavior—self-annihilation—a norm that is now accepted as a valid discharge of religious obligation under the law of the military Jihad.[4]

Al-Qaradawi is a leading figure in online terrorist fatwas and a good example for the use of fatwas to legitimize terrorist actions. Here is a typical question and answer form of fatwa issued by al-Qaradawi:

> The martyr operation is the greatest of all sorts of jihad in the cause of Allah. A martyr operation is carried out by a person who sacrifices himself, deeming his life [of] less value than striving in the cause of Allah, in the cause of restoring the land and preserving the dignity. To such a valorous attitude applies the following Qur'anic verse: 'And of mankind is he who would sell himself, seeking the pleasure of Allah; and Allah hath compassion on (His) bondmen'. (Qur'an, 2: 207). But a clear distinction has to be made here between martyrdom and suicide. Suicide is an act or instance of killing oneself intentionally out of despair, and finding no outlet except putting an end to one's life. On the other hand, martyrdom is a heroic act of choosing to suffer death in the cause of Allah, and that's why it's considered by most Muslim scholars as one of the greatest forms of jihad. When jihad becomes an individual duty, as when the enemy seizes the Muslim territory, a woman becomes entitled to take part in it alongside men. Jurists maintained that when the enemy assaults a given Muslim territory, it becomes incumbent upon all its residents to fight against them to the extent that a woman should go out even without the consent of her husband, a son can go too without the permission of his parent, a slave without the approval of his master, and the employee without the leave of his employer. This is a case where obedience should not be given to anyone in something that involves disobedience to Allah, according to a famous juristic rule.[5]

Qaradawi's fatwa on suicide was echoed in numerous other terrorist fatwas. For example, Dr. Abd al-Aziz al-Rantisi, then one of the leaders of Hamas, argued that:

> suicide depends on volition. If the martyr intends to kill himself, because he is tired of life—it is suicide. However, if he wants to sacrifice his soul in order to strike the enemy and to be rewarded by Allah—he is considered a martyr (*shahid*). We have no doubt that those carrying out these operations are martyrs.

Rantisi based the distinction on Qaradawi's fatwa.

Jannah: The Promise of Afterlife

A somewhat different approach to reach the same imbalance is represented by the ISIS notion of afterlife. Islamic State's taunt that 'we love death more than you love life' was always a threat as well as a fact. Death, in the minds of certain Jihadists, is not the end of life. It's the way to an eternal life of bliss in Heaven (*Jannah*) to a Paradise depicted in the most extravagant, sensual terms in certain Muslim writings that are accepted as literal truth. According to Jihadi perception and presentations, Heaven is a real place, a Garden full of sensual delights. For example, the promise of 72 virgins. The promise of virgins upon martyrdom has become a familiar notion when discussing Jihadi perception of afterlife, but in fact, the conceptualization of afterlife is more complex and multifaceted.

Regarding the motivational imbalance, Heaven in Jihadi propaganda is not a diluted doctrine as it has become for many Christians. Heaven in jihadi belief is an operational concept, especially when targeting certain audiences. It outweighs all other motives, needs, aims, and values. For example, on July 16, 2019, Islamic State fighters in Tunisia renewed their pledge of allegiance to group leader Abu Bakr al-Baghdadi in a video, and incited Muslims to support and participate in jihad, and to take action in their own countries and attack tourists from among 'citizens of the Cross'. The video features two speaking segments, with one of them declaring that jihad is an obligation, and telling those Muslims who abstain from it:

> Are you satisfied with this life more than the hereafter? The pleasure of this world is small compared to the pleasures of the hereafter. You will stand before Allah the Almighty and you will be asked about your support of this religion. Did you support this religion or let it down? Whoever abandons jihad, due the lack of ammunition of the mujahideen, we say to him: go back to your religion and meditate on the traditions of the Prophet, Allah's peace and blessings be upon him. Then you will see that you are in a great fault. So rise up O brother in the doctrine, and support this religion and back your brothers the mujahideen. Jihad is one of the doors of Paradise whereby Allah takes away gloom and distress. Join your brothers the mujahideen. He who is unable to do this, fight the enemies of Allah. Turn their safety unto terror, and their joy unto sadness. Fight the citizens of the Cross in your country. Strangle them and turn their tourism into mourning.

As noted by Tiersky (2016), 'ISIS's deadliest weapon is the idea of Heaven'. Indeed, an analysis of ISIS online message to foreign fighters (Weimann, 2016) reveals the frequent use of the afterlife, the promised paradise as a key persuasive element. In 2014, ISIS released a video inciting Muslims to come participate in jihad, featuring a German chant with an English translation. The 5-minute, 26-second video, entitled, 'Haya Alal Jihad' (Let's Go for Jihad), was produced by the ISIS' al-Hayat Media Center and was posted on its Twitter account on June 15, 2014. Footage shows operations by ISIS fighters while the voiceover (of the German jihadist Denis Cuspert, also knowns as Abu Talha al-Almani) calls on Muslims to join the jihad and seek martyrdom:

> Brothers, it's time to rise/Set forth for the battle if you are truthful/Either you get victory or the shahadah [martyrdom]/Do you fear death?/There is no escape!/Get dignity!/And shahadah fi sabilillah [martyrdom in the cause of Allah] is the entrance to Paradise.

As ISIS' self-declared caliphate crumbled with heavy losses in territories, ISIS has refocused its media narrative from its triumphs to its commitment to a 'long war' against its enemies (Munoz, 2018). In this new phase, the ISIS propaganda tells fighters that true victory lies in attaining paradise through martyrdom rather than controlling lands and cities. Thus, for example, the 13th issue of *Rumiyah*, the online magazine of ISIS published in 2017, featured the third part of an essay written by the late al-Qaeda in Iraq leader Abu Musab al-Zarqawi titled 'Important Advice for the Mujahidin', which emphasized that the mujahideen have sold their lives to God in exchange for paradise and that they must 'favor aqidah (creed) over life'.

Seducing Women

A perfect combination of 'narrowcasted motivational imbalance' is the targeting of women. Since ISIS leader Abu Bakr al-Baghdadi called on women to travel to the proclaimed Islamic State on June 29, 2014, thousands of women responded including about a thousand Western women and teenage girls who moved to Syria and Iraq to join the ISIS. According to some estimates, 13% (4,761) of the total 41,490 travelers to the Islamic State were women. This was an unprecedented success in attracting women from the West that no other jihadist group had before. How has ISIS lured women from the West, what motivated women to join such a violent and notorious terrorist group, famous for its mistreatment and enslavement of women? One of the explanations provided by studies reveals the notion of priming a motive for selected groups of females (or narrowcasting based on combination of gender and motives).

Tarras-Wahlberg (2017) conducted a study of 'The Promises of ISIS to its Female Recruits' based on analysis of over more than 1,000 pages of official ISIS online propaganda, most produced in its online magazine, *Dabiq* complemented by an analysis of 11 statements made by official spokesmen of ISIS. The analysis revealed that women are promised seven things: the possibility to fulfill their religious duty, become important state builders, experience deep and meaningful belonging and sisterhood, live an exciting adventure in which they can find true romance, as well as being increasingly influential. However, according to our suggested combination of narrowcasting and motivational imbalance, the motives are directed, differentially, to specific groups. Thus, for example, there are women in the West who are descendants of Muslim immigrants. They often expressed their frustration with their status in the West and face an identity crisis. For them, the appealing message is a sense of belonging and recognition within the newly proclaimed Muslim Caliphate. Tarras-Wahlberg reports that the call for Western Muslim women to migrate and join ISIS is portrayed as an obligation for all pious Muslims in both *Dabiq* and the official ISIS statements. In a *Dabiq* article by Umm Sumayyah al-Muhajirāh, women are portrayed as the 'twin halves' of men when it comes to the subject of migration. There is no difference between the sexes in relation to the duty of *hijrah*. Al-Muhajirāh writes: This ruling [of migration] is an obligation upon women just as it is upon men'.

Similarly, Perešin (2018) who studied 'Why Women from the West are Joining ISIS' argued that:

> Personal reasons that lead women to move to the Caliphate are dissatisfaction or disappointment with different aspects of their lives, boredom, desire for adventure and alternatives to their current life, adolescent rebellion, troubling family relationships and traumatic experiences, sexual abuse or honor-related violence, etc.

The ISIS online recruitment campaign targets women according to these motives, offering each sub-category a special motive or reward while downloading others. These include idealistic goals of religious duty to help build a utopian Muslim Caliphate; new meaningful life that offers a sense of belonging and sisterhood; romantic and adventurous experience of life, taking part as equal partners in the state-building process and be given an important role in creating the new ideologically pure state, in contrast with the imperfections of the infidel Western society.

However, the seductive rewards to be received are not limited to idealistic or religious ones. For some target groups, there are earthly rewards such as a husband and family. As Viano (2018) argues:

> Foremost among female supporters of ISIS is the romantic idea of marrying a brave and noble warrior. Marriage for many women represents the passage from childhood into adulthood, especially in cultures that tightly control women's lives. Thus, it is a core factor in spurring female migration to ISIS land. Women traveling to Syria and Iraq and expecting to remain single and independent do not fit well there. They are out of place and quite vulnerable.

Indeed, for certain female groups, the romantic pull as well as the notion of jihadi brides became an attractive instrument for ISIS recruitment. This corresponds mainly to the findings of Astrid Gorris, the mother of Tatiana Wielandt (a Belgian teenager who left for Syria to join her husband and IS-fighter Mujahid).

She testified that young girls are lured into marrying a jihadi fighter by painting a false picture of a romantic adventure within the boundaries of Islam. 'I think she left with a very romantic idea of a rebellious struggle, but the only reason why ISIS is bringing girls into Syria is to produce offspring. More fighters for the cause'. Thus, for some groups, the romantic appeal was joined by the role of motherhood and family. ISIS used also its female-focused efforts spreading the message that women are valued, not as sexual objects, but as mothers to the next generation. On January 23, 2015, a manifesto on women was released by the Al-Khanssaa Brigade. This document was clearly designed as a means of drawing in Muslim women from countries in the region. The emphasis throughout the manifesto lies in the importance of motherhood and family support.

Revenge: Channeling Alienation, Frustration, and Anger

Finally, the use of motivational imbalance and narrowcasting is evident in the targeting of frustrated, alienated, and angry groups. In May 2016, a video labeled 'Army of Orphans' emerged online. Numerous orphans who lost their parents due to foreign anti-terror attacks in Syria are presented as 'thirsty for revenge'. The video, created by the jihadist group's media arm, Alhayat Center, shows them as a squad of well-equipped child soldiers preparing for battle and walking across the ruins of a city destroyed during the Syrian conflict. Amid the scenes of destruction, the faces of Western leaders and politicians blend in with carefully drafted propaganda messages calling for revenge. Revenge, promoted to a main motive, has been used by ISIS when targeting specific audiences. These audiences were defined by their preexisting anger and frustration caused by social and political alienation.

Deprivation, marginalization, frustration, poor economic conditions, and grievances of Muslim populations were the main causes leading individuals to join violent Jihadists' groups (Rocca, 2017). Jihadi propaganda attempted to channel these feelings into one direction: taking revenge by fighting back, by making 'them' pay for their wrongdoings, sins, and crimes. Materializing grievances as a motive for recruitment can also be personal or collective. McCauley and Moskalenko (2008) found that suicide terrorists are often driven by a sense of personal victimization and revenge. Nevertheless, those personal rationales are 'unlikely to account for group sacrifice unless the personal [grievance] is framed and interpreted as representative of group grievance' (Ibid., 419). In the context of Jihadism, those grievances are embedded in the perceived oppression and alienation of Muslims (Brachman, 2009, p. 11).

ISIS was born in Iraq, as a consequence of the collapse of the Sunni-dominated Baath government. The Sunni segment of Iraq's population fell into a social, religious, and political vacuum with no united leadership. Some aligned themselves with tribal leaders and some with the Islamist parties that slowly began to emerge, while others depended on Sunni Islamist insurgents. Sunni insurgents were using grievances against Shia-dominated executives, and their discrimination against the Sunnis, which they perceived as part of a systemic disempowering of their community. ISIS is in fact a symptom of the frustrated Sunnis in Iraq and later in Syria. In the political vacuum in these countries, ISIS recruited Iraqi Sunnis, many of whom were former career military and intelligence officers under Saddam Hussein. The frustration, alienation, and perceived weakness were channeled by ISIS to a single motive: revenge.

Social grievances used for channeling anger to revenge included Islamophobic attacks, a perceived lack of belonging and failed integration. Following the attack in Christchurch, New Zealand, that left 50 people dead at two mosques during midday prayers, ISIS launched a revenge campaign. In its online messages, ISIS has called on its followers to retaliate. The group's spokesman, Abu Hassan al-Muhajir, broke a 6-month silence to call on ISIS supporters to 'take vengeance for their religion' in a 44-minute audio recording distributed by Al Furqan, a media organization operated by ISIS:

This slaughter in those two mosques is no more than another tragedy among past and coming tragedies, which will be followed by scenes of force that reach all who were tricked to living among the polytheist. We'll tell you what's true. You can form your own view.

> From 15p €0.18 $0.18 USD 0.27 a day, more exclusives, analysis, and extras. Subscribe now.
> The scenes of death in the two mosques are enough to wake the sleep and incite the supporters of the caliphate who live there, to take vengeance for their religion and for sons of their Ummah, who are killed everywhere in the world.

Recently, following ISIS heavy losses, the priming of revenge has become more evident. On April 8, 2019, ISIS launched what it called 'the Campaign of Vengeance for the blessed al-Sham Province', calling attacks by its affiliates around the world under this banner. The notion of 'Vengeance for Sham' appears to be a coordinated campaign in an attempt to restate the group's capabilities following the fall of its self-declared caliphate. In April 2019, ISIS' leader Abu Bakr al-Baghdadi threatened a wave of attacks worldwide in revenge for the defeat of his militant group in its Iraq and Syria heartlands. He hailed the jihadists who he said had fought to the death to defend Baghuz, ISIS last stronghold, where the self-proclaimed caliphate came to an end: 'The battle of Baghouz is over. But it did show the savagery, brutality and ill intentions of the Christians towards the Muslim community'. He added that Isis would seek revenge for the killing and imprisonment of its militants, and that his group was in a 'battle of attrition' a mission that now outweighs all other motives.

Evidence from Recruits

The recruits themselves often testify on the process of their radicalization and online recruitment. These reports reveal the combined use of online platforms, motivational imbalance, and narrowcasting. Speckhard and her colleagues at the ICSVE *Breaking the ISIS Brand Counter Narrative Project* conducted interviews with a sample of more than 100 ISIS and 16 al Shabaab defectors, prisoners, and returnees. The interviews reveal how these terrorist groups' strong social media and Internet presence allowed them to attract attention of many, causing key individuals to become interested in violent extremism and ultimately to make contacts with recruiters and facilitators for travel to their terrorist havens in Iraq and Syria or, in the case of al Shabaab, into Somalia. Moreover, these interviews reveal that in the process of seduction, a dominant motive is highlighted when targeting specific subgroups (www.icsve.org/project/breaking-the-isis-brand/).

Let us use illustrative examples from the numerous personal accounts of such recruitment. As noted earlier, the belief in an afterlife and the fear of eternal damnation is one way to manipulate a potential recruit. Such manipulative tactic worked on Hoda Muthana, a then 20-year-old college student, who in 2014 lived between two worlds: her strict Yemeni parents and as an average teenage girl growing up in the West. Hoda joined ISIS and spent 5 years inside the ISIS Caliphate. In a personal interview (Speckhard & Shajkovci, 2019), Huda recalled ISIS' competency and exploitation on social media platforms. She described how the group succeeded in making inroads into her personal life: first luring and later fully integrating her into the subcultures of the Islamic State's strong Twitter presence. Reflecting on her social media experience 6 years ago, Hoda vividly described:

> There's basically a Twittersphere of Muslims. That's what they used to call it. And in that group, at the time, we were all not very practicing [Muslims]. We were just being our normal selves. We just all got this account and we all started following each other. We got to know each other and stuff and then suddenly I think some people in the group became practicing and they started inviting us to become practicing as well.

Referring to the 'Twittersphere', she clarified:

> In the beginning it [the verses on Twitter], was just to get you practicing basically. Verses about, I think, hellfire, heaven, things that the Prophet Mohammed would do, maybe to get you closer to heaven, because if you follow his way, you know this is what we believe, if you follow his way. It was just basic things. Nothing radical yet.

But soon the messages were more threatening, pressuring, and demanding:

> So, they were interpreting things, verses. Giving out quotes and verses. Verses that if you do read them, they are very black and white. Especially about the immigration [*hijrah*] that it's obligatory on you to immigrate to Islamic State, and if you don't, you are doomed basically to hellfire, and this is what scared me the most. They would, yeah, they started mentioning stuff [to describe hell on Twitter]. I started looking up stuff myself and I would get even more scared. I wanted to do anything obligatory on me really to avoid going to hell.

When asked how many of them were in her [online] group, she replies, 'Thousands', adding, 'I would say most of the youth who were in ISIS came [to Syria] because of Twitter'. In 2014, when ISIS announced the so-called caliphate, Muthana said, she decided to join ISIS and she did so because she was driven by a fear of God and a fear of doing the wrong thing, so she traveled to Syria without thinking of the consequences.

The powerful net that Jihadi groups like ISIS cast within the Internet to catch vulnerable individuals is using other motives too as revealed by Abu Ayad, an Iraqi who joined ISIS in 2016, and told ICSVE researchers in 2017:

> [I watched their videos on] Twitter and Facebook. They affected me because I was poor. I wanted to improve my living condition [for] marriage. I wanted us to have a house and for me to get married, and to have money. [We were living] in [Baghdad], in Dora. [My family] was good, but we were poor.

'[I watched] about 10 videos, approximately. I used to listen only to what was said, and the videos that had killing or blood, I fast-forwarded through them. I didn't see them. I didn't agree on killing people and things', he added. Yet, he found the group compelling and thought they could address his needs: '[After watching, I contacted] the one who publishes the video or the pages where you find these videos on Twitter and Facebook'. He went on to join and even, according to Iraqi security officials, volunteered to become a suicide bomber.

Abdirizak Warsame, a convicted Jihadi from Minnesota, told the media that he was so brainwashed that he would have had no qualms about carrying out brutal ISIS executions. Warsame, a Somali immigrant from Minneapolis who came to the United States at age 10 was 19 in 2014 when he helped two pals get passports and hooked them up with ISIS contacts in Syria. His two pals reached Syria via Turkey and were soon killed in action. He planned to join the terror group, where he said he intended to commit the kind of atrocities that have made the organization so notorious, such as beheadings, drownings, and burning prisoners alive. 'Yeah, I was going to be, I was going to be participating in those activities', Warsame told '60 Minutes'. Warsame explained that he turned to ISIS because he felt lost between the traditional Muslim household of his parents and his new, confusing American life. He was introduced to terrorism through online videos of Anwar al-Awlaki, the radical American-born cleric who was killed by a US drone in Yemen in 2011. 'He explained how Islam was, you know, like my calling', Warsame described. 'It was almost like he was talking to you. And, like, it made you feel like you were special, and like you're the chosen one'. Watching al-Awlaki preaching on YouTube became his reality: 'It kind of takes control of you', Warsame said. 'Most of the videos would talk about how, if you would engage in jihad, you would be doing your family favor. And that you would be saving their lives from eternal hellfire'.[6]

Israfil Yilmaz, a Dutch citizen of a Turkish descent, joined ISIS in Syria. Before that he served 15 months in the Turkish Armed Forces and then, in 2010, he enlisted in the Royal Netherlands Army to join a commando group but did not meet the requirements to do so. He nevertheless finished basic training and spent a short time in an infantry unit. Arriving in war-torn Syria in late 2012 or early 2013, Yilmaz joined various Jihadi groups until 2014 when he joined the ranks of ISIS and achieved his ultimate notability. Highly active on social media outlets such as Instagram, ASKfm, Twitter, and Tumblr, Yilmaz garnered media notice. His popularity among the ranks of ISIS was such that he received an astonishing 10,000 marriage requests during his time as a jihadi fighter up until his marriage. On one post, before the account was closed down by Tumblr, he posted a picture of himself holding a kitten and another showed him holding a newborn child aloft.

In January 2014, the Dutch television program *Nieuwsuur* aired an interview with Yilmaz and showed him training a small group of foreign fighters. In a later televised interview aired by CBS on October 7, 2014, Yilmaz explained his motivation to join Jihad and ISIS:

> We don't want you, we want our own laws, the Islamic law, this is the only solution . . . This fight never ends, this is our religion. This is our faith, this is what we believe in . . . I will fight anyone . . . even if it was my father, I will fight him and kill him myself . . . We left everything behind. When we migrated here, we left everything behind. Everything. Our families, our friends, our future.[7]

Israfil Yilmaz is believed to have been killed by a coalition airstrike on Raqqa, Syria on September 5, 2016.

Conclusions

What can we learn from the use of psychological and social needs to understand the various forms of extremism? This chapter suggests the combination of two concepts, motivational imbalance and narrowcasting when trying to explain Jihadi online seductive appeals. In fact, we argue that the unique combination of medium (online platforms), content (motivational imbalance), and selected audiences (narrowcasting) is the key to the success of Jihadi persuasive campaigns. This medium–content–audience triangle relies on the compatibility of its elements: the use of online social media, for example, allows for a perfect narrowcasting while effective narrowcasting contributes to the effectiveness of tailoring motivational imbalance for specific groups and individuals. Our illustrative examples of the appeals used to create, promote, and maintain motivational imbalance show that online recruitment attempts to address certain needs and promote extreme means usually perceived to be more instrumental for addressing those needs. Moreover, combining narrowcasting with motivational imbalance increases the effectiveness of recruitment because it highlights people's unique needs and vulnerabilities and promises to address them. Narrowcasting allows for more effective recruitment because of the ability to create the motivational imbalance. The evidence from the recruits themselves illustrates the success of this combination.

However, this is only the initial step of linking the concepts of motivational balance to radicalization and recruitment by political violence entities. There is certainly a need for developing and validating scales or measures for studying motivational imbalance and its application to online contents and their audiences. With the growing popularity of social media among radical groups and extremists, this direction is more than promising.

Notes

1. The database is the digitized archive on online terrorism. When our data collection began in the late 1990s, there were merely a dozen terrorist websites (Tsfati and Weimann 2002); by 2000, virtually all terrorist groups had established their presence on the Internet, and in 2003, there were more than 2,600 terrorist websites (Weimann 2005). The number rose dramatically, and by October 2019, the project archive contained more than 9,800 websites serving terrorists and their supporters. On top of websites, the archive contains postings from all social media used by terrorist groups (Weimann, 2016). The project enjoyed funding from various academic foundations, including the United States Institute of Peace (USIP), the National Institute of Justice (NIJ), The Woodrow Wilson Center, Australian Research Council (ARC), and more.
2. *Al-Riyadh* (Saudi Arabia), October 10, 2006.
3. See, for example, Sheikh Yusuf Qaradawi, "Fatwa: Whether Suicide Bombings are a Form of Martyrdom", Nov. 8, 2003 available at http://qaradawi.net (in Arabic).
4. Bernard Freamon, "Martyrdom, Suicide and the Islamic Law of War", *Fordham International Law Journal (*27, 2003), p. 300.
5. At: www.meforum.org/646/the-qaradawi-fatwas
6. "60 Minutes' Interviews Minnesota Man Linked To ISIS", October 27, 2016, https://minnesota.cbslocal.com/2016/10/27/60-minutes-terror-suspect/
7. "Western jihadist on why he fights". CBS Evening News, October 7, 2014.

References

Ahmad, A. (2014). The role of social networks in the recruitment of youth in an Islamist organization in Pakistan. *Social Spectrum, 34*(6), 469–488.

Berger, J. M. (2015). Tailored online interventions: The Islamic State's recruitment strategy. *CTC Sentinel, 8*(10), 19–23.

Berger, J. M., & Morgan, J. (2015). *The ISIS Twitter census*. Washington, DC: The Brookings Institution. Retrieved from www.brookings.edu/wp-content/uploads/2016/06/isis_twitter_census_berger_morgan.pdf

Blackwell, E. (2016). Radical approach: How Islamic state exploits young people—How Islamic State exploits Maslow's hierarchy of needs to exploit you. *Huffington Post*. Retrieved from www.huffingtonpost.com.au/2016/05/22/radical-approach-how-islamic-state-exploits-young-people_a_21381369/

Brachman, J. (2009). *Global jihadism: Theory and practice*. London: Routledge.

Carter, J., Maher, S., & Neumann, P. (2014). *#Greenbirds: Measuring importance and influence in Syrian Foreign fighter networks*. London: International Centre for the Study of Radicalisation.

Ducol, B. (2012). Uncovering the French-speaking Jihadisphere: An exploratory analysis. *Media, War and Conflict, 5*(1), 51–70.

GerDenning, D. (2010). Terror's web: How the Internet is transforming terrorism. In Y. Jewkes, & M. Yar (Eds.), *Handbook of internet crime* (pp. 194–213). Cullompton: Willan Publishing.

Gerwehr, S., & Daly, S. (2006). Al-Qaida: Terrorist selection and recruitment. In D. Kamien (Ed.), *The McGraw-Hill Homeland security handbook*. New York: McGraw-Hill.

Hegghammer, T. (2014). Interpersonal trust on Jihadi internet forums. *Norwegian Defence Research Establishment, 5*, 1–43.

Just, J. (2015). *Jihad 2.0: The impact of social media on the Salafist scene and the nature of terrorism*. Hamburg: Anchor Academic Publishing.

Kruglanski, A. W. (2018). Violent radicalism and the psychology of prepossession. *Social Psychological Bulletin, 13*(4), 1–18.

Kruglanski, A. W., Fernandez, J. R., Factor, A. R., & Szumowska, E. (2019). Cognitive mechanisms in violent extremism. *Cognition, 188*, 116–123.

Kruglanski, A. W., Gelfand, M. J., Bélanger, J. J., Sheveland, A., Hetiarachchi, M., & Gunaratna, R. (2014). The psychology of radicalization and deradicalization: How significance quest impacts violent extremism. *Political Psychology, 35*, 69–93.

Kruglanski, A. W., Jasko, K., Chernikova, M., Dugas, M., & Webber, D. (2017). To the fringe and back: Violent extremism and the psychology of deviance. *American Psychologist, 72*(3), 217–230.

Mahood, S., & Rane, H. (2017). Islamist narratives in ISIS recruitment propaganda. *The Journal of International Communication, 23*(1), 15–35.

McCauley, C., & Moskalenko, S. (2008). Mechanisms of political radicalization: Pathways toward terrorism. *Terrorism and Political Violence, 20*(3), 415–433.

Munoz, M. (2018). Selling the long war: Islamic state propaganda after the caliphate. *CTC Sentinel, 11*(10). Retrieved from https://ctc.usma.edu/selling-long-war-islamic-state-propaganda-caliphate/

Neumann, P. (2012). *Countering online radicalization in America*. Washington, DC: Bipartisan Policy Centre, Homeland Security Project.

Perešin, A. (2018). Why women from the west are joining ISIS. *International Annals of Criminology, 56*(1–2), 32–42.

Ramsay, G. (2008). Conceptualising online terrorism. *Perspectives on Terrorism*, *2*(7), 3–10.

Rocca, N. M. (2017). Mobilization and radicalization through persuasion: Manipulative techniques in ISIS' propaganda. *International Relations and Diplomacy*, *5*(11), 660–670.

Shah, J. Y., Friedman, R. S., & Kruglanski, A. W. (2002). Forgetting all else: On the antecedents and consequences of goal shielding. *Journal of Personality and Social Psychology*, *83*, 1261–1280.

Speckhard, A., & Shajkovci, A. (2019, April 23). Born in America Hoda Muthana shares her story of joining and escaping from the Islamic state caliphate. *Homeland Security Today*. Retrieved from www.hstoday.us/subject-matter-areas/terrorism-study/american-born-hoda-muthana-tells-all-about-joining-isis-and-escaping-the-caliphate/

Tarras-Wahlberg, L. (2017). *Seven promises of ISIS to its female recruits*. Retrieved from www.icsve.org/research-reports/seven-promises-of-isis-to-its-femalerecruits

Tiersky, R. (2016). ISIS's deadliest weapon is the idea of Heaven. *Real Clear World*, 19.

Viano, E. (2018). Introduction to the special issue on female migration to ISIS. *International Annals of Criminology*, *56*(1–2), 1–10.

Weimann, G. (2006). *Terror on the internet: The new arena, the new challenges*. Washington, DC: US Institute of Peace Press.

Weimann, G. (2007). Using the internet for terrorist recruitment and mobilization. *Hypermedia Seduction for Terrorist Recruiting, NATO Science for Peace and Security Series*, *25*, 47–58.

Weimann, G. (2010). Terror on Facebook, Twitter and YouTube. *Brown Journal of World Affairs*, *16*(2), 45–54.

Weimann, G. (2011). Cyber-fatwas and terrorism. *Studies in Conflict and Terrorism*, *34*, 1–17.

Weimann, G. (2014). *New terrorism and new media*. Washington, DC: Woodrow Wilson International Center for Scholars. Retrieved from https://preventviolentextremism.info/sites/default/files/New%20Terrorism%20and%20New%20Media.pdf

Weimann, G. (2015). *Terrorism in cyberspace: The next generation*. New York: Columbia University Press.

Weimann, G. (2016). The emerging role of social media in the recruitment of foreign fighters. In *Foreign fighters under international law and beyond* (pp. 77–95). The Hague: TMC Asser Press.

Zelin, A. (2013). *The state of global jihad online*. Washington, DC: New America Foundation.

INDEX

Note: Page numbers in *italic* indicate a figure and page numbers in **bold** indicate a table on the corresponding page.

addiction 4–7, 19–22, 77–79, 184
afterlife *see Jannah*
agentic, feeling 262–263
aggression 25–33, 74–75, 118, 264–265
agreement with society 42–43, *43*
alienation 296–297
ambidexterity 143–144; *see also* tight–loose ambidexterity
anger 296–297
appeals: the highest demand 287–289; *Jannah* 292–294; legitimizing and promoting suicide 291–292; the power of fatwa 289–291; revenge 296–297; seducing women 294–296
attitude–behavior consistency 52, 54–57, *56*
attitudes: determinants of attitude certainty 51; determinants of attitude unusualness 51–52; polarized, confident, and unusual 44–46, *45*; properties that increase attributions of extremism 41–44; properties that predict extreme behavior 52–54
attitude strength (certainty) 6, 35–38, 52–53, 56–57, 163
attitudinal extremism: candidates for inclusion in a model of 37–41; developing a model of 34–37; information exposure and processing 46–47; mere thought 47–48; metacognitive validation 48–49; normative influence 49–52; processes that produce polarized, confident, and unusual attitudes 44–46; properties of attitudes that increase attributions of extremism 41–44; properties of attitudes that predict extreme behavior 52–54; threat as a moderator of compensation effects 54–57
attitudinal imbalance 39–41
authentic extremism 34–36, 41

balance 2, 242–244
balancing *see* tight–loose ambidexterity
BASE jumping 183–184, 186, 191, 195
Baumgartner, Hannah M. xiii, 5, 6
behavior *see* attitude–behavior consistency; extreme behavior; interpersonal behavior; intrapersonal behavior; paradoxical behaviors
Berridge, Kent C. xiii, 5, 6
boredom susceptibility (BS) 185
Bouchat, Pierre xiii, 6, 8
Briñol, Pablo xiii–xiv, 5, 6
Brymer, Eric xiv, 6, 8, 193–195
burnout 80–81, 85, 123

CEOs 204, 206, 218, 221–223, 225
certainty 35–40, 43–45, *45*, 51–57, *56*, 101–102
charismatic groups 104–106; *see also* extreme groups

Choi, Virginia K. xiv, 5, 7
cognitive closure 47, 50, 102, 122, 266
commitment signaling 167, 169, 174
compensation effects: threat as a moderator of 54–57
confident attitudes 44–46
COVID-19 pandemic 42–44, 49, 54–57, 260
cults 104–105, 121, 128
culturally approved violence 264–265
cultural tightness–looseness 144, *151*
curvilinear relationship 147–148, 150, *152*

'dangerous desire' 21–25
decision utility 15–21, 23, 25–27
dehumanization 74–75, 108–109, 124
desirability of the morally exceptional 234
determinants 66–67
developmental history 238
different balance 242–244
disinhibition (DIS) 185
domains of human endeavor 5, 8–10
dopamine *see* mesolimbic dopamine system
Dualistic Model of Passion 67–72

economic grievances 266–267
edgework 187–188
environmental factors 165–166
exceptional 249–252; *see also* morally exceptional
experienced utility 16–17, 19–21, 23–25
experience seeking (ES) 185–186
exploitation 142–144
exploration 142–144
extraordinary experience 195–197
extreme aggression 25–27
extreme behavior 85–89; and determinants of harmonious and obsessive passion 83–85; and a dualistic model of passion 67–72; impersonal 72–77; intrapersonal 77–83; properties that predict 52–54; the role of passion in 66–67, 72–83; and the study of the morally exceptional 231, 234–237, 241–242, 245, 252
extreme groups 96–98, 128; and narratives 106–113; and networks 113–127; and the quest for significance 98–106
extremely moral 235–244
extreme motivation 15–16; attribution of incentive salience 18–21; implications in extreme aggression 25–27; 'irrational miswanting' and 'dangerous desire' 21–25; reward utilities and 'wanting' 16–17

extremes of innovation 140–142, 151–155; cultural tightness–looseness 144; exploration and exploitation 142–144; exploring looseness, exploiting tightness 144–147; the Goldilocks principle 147–151
extreme sports 183–184; positive psychology perspectives 192–197; beyond the risk-taking narrative 191–192; traditional perspectives on the psychology of 184–190
extremism: attributions of 41–44, *43*, *45*; conceptualizing 39–41; evolution of 163–176; function of 167–168; and greed 206–208, 222–224; and imbalance 2–4; morality's attenuation and exacerbation of 171–173; as a motivational construct 2–4; unique to humans 168–170; *see also* attitudinal extremism; violent extremism

fatwa 289–291
Fleeson, William xiv, 6, 9
flow 192–194, **193**
Fox, Nadia xiv, 5–6, 8
frustration 296–297
Furr, Mike xiv, 6, 9

Gelfand, Michele J. xv, 5, 7, 145, 148, 156
genetic factors 20, 164–166, 168, 173
genome-wide association studies (GWAS) 164–165
Goldilocks principle 147–151
greed **209–215**; defined 206–208; history of 208, 216–218; mitigating the problem 224–226; multilevel model of 222–224; narratives 203–205; in the social sciences 218–222
group development 125–127
group socialization 109–111, 113–125, *115*; investigation phase 114–118; maintenance phase 120–122; resocialization phase 122–125; socialization phase 119–120

harmonious passion 68–71, 79, 82; determinants of 83–85
Haynes, Katalin Takacs xv, 6, 8–9, 221–224
highest demand 287–289
Holocaust rescuers 235–236, 241–244, 248–250
humanitarianism 79–80, 238, 240

identity 68–72, 99, 101–103, 264, 266–268; as motivation 240–241
identity fusion 53–54, 103–104, 107
imbalance 2–4, 242–244; *see also* attitudinal imbalance; motivational imbalance
immoral behaviors 74, 239–240
incentive salience 15–16; attribution of 18–21; implications in extreme aggression 25–27; 'irrational miswanting' and 'dangerous desire' 21–25; reward utilities and 'wanting' 16–17
inequality 266–267
information exposure 46–47
information processing 46–47
ingroup identification 263–264
innovation 125–126; *see also* extremes of innovation
intergroup conflict 103, 166
interpersonal behavior 72–77
intrapersonal behavior 77–83
investigation phase 114–118
irrational miswanting 21–25

Jannah 292–294
Jasko, Katarzyna xv, 6, 9, 270
Jayawickreme, Eranda xv–xvi, 6, 9
Jihadi online recruitment 280–284, 300–301; appeals used 287–297; evidence from recruits 298–300; and the highest demand 287–289; and *Jannah* 292–294; legitimizing and promoting suicide 291–292; and narrowcasting 286–287; and the notion of motivational imbalance 284–286; and the power of fatwa 289–291; and revenge 296–297; seducing women 294–296; using motivational imbalance in 287–297

Knobel, Angela xvi, 6, 9
Kopetz, Catalina xvi, 3, 66
Kruglanski, Arie W. xvi, 5–6, 7, 9; and attitudinal extremism 36, 39; and the evolution of extremism 163; and the extreme group 97, 102, 128; and harmonious and obsessive passion 66, 77, 84, 86–87, 89; and Jihadi online recruitment 287; and the morally exceptional 242–243; and the psychology of extreme sports 196; and the psychology of greed 207

leadership aspirations 169
legitimizing suicide 291–292
levels of analysis 5–10
Levine, John M. xvi–xvii, 5, 7, 124
lived experience 183, 191–193
looseness 144–147; *see also* tight–loose ambidexterity

maintenance phase 120–122
managers 204, 221, 224–225
Maslow, A. *see* peak experience
material wealth 204, 206–208, 216–217, 220–226
mental health 80, 83
mere thought 45–48
mesolimbic dopamine system 18–22, 26
metacognitive validation 48–49
Miller, Christian B. xvii, 6, 9
mitigation 223–226
moderate position 37, 42
moderating effects *56*
moderation 2
Molinario, Erica xvii, 6, 9, 100
moral 230–235, 245–246, 249–252; what counts as 247–249; why some people are extremely moral 235–244
moral disengagement 74–75, 108–109
morality 171–173; study of 252–253
morally exceptional *see* positive moral exceptionality
morally good 247–249
moral reasoning 239–240
motivation: identity as 240–241; *see also* extreme motivation
motivational balance 2–3, 230, 234, 242, 268, 285, 301
motivational construct 2–4
motivational imbalance 2–4, 39–41, 280–281, 300; consequences of 4–5; across domains and levels of analysis 5–10; in Jihadi campaigns 287–300; the notion of 284–286
multilevel model 205; greed as extremism 222–224
multilevel regression models **149**

Naffziger, Erin E. xvii, 5, 6
narratives 106–113, 269–270; greed narratives 203–205; risk narrative 189–192
'narrowcasting' 286–287
national outputs 149–150, 157
nation-level tightness *152*

nations 140–145, 148–152, 154, 156–157
nature of work 148–149, 156–157
needs 267–269
networks 113, 270–271; group development 125–127; group socialization 113–125; *see also* extreme groups
Nguyen, David xvii, 5, 6
normative influence 45–46, 49–52, 98, 111
norms 50–51, 122–123, 140–148, 152–155, 156–157, 168–171, 262–267
noticed, feeling 262–263

obsessive passion 68–70, 76, 79, 81–82, 86–87; determinants of 83–85
online recruitment 281–284; *see also* Jihadi online recruitment
organ donors 231, 235, 241, 248
outcomes 66–73, 76–81, 83–86, 89

Paquette, Virginie xvii, 5, 7
paradoxical behaviors *147*
partial regression plots 150, *151*
passion 85–89; determinants of harmonious and obsessive passion 83–85; a dualistic model 67–72; and extreme impersonal behavior 72–77; and extreme intrapersonal behavior 77–83; role in extreme behavior 66–67, 72–83
peak experience 192–195, **194**
perceived extremism 34–35, 44
personality traits 238–239
Petty, Richard E. xvii–xviii, 5, 6
polarization *43*, *45*; determinants of 45–46
polarized attitudes 44–46
political discrimination 267
political groups, single-issue 96, 98, 128
positive 230, 242
positive moral exceptionality 9, 230–231; empirical findings 235–244; existence of morally exceptional people 232–233; the promise of studying 231; theoretical, philosophical, and theological accounts of 233–235; three difficulties for the study of 246–253; what is not known 244–246
positive psychology perspectives 192–197
predicted utility 16–21, 23–27
Prokopowicz, Piotr xviii, 5, 7
promoting suicide 291–292

proximate situational causes 237–238
proximaty flying 184

radicalization 112, 261–262, 270, 281–285, 288, 298, 301
recruits 298–300; *see also* Jihadi online recruitment
relapse 19–21, 26
relative deprivation 266–267
religious orders 96, 98, 128
remembered utility 15–17, 24–26
resocialization phase 122–125
revenge 296–297
reversal theory 188–189
reward cues 18–20, 22
reward utilities 16–17
risk narrative 189–192; *see also* narratives
road rage 75, 77, 86

sacrifice 103–104
seducing women 294–296
self-uncertainty 101–102
sensation seeking 185–187
Siev, Joseph J. xviii, 5, 6
significance, need and quest for 98–107, 113–118, 267–271, 284–285
significance gain 100–101, 268
significance loss 99–100, 101, 269
Significance Quest Theory 98–106, 271
single-issue political groups 96, 98, 128
social disapproval 35–36, 39, 42
social identity 99, 101–103, 264, 267–268
socialization phase 119–120
social marginalization 267
social media 281–284, 286–287, 298–301
social networks *see* networks
social psychology 259–260; culturally approved violence 264–265; feeling noticed and agentic 262–263; ingroup identification 263–264; perceived efficacy of violence 260–262; relative deprivation and inequality 266–267; the 3N framework 267–271; violence as a clear response 265–266
social sciences 157, 205; greed in 218–222
social structures 237
societal position 42, *43*
sports *see* extreme sports
stalking 75, 77, 86
suicide 291–292
surfing 183–184, 191
Szumowska, Ewa xviii, 66, 163

targets 42–45, *43*, *45*
threat 47; as a moderator of compensation effects 54–57, *56*; uncertainty threat 101–102
3N framework 267–271
thrill and adventure seeking (TAS) 185–186
tight–loose ambidexterity 140–142, 151–155; cultural tightness–looseness 144; exploration and exploitation 142–144; exploring looseness, exploiting tightness 144–147; the Goldilocks principle 147–151
tight–loose trade-off 144–146, *145*
tightness 144–147; *see also* tight–loose ambidexterity
tightness–looseness 156; *see also* tight–loose ambidexterity
traditional perspectives 184–190, 196
transcendent experience 195–197
tribalism 173

uncertainty: desire to reduce uncertainty 101–103
unusual attitudes 44–46; determinants of 51–52
unusualness (infrequency) 35–36, 38–42, 44–45, 51–52

Vallerand, Robert J. xviii–xix, 5, 7, 66–67, 76, 78–81, 85

violence: as a clear response 265–266; culturally approved 264–265; perceived efficacy of 260–262
violent extremism 259–260; culturally approved violence 264–265; feeling noticed and agentic 262–263; ingroup identification 263–264; perceived efficacy of violence 260–262; relative deprivation and inequality 266–267; the 3N framework 267–271; violence as a clear response 265–266; *see also* Jihadi online recruitment
virtues: and the morally exceptional 234–235
virtue signaling 167, 169, 174
von Hippel, William von xv, 5–6, 8

'wanting' 16–17
wealth *see* material wealth
Webber, David xix, 6, 7, 9, 102
Weimann, Prof. Gabriel xix, 6, 9, 285, 287
well-being, psychological 67, 78, 81–83, 87–88
whistleblowing 231, 237–239, 241, 243, 247
women 172, 196–197, 286–287, 294–296
work tasks 148–149, **149**, 156–157

youth 286, 299

For Product Safety Concerns and Information please contact our EU
representative GPSR@taylorandfrancis.com
Taylor & Francis Verlag GmbH, Kaufingerstraße 24, 80331 München, Germany

www.ingramcontent.com/pod-product-compliance
Ingram Content Group UK Ltd.
Pitfield, Milton Keynes, MK11 3LW, UK
UKHW021450080625
459435UK00012B/437